The Soybean Through World History

This book examines the changing roles and functions of the soybean throughout world history and discusses how this reflects the complex processes of agrofood globalization.

The book uses a historical lens to analyze the processes and features that brought us to the current global configuration of the soybean commodity chain. From its origins as a peasant food in ancient China, today the protein-rich soybean is by far the most cultivated biotech crop on Earth; used to make a huge variety of food and industrial products, including animal feed, tofu, cooking oil, soy sauce, biodiesel and soap. While there is a burgeoning amount of literature on how the contemporary global soy web affects large tracts of our planet's social-ecological systems, little attention has been given to the questions of how we got here and what alternative roles the soybean has played in the past. This book fills this gap and demonstrates that it is impossible to properly comprehend the contemporary global soybean chain, or the wider agrofood system of which it is a part, without looking at both their long and short historical development. However, a history of the soybean and its changing roles within equally changing agrofood systems is inexorably a history about globalization. Not only does this book map out where soybeans are produced, but also who governs, wields power and accumulates capital in the entire commodity chain from inputs in production to consumption, as well as identifying the institutional context the global commodity chain operates within. The book concludes with a discussion of the main challenges and contradictions of the current soy regime that could trigger its rupture and end.

This book is essential reading for students, practitioners and scholars interested in agriculture and food systems, global commodity chains, globalization, environmental history, economic history and social-ecological systems.

Matilda Baraibar Norberg is Associate Professor at the Department of Economic History and International Relations, Stockholm University, Sweden.

Lisa Deutsch is Assistant Professor and Senior Lecturer at the Stockholm Resilience Centre, Stockholm University, Sweden.

Routledge Studies in Food, Society and the Environment

The Vegan Evolution
Transforming Diets and Agriculture
Gregory F. Tague

Food Loss and Waste Policy
From Theory to Practice
Edited by Simone Busetti and Noemi Pace

Rewilding Food and the Self
Critical Conversations from Europe
Edited by Tristan Fournier and Sébastien Dalgalarrondo

Critical Mapping for Sustainable Food Design
Food Security, Equity and Justice
Audrey G. Bennett and Jennifer A. Vokoun

Community Food Initiatives
A Critical Reparative Approach
Edited by Stefan Wahlen, Esther Veen, and Oona Morrow

Food Futures in Education and Society
Edited by Gurpinder Singh Lalli, Angela Turner, and Marion Rutland

The Soybean Through World History
Lessons for Sustainable Agrofood Systems
Matilda Baraibar Norberg and Lisa Deutsch

For more information about this series, please visit: www.routledge.com/books/series/RSFSE

"*The Soybean through World History* is an essential guide to understanding how soy has come to play such a central role in the world food economy today. In this innovative and well-written volume, Baraibar Norberg and Deutsch provide a fascinating look at the *longue durée* of soy, mapping the long historical cycles through which its production has fueled vast trade networks as well as complex ecological consequences."

Jennifer Clapp, *Political Economist Professor and Canada Research Chair in Global Food Security and Sustainability at the University of Waterloo*

"This is not the first book on the history of soy – but it's the first truly global and long-term account combining politico-economic and socio-ecological perspectives. Following soy's pathways from ancient China to modern globalization, the book explains how this commodity has become so central in the current agro-food system, including its burden for society and nature."

Ernst Langthaler, *Economic Historian Professor and Head of the Department of Social and Economic History at the Johannes Kepler University Linz as well as Head of the Institute of Rural History in St. Pölten*

"From the meticulous examination of ancient texts and humble contexts to the investigation of today's conglomerates, the authors trace the successive historical transformations of soy, now a key player in the planet's increasingly vulnerable and unsustainable agrofood system. Soy's social, economic, and political history, fraught as it has become, nonetheless offers a means by which the next transformation, more stable and equitable, can still take place."

Carole L. Crumley, *Anthropology Professor at University of North Carolina, Chapel Hill and Executive Director of IHOPE at Uppsala University*

"In line with revolutionizing studies of sugar, coffee, salt, cod and other transformative global commodities, Baribar Norberg and Deutsch have crafted a comprehensive, yet convincing and accessible world history of soy – this ancient, ever-changing bean, arguably the most expansive element of current unsustainable Anthropocene food-chains, but also with untapped potential to support a resilient future."

Sverker Sörlin, *Environmental Historian Professor, KTH Royal Institute of Technology, Stockholm and co-founder KTH Environmental Humanities Laboratory*

The Soybean Through World History

Lessons for Sustainable Agrofood Systems

Matilda Baraibar Norberg
and Lisa Deutsch

Routledge
Taylor & Francis Group
LONDON AND NEW YORK
earthscan
from Routledge

First published 2023
by Routledge
4 Park Square, Milton Park, Abingdon, Oxon OX14 4RN

and by Routledge
605 Third Avenue, New York, NY 10158

Routledge is an imprint of the Taylor & Francis Group, an informa business

British Library Cataloguing-in-Publication Data
A catalogue record for this book is available from the British Library

ISBN: 978-0-367-40631-8 (hbk)
ISBN: 978-1-032-50935-8 (pbk)
ISBN: 978-0-367-82286-6 (ebk)

DOI: 10.4324/9780367822866

Typeset in Goudy
by codeMantra

Contents

Acknowledgments ix

1 Combining insights from political economy and environmental history: what can the soybean tell us about changes in the global agrofood system? 1

2 The first soybean cycle (domestication to 900 CE) 22

3 The second soybean cycle (1000–1860) 57

4 The roots of the third soybean cycle (1860–1949) 115

5 The regime of the third soybean cycle (1950–today) 155

6 Historicizing soy: toward a new rupture? 213

Index 243

Acknowledgments

We would like to give special thanks and acknowledge those who have supported and contributed to this book in different ways. First, and foremost, we would like to thank the Economic History Professor Ulf Jonsson and Agrarian History Professor Janken Myrdal, who are special contributors to this book. With support from the Swedish Research Council (Vetenskapsrådet - project number 2018-02051), the four of us have worked jointly on a project about the "Miracle Bean" throughout world history. As we mention briefly in the methods section of Chapter 1, we have been in constant dialogue with these experienced and well-read project colleagues, and it is impossible to overstate their importance for this book. While both have been central sparring partners and close readers of earlier drafts, Prof. Myrdal has been particularly important as an antidote to the danger of soy-centrism, offering his advice and comprehensive knowledge on matters of early Asian agrarian systems. He has done this both personally and in print, with his texts *A Theory Kit for World History* (2022) and the forthcoming *The Agricultural History of East and Southeast Asia*. Meanwhile, Prof. Jonsson has been particularly helpful in exploring the roles of soy and other fats and oils in the burgeoning oil and margarine industries in Europe during the first half of the 20th century. Among other things, Ulf examined the historical production and trade reports in the FAO's Rome library during March 2019 before it was completely closed due to the pandemic. While an increasing part of the vast material of the FAO library is digitized, important older sources still require long days of archival work before they can be found and scanned. The combined contribution of Profs. Jonsson and Myrdal have provided us with a solid sense of soy's role in its wider context.

We have also discussed sources and interpretations with other scholars and experts in specific historical periods as well as thematic and geographical points of attention. In this respect, we are grateful for critical comments on earlier drafts of this book provided by colleagues at the Department of Economic History and International Relations, Stockholm University. We are particularly thankful for the comments and other input provided by Gonzalo Pozo Martin, Rodney Edvinsson, Paulina de los Reyes, Olov Lund, Klara Arnberg, Alf Sjöblom, Sven Hellroth, Lars Ahnland, Karina Shyrokykh, Daniel Berg, Daniel Stridh, Rasmus Fleischer, Gustav Ingman and by the late Johan Söderberg.

Thanks also to colleagues at Stockholm Resilience Centre, Stockholm University for their support of the original ideas behind this work. We are particularly indebted to Oonsie Biggs, Line Gordon and Max Troell. We have also received valuable input on an early draft from colleagues at the Nordic Institute of Latin American Studies (NILAS) at Stockholm University. We are particularly grateful to our discussant Magnus Lembke. Thanks also for important soy-discussion with colleagues and friends at the South American Institute for Resilience and Sustainability Studies (SARAS), particularly to Nestor Mazzeo, Esteban Jóbbagy and Cristina Zurbriggen. We are thankful for fruitful conversations with scholars linked to the Soybean and Globalization Network and *Soy in the Anthropocene*, particularly Ernst Langthaler at the University of Vienna, Claudio de Majo at Ludwig-Maximilians-University of Munich and Claiton Marcio da Silva at Universidade Federal da Fronteira Sul. Thanks also to Karin Limburg at SUNY, to Dennis Swaney at Cornell and to Janne Bengtsson at the Swedish University of Agricultural Sciences (SLU) for discussions along the way.

Without the support of amazing librarians, we would not have been able to do this work: Anna Cigwald Lööf at the Swedish Agricultural University library, librarians at KSLA and the many librarians at our Stockholm University library and Olof Rodriguez-Åkerstedt at NILAS. We are also indebted to the SoyInfo Center, particularly Bill Shurtleff. Thanks for your invaluable assistance.

Last, but not least, we would also like to thank our highly competent, systematic and hard-working research assistants Huai-Tse Yang, Mateusz Pozar, Enrique Mejía and Torkel Gure. We are deeply indebted to all of them.

1 Combining insights from political economy and environmental history

What can the soybean tell us about changes in the global agrofood system?

The present agrofood system is capable of providing calories for billions of humans. In fact, the world is experiencing a crisis of obesity, with more people dying from health issues related to this condition than from lack of food: at the same time, hundreds of millions remain hungry (FAO et al. 2021). The agrofood system is also a major driver of biodiversity loss, deforestation, pollution, land degradation and significant greenhouse gas emissions (Ortiz et al. 2021). These are just a few of the extreme contradictions that characterize today's global food order. Many more could be listed. The questions arise: how did we get here, and what alternatives were missed along the way? In this book, we approach these questions through a historical lens and focus on one of the most dominant crops in the current agrofood system: the soybean.

The soybean – *Glycine max* – has become one of the most widely cultivated plants in the world – right after wheat, maize and rice (AMIS 2022). Over a period of just 70 years, world soybean acreage exploded; increasing almost eightfold, from 16.5 million hectares (Mha) in 1950 to 127 Mha in 2020 (FAOSTAT 2022). Soy has expanded over existing agricultural lands and pastures, replacing other crops and grass-fed livestock, and it has also expanded over frontier areas – displacing forests, wetlands, natural grasslands and savannahs (Baraibar Norberg 2020b). The soybean is by far the world's most widely planted genetically engineered (GE) crop by acreage: almost half of all GE crops in 2019 were soybeans. In fact, of a total of over 190 Mha of biotech crops, 92 Mha are GE soybeans – roughly the area of Nigeria (ISAAA 2019). Soy and its derivatives are also among the most traded agricultural commodities – only exceeded by coffee and cacao. In 2019, every other soybean was traded on the international market (AMIS 2022).

Soy is ubiquitous. Nearly all of us consume the bean in some form on a daily basis. A common way to conceptualize its accelerating production and consumption has been as a mere natural response to a century-long steady increase in world demand. Likewise, the standard narrative about the supply side (the rapid expansion of soybean, referred to in Latin America as *sojización*) is that it comes simply as the natural consequence of "rational", profit-seeking farmers and traders responding to new price relations, where soybeans yield the highest annual returns (Baraibar Norberg 2022). While partly demand-driven, this growth is not as straightforward a process as is often proposed. For example, the rise of

DOI: 10.4324/9780367822866-1

genetically engineered soybeans did not come as a direct response to consumer wishes or preferences. In fact, few consumer products have generated as much opposition as genetically engineered crops, in general, and soybeans, in particular, given the widespread (and extremely contested) use of the associated weed killer glyphosate (originally patented as *Roundup* by Monsanto). Moreover, the loss of natural forests and grasslands in Latin America associated with soybean expansion has been a hot point of protest by consumers. Attributing end consumers with ultimate responsibility for the increasing supply of soy is, therefore, partial at best.

This contradiction can, to a great extent, be explained by the fact that most soy consumption is indirect: end consumers are often not even aware that they are eating or drinking soy-based products. The invisibility of soy is due to the fact that the lion's share is first ingested by animals as feed. Most end users never consider how their chicken, pork, beef, fish and dairy products are produced, which can, in turn, be partly explained by the vast geographic and social distances separating consumers, on the one hand, and the people and places where their food is made, on the other. Globally, soybeans are second only to palm trees as the largest source of vegetable oil (Du Bois 2018, 8). In the United States, by contrast, soy oil is the most consumed edible fat and a key ingredient in many products. Yet the word "soybean" seldom appears on the front label of salad dressings, margarine, ready-made meals and many other foodstuffs (Mintz, Tan, and Du Bois 2008, 5). Likewise, many will not know that what they eat was transported in a vehicle running on soy-based biodiesel. In addition to the feed, food and fuel industries, the soybean is also central to other sectors: biotech, agrochemicals and commodity trading (Figure 1.1). Plainly, as an element of the modern agrofood system, soy is everywhere, and yet it remains mostly invisible, or even hidden, from end consumers.

This impressive increase in the production and trade of soy has thus also responded to the interests of powerful actors throughout the global agrofood system, and not just to a "natural" rise in demand among consumers at the end of the chain. In addition, a plethora of political decisions and regulatory shifts have created the necessary conditions that allowed soy to become what it is today. Instead of naturalizing, de-politicizing and reducing the *Great Acceleration* of world production, trade and consumption of soy into a mere tale of shifts in supply and demand, we ask: how and when did this demand emerge? If it was constructed, how and by whom? What other roles and functions has soy had throughout its long history? In which ways are wider changes and continuities in the agrofood system reflected throughout the historical journey of the soybean?

To respond to these questions, we use a historical lens and analyze the processes and features that brought us to the current global configuration of soy. One necessary element in order to understand how systems change is to consider their heritage or historical background (Crumley 2006). While there is a burgeoning amount of literature on how the contemporary global soy web affects large tracts of our planet's social-ecological systems, little attention is given to the questions of how we got here and what alternative roles the soybean has

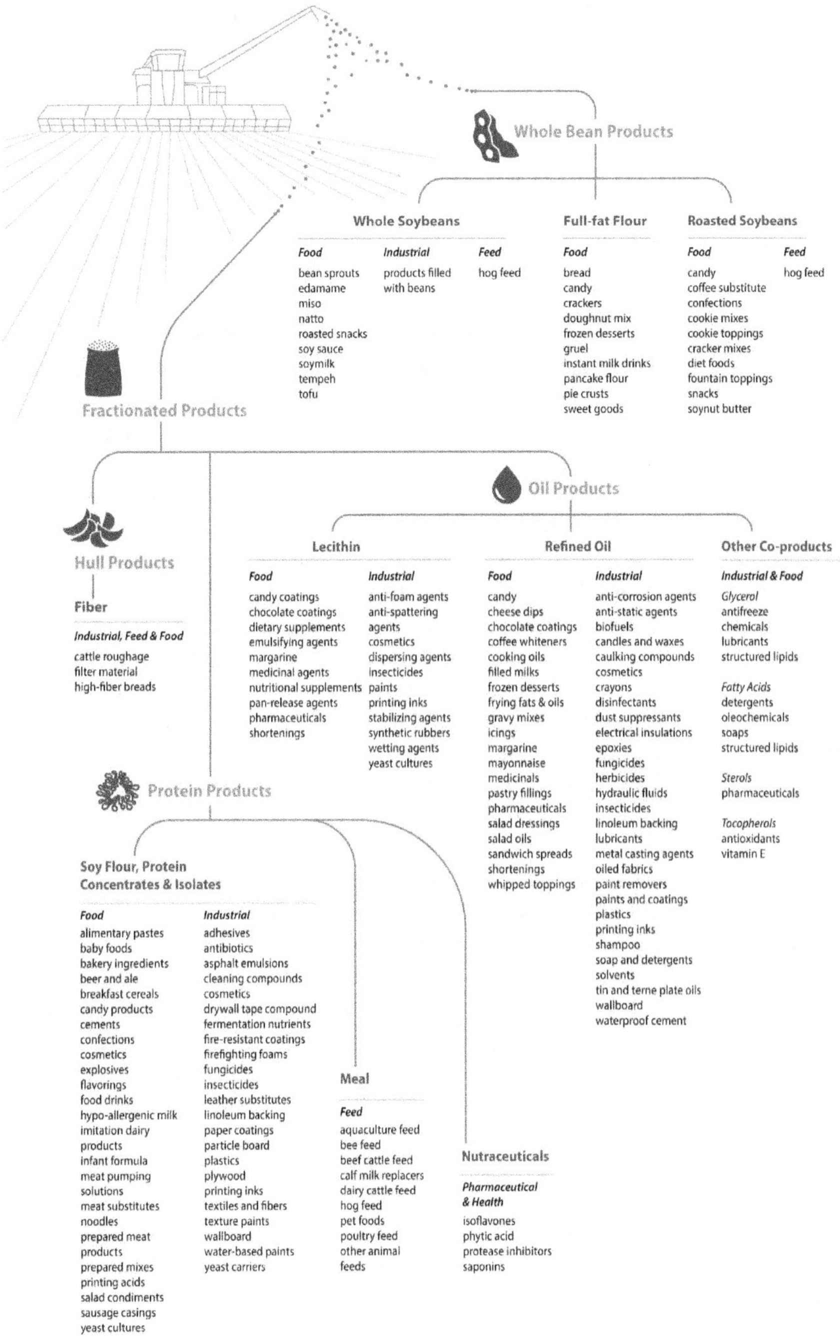

Figure 1.1 Soybeans – one crop, many uses.

played in the past. With our book, we attempt to fill this gap. It is impossible to properly comprehend the contemporary global soybean chain, or the wider agrofood system of which it is a part, without looking at both their long and short historical development (Oliver et al. 2018).

A world history of soy: agrofood globalization through the lens of the soybean

This book is primarily conceived as a contribution to the expanding research field of the world history of commodities: indeed, the changing roles and functions of the soybean offer a fruitful entry point into the complex and uneven processes of agrofood globalization. World history represents an "intellectual conceptualization of the planet as a single entity with shared problems, and an insight that these are a consequence of the long history" (Myrdal 2022, 16). The foundation of world history rests upon the ways in which world spatialities and their historical formations are conceived and structured (Drayton and Motadel 2018).

The book seeks to retrace the soybean's deepest historical origins as it became domesticated and spread throughout Southeast Asia at least 7,000 years ago. The English word "soy" – like its Spanish and Portuguese cognate, *soja* – derives from the Early Chinese word *shu*, reflecting the fact that soybeans were an integral part of the Asian diet for more than a millennium before they were "discovered" by the West in the late 17th century. We thus begin our study in China. Ancient Chinese written records show that the soybean was both an important starvation crop and a food staple for the military. The soybean served both to save people from drought and to prevent soils from degradation in the intensive cultivation system of the Yellow River valley under the unified Han Dynasty (200 BCE to 200 CE). While praised for its robustness, it was also disdained for being difficult to digest and for its beany flavor. With the abdication of the last Han emperor, Hsien-ti, however, the Northern Chinese agricultural system along the Yellow River disintegrated, and the center of agricultural production and commercialization gradually shifted southwards to the Yangtze valley. While the population gravitated to the South, where a high-yielding wet-rice system emerged, grains and beans remained mainly a product in the North. This geographic specialization spurred trade. Moreover, soybeans went from mostly being eaten boiled or steamed, to becoming the basis of increasingly sophisticated processed products, resulting from a gradual, centuries-long development through new food processing and preparation techniques such as fermentation.

Indeed, soyfoods were well known for millennia, but they did not become a fundamental ingredient of people's everyday eating habits until the mid-Ming Dynasty (1368–1644). Reflecting the hegemonic position of China in the region, the intensity in the exchange and spread of cultivation traditions in neighboring countries increased during the early-modern period. It should be mentioned however that monks and merchants had already disseminated seeds and food traditions in the South Asiatic region since ancient times, particularly to Japan and Korea. By the 17th century, an increasing number were purchasing soybeans or other soyfoods on the market and soy sauce had become a fixture in most meals throughout Southeast Asia. Aside from providing savory flavoring and being sought out as food in itself, the residues from oil extraction or tofu production (soybean cake and meal) also provided fertilizer and feed, while soy oil became ever more important in cooking. This period also saw a long-term reversal of fortunes with the rise of the West and the relative decline of the East. The shifts are faithfully reflected in the great amount of texts written during this time, not least by European travelers, missionaries and merchants recording the wide use of soybeans and soyfoods in East Asia. The Western documentary history of soy also emerges in the trade lists of the European merchant companies taking over a growing share of intra-Asiatic commerce (including soy sauce). By the end of this period, it begins featuring in European cookbooks for the upper classes celebrating soy sauce as a new exotic and tasty ingredient. Soy also became the stuff of scientific "discovery" by enthusiastic botanists and agronomists in the West, who would begin producing lengthy reports about the bean's many qualities and advocate its cultivation and extension outside Asia. Yields, alas, proved disappointing, so the soybean remained confined to field trials for quite some time.

The historical trajectory of soy in Asia did not stay static. China's power began to diminish, and its rulers failed to translate their early dominance of the world economic system into a successful strategy of sustained economic development. The mid-19th century marked the end of Sinocentric soybean history. As world agriculture became increasingly industrialized, as transport costs for bulky goods began falling in the late 19th century and as growing swathes of new frontier land were incorporated into soybean production in Manchuria, soybean imports became a key input in the burgeoning margarine and dairy industries of Europe, North America and Japan. Manchuria became the main global supplier of soybeans, but it would do this under the control of Japanese (and to a lesser extent Russian) merchants. This situation lasted until the mid-1940s, when World War II finally cut off the Asian trade and the epicenter for production and exports shifted to the United States. This shift marked the emergence of a new regime which consolidated after 1945, the era of the Great Acceleration (Steffen et al. 2015), during which soybeans quickly assumed their present role as a key ingredient in animal feed – providing the protein that fueled the mass production of cheap meat, under US hegemony and multilateralism.

Soybean production here underwent a period of rapid growth and technological innovation (Mintz, Tan, and Du Bois 2008, 5). The role of soybeans as cheap feed would quickly become global with the liberalization of agricultural trade and the *Gene Revolution* of the post-1990s global economy. As the pace of production and consumption accelerated further, the soybean became the most important oilseed in the world and, as mentioned above, the most widely cultivated genetically engineered crop. In recent years, the expansion of soybean cultivation has primarily taken place in Latin America – today, the region is the biggest global soybean producer and exporter (Baraibar Norberg 2022). At the same time, increased demand for animal products and the deregulation of soybean imports in China made the erstwhile birthplace of soybeans the world's largest buyer and consumer.

This illustrates a wider shift in the geopolitical map of food flows. Under the era of the Great Acceleration, soy has not only become a key commodity but it has also been industrialized and financialized. It is, as intimated above, a key driver of deforestation and land degradation and as the main ingredient behind "cheap meat", it also plays a central role in the rising consumption of animal products associated, in turn, with cardiovascular diseases and obesity.

The historical evolution of the soybean exhibits a spatial reconfiguration of production – from its origins in North China, to its expansion into the Manchurian frontier, its move west to the Corn Belt in the US Midwest and, finally, its expansion south to Latin America – where most soybeans are sourced now. At the same time, the emerging economies of Asia, with China in the lead, have become the largest buyers of soybeans. Meanwhile, the owners of seed technology and agrochemicals, along with the main international traders and crushers, have been concentrated in the United States since the turn of the 20th century. However, China is now entering these markets in leaps and bounds (Baraibar Norberg 2020a, 130–31). Our examination of the changing locations, modes of production, trade routes and markets for soybeans not only illuminates the myriad of ways through which very distant communities were entangled in increasingly complex and wide-ranging economic networks, but also illuminates the concomitant social, cultural, economic and ecological consequences brought by this globalization of agriculture and trade. More concretely, through an analysis of the shifting geographies, uses, practices and values of the soybean throughout world history, we explore the widely shifting roles it has played within agrofood systems.

A history of the soybean is inexorably a history about globalization. A common way to conceptualize globalization is as "a process in which the network of human interaction gradually widens and takes on new and more complex forms" (Jarrick, Myrdal, and Wallenberg Bondesson 2016, 6). While we agree with this definition, we also want to emphasize that alongside the broad trend of increasingly complex networks of human interaction, there is an equally strong movement toward ecological simplification, or, in other words, the loss of landscape complexity, integrity and niche diversity in ecosystems (Peipoch et al. 2015, 1057). Agriculture is one of the main forces behind biodiversity loss, since, by definition, it favors a few specific crops at the expense of all other organisms and non-crop habitats. Today's high-tech, monocultural soybean production, centered around genetically modified seeds, represents an extreme case of ecological simplification. The way genetically engineered soybeans are produced and used further results in a loss of traditional agricultural practices. In this way, the history of the soybean is the story of a double movement, creating increasingly complex forms of human trade networks, while simultaneously impoverishing the variety and richness of local and regional social-ecological systems (Baraibar Norberg 2022).

In the pages that follow, then, we have traced the long social-ecological history of the soybean using a world historical approach, paying attention both to the socioeconomic complexity it generates as well as to the ecological simplification it imposes. Besides pointing to clear world historical shifts, we have also explored patterns of long-term inertia and continuity. We have taken, in other words, inspiration from the *longue durée* perspective, coined by the French historian Fernand Braudel and the *Annales* school. This approach to history moves beyond brief spans of time, beyond the episodic and even beyond the cyclical, and fixes its attention on the very long duration (Braudel and Wallerstein 2009). The idea is to discern structures which change only slowly – so much so, that they are often imperceptible to those who experience them – but which nevertheless provide the main foundations (pillars) and limitations (obstacles) on societies and people. Structure here is meant in a purely Braudelian sense, i.e.

> an assemblage, an architecture, but even more it is a reality that time can only slowly erode, one that goes on for a long time. Certain structures, in their long life, become the stable elements of an infinity of generations. They encumber history and restrict it, and hence control its flow. Other structures crumble more quickly. But all structures are simultaneously pillars and obstacles.
>
> (Braudel and Wallerstein 2009, 178, 196)

From this perspective, and through the lens of soy, we explore longer-term (slow) global trends, such as agrarian frontier expansion, falling transport costs, rising population, urbanization, increased social complexity and ecological simplification. We also consider how the longer-term structures interact with events and shorter-term shifts.

Combining insights between political economy and environmental history

The environmental historian J.R. McNeill argued that the modern ecological history of the planet and the socioeconomic history of humanity make full sense only if seen together (McNeill 2001). We agree with McNeill and have therefore worked to move beyond discipline-specific approaches to integrate different sciences (e.g. economic history, ecology, environmental or agrarian history), hopefully transcending each of their traditional boundaries (Westley and Miller 2013; Nicolescu 2014). We combine methods and approaches from Global Commodity Chain (GCC) analysis and food regimes – both with intellectual roots in world systems theory – with resilience thinking (Holling 1973; Levin 1999). This allows us to take a fresh look at changes and (dis)continuities in the agrofood system over millennia, through the lens of soy. Despite huge technological advances, people are still ultimately dependent on nature for life support, often referred to as "ecosystem services" (Sarukhán et al. 2005). These ecosystem services are provided by functioning ecosystems, i.e. ecological processes and structures, where humans are embedded parts of the biosphere. We apply resilience thinking to keep our attention on the features of the social-ecological system that underlie its (in)capacity to deal with change through persistence, adaptation and/or transformation (Biggs, Schluter, and Schoon 2015).

While both resilience thinking and world system theory are structural, long-term and transdisciplinary approaches, the main focus of the latter is unequal power relations. The world is seen to be characterized by center-periphery (and hinterland) complexes, driven by the process of capitalist accumulation, and dependent on the global division of labor (Frank and Gills 2000, 4–5). One way of empirically investigating these relations is by mapping and analyzing the "commodity chain". The term was coined by Terrence Hopkins and Immanuel Wallerstein in an article from 1977, where the authors called on researchers to take on a radically new way of conceptualizing the world system:

> Let us conceive of something we should call, for want of a better conventional term, "commodity chains". What we mean by such chains is the following: take an ultimate consumable item and trace back the set of inputs that culminated in this item – the prior transformations, the raw materials, the transportation mechanisms, the labor input into each of the material processes, the food inputs into the labor. This linked set of processes we call a commodity chain. If the ultimate consumable were, say, clothing, the chain would include the manufacture of the cloth, the yarn, [...] the cultivation of the cotton, as well as the reproduction of the labor forces involved in these productive activities.
>
> (1977, 128)

Commodity chain analysis is thus about tracing the path of a commodity by illuminating how actors and activities are connected to each other as constituent links of the chain – from inputs to production to end consumption – moving beyond the territorial confines of specific localities or national economies

> Studying commodity chains is for the political economist something like looking through the Hubble telescope for the cosmologist. We are measuring indirectly and imperfectly a total phenomenon that we cannot see directly no matter what we do. The point however is to figure out how this total phenomenon operates, what are its rules, what are its trends, what are its coming and inevitable disequilibria and bifurcations. It requires imagination and audacity along with rigor and patience. The only thing we have to fear is looking too narrowly.
>
> (Wallerstein 2009, 89)

The commodity chain literature has expanded and taken many new forms since Hopkins and Wallerstein first coined the concept (Bair et al. 2021). While we remain close to the world systems tradition, we also draw on the Global Commodity Chain (GCC)/Global Value Chain (GVC) approaches that developed in the 1990s (Gereffi, Korzeniewicz, and Korzeniewicz 1994, 9; Fernandez-Stark and Gereffi 2019; Ponte, Gereffi, and Raj-Reichert 2019; Bair et al. 2021).[1] These provide a structured approach to mapping the set of interlinked actors, assets and activities involved at different stages of the global soybean productive and commercial chain (Wallerstein 1974; Gereffi 1994). Leaning on these traditions, we have thus explored the full spectrum of whos, whens, wheres and whys of soybean production, trade and consumption over history.

GCC further assumes that it is necessary to comprehend the full range of entities – natural, social, technological and economic – found in a commodity chain, including informal or formal contracts (Gereffi, Korzeniewicz, and Korzeniewicz 1994, 9; Gutman et al. 2006, 255; Morgan, Marsden, and Murdoch 2006, 18; Bair 2009). This also means asking who governs, wields power and accumulates wealth in the entire commodity chain – the so-called governance structure of the chain – as well as identifying the wider institutional context in which the chain operates (Gereffi, Korzeniewicz, and Korzeniewicz 1994; Bair 2005; 2009, 8). Thus, the chain is in fact more of a network than a chain, and calls for a systemic and structuralist perspective in International Political Economy. Notwithstanding this structuralist approach, power is mainly conceptualized in the form of the strategic behavior of lead economic actors, which under a capitalist mode of production are leading firms acting as chain drivers (Raikes, Jensen, and Ponte 2000, 395). This allows us to eschew many of the classical pitfalls of structuralist approaches, which are too frequently blind to agency within a particular system, favoring only deterministic explanations. By definition, however, the application of these GCC insights only becomes fully meaningful after the advent of extended capitalist relations of accumulation (wherever we locate their temporal origin). All the same, we invoke the form of some key GCC concepts to understand the evolving role of soybeans throughout all the period before the actual rise of capitalism. To wit, in the absence of capitalist relations strictly speaking, we still retain Gereffi's four "dimensions of analysis": (1) Territorial or geographical configuration of the chain, involving the spatial concentration or dispersion of production networks. (2) Input-output structure, describing the process where actors, products and services are linked together into final production. (3) Governance structure, to illuminate the nature of power relations in the chain by showing how particular players (often leading firms or chain drivers) exert control over other participants. (4) Institutional structure, focusing on the wider context in which the productive and commercial chain is embedded (Gereffi 1994, 96; 2014; Hamilton and Gereffi 2009, 140).

Soy history through regimes

Since the influential idea of "paradigm shift" in Thomas Kuhn's seminal book *The Structure of Scientific Revolutions* (1970), a broad array of perspectives both within the natural and social sciences have imagined different frames based on successions of cycles involving regimes (stability), crises and reorganization. In resilience thinking, for instance, this has been theorized as the adaptive cycle, focusing upon processes of destruction and reorganization (Gunderson 2001) in which periods of relative stability alternate with periods of crisis and opportunities for innovation. The adaptative cycle rests on the notion that a specific social-ecological system (SES) can have multiple stable states and thresholds. Within such states, different relations and feedback mechanisms shape their features and dynamics. In turn, SESs are complex but adaptive systems, characterized by non-equilibrium and nonlinear behavior (Levin et al. 2013). The capacity of SESs to deal with change, while still retaining its essential structure and functions, is precisely the definition of social-ecological resilience (Carpenter et al. 2001; Walker et al. 2004). Changes in the feedback flows, diversity and connectivity of the main components of the system are at the center of the analysis. A system can cope with change by either resisting or adapting so as to remain in the same regime, but if the capacity to deal with perturbations is eroded, change can be abrupt and often nonlinear, with small changes leading to seemingly larger effects. Eventually, the regime falls apart and/or transforms into a new regime, leading to a regime shift.

At a more concrete level, however, we have proceeded by way of critical dialogue with the food regime approach, formulated by the agrarian sociologists Harriet Friedmann and Philip McMichael (Friedmann and McMichael 1989; McMichael 2009). This framework is particularly concerned with the geopolitical history of industrial capitalism through an agrofood lens (McMichael 2021, 218). In McMichael's words:

> Food is intrinsic to capital's global value relations since it is central to the reproduction of commodity producing labor, and the food regime characterizes the political mechanisms by which capital can reduce cost by cheapening wage foods. How this is accomplished in particular episodes depends on geo-political arrangements governing and governed by forms of capital accumulation
>
> (2015-307-308)

The food regime approach emphasizes the nonlinearity of capitalist development across an unequal state system and sees successive political-economic hegemonic orders as the foundation of subsequent food regimes. Like GCC analysis, this framework has its intellectual roots in world systems theory, and both pose a stark contrast to the widely spread assumptions of modernization theory, in which societies are typically seen to transition progressively from traditional to modern. In our view, rather, world history has been characterized by periods of relative stability – regimes with temporary constellations of formal and informal rules and relations – followed by periods of crisis, change, reconfiguration and transition.

Another key feature common to GCC analysis and food regimes is that both draw on "conventions theory", resting on the assumption that any form of coordination in economic, political and social life requires agreement of some kind – non-codified traditions and ways of doing things – among its agents. A regime thus depends on a certain amount of agreement among actors, on implicit rules creating predictability in the system, and crucially, smoothing out the internal contradictions that can beset it. However, when and if these contradictions intensify, the regime starts to erode and eventually falls apart in a crisis of structures, logics and forms of accumulation (Williams 2014, 406). Eventually, all regimes fall apart and/or morph into new ones.

Regimes, then, are understood as a phase with a relatively stable institutional structure, shaped by a particular global division of labor and sets of power relations between farmers, firms and governments (McMichael 2021; 2016; 2009, 140). This said, there is an important strand of food regime scholars focusing on the prevalence of challengers or counterpoints which eventually can bring down the regime (Carton 2018; Jeff 2009; McMichael 2008). While we focus on the mainstream features and trends of the soybean, we also bring forth some of the alternative examples of soybean production and consumption; we pay some attention to the role of challengers and counterpoints in the disintegration of the old regime and formation of a new one.

Regime analysis, then, offers a fruitful tool with which to examine the political, economic and – we argue – ecological relations governing the global production and circulation of food from a historical perspective.

Regimes operationalized: our soy periodization

McMichael and other scholars have famously periodized the history of global food and agricultural systems into three differentiated food regimes. He places the first regime under the period of British hegemony (1860–1914) and characterizes it by its combination of free trade policies, falling transport costs and the Gold Standard. The massive import of temperate foods (such as grain and meat) from the new settler frontiers and other colonial tropical products provided cheap supplies to an industrializing Europe (McMichael 2021). The second regime took shape under Washington's hegemonic oversight (1940s–70s): here, the United States managed to impose a particular national capitalist development model on the rest of the world, marked by protectionism, an encouragement of surpluses, "meatification", standardization, industrialization of agriculture (Green Revolution), food aid and selective "free" trade. The third food regime, the neoliberal, under corporate hegemony (1980s–the present), is premised on the dismantlement of economic nationalism in favor of transnationalization, the Gene Revolution, the dispossession of small farmers and the phenomenon of "agriculture from nowhere" (McMichael 2021, 218–19). A few caveats are nevertheless necessary here. While we have taken inspiration in McMichael's clear periodization, we argue that the general patterns exposed in these international regimes do not conform to all food production and consumption everywhere (Wilkinson and Goodman 2018). Some general features of agrofood globalization are, of course, mirrored in the history of the soybean, but as we also show, the soybean has important particularities and does not always follow the path of other agro-commodities. In addition, and by way of a substantive contrast, instead of using the periodization of the food regime approach, we have followed the soybean *since before* the emergence of modern capitalism. We thus extend the food regime framework back in time to explore additional key markers of periodization depending on the roles played by the soybean in production, farming, trade and consumption (both domestic and regional) in earlier historical periods. We have adopted an inductive research design, basing our own periodization on an empirical analysis of the historical records. By tracing and mapping the shifting roles, functions and meanings of soy, in production, consumption and trade, and building our analysis from there, we have identified three main "soy cycles", with periods of reorganization (*roots* and *ruptures*) in between regimes. Taken together the world history of soy has the following three cycles of change.

The first soy cycle (domestication to 900 CE)

The most striking difference between our periodization and that of the influential food regime approach is that our study begins in ancient China, long before "the first food regime", characterized by massive global agrofood trade, whose origins McMichael and other food regimes scholars locate in the 1870s. By contrast, we argue, soybeans emerged as one among several crops quite early on, in the highly stratified early agricultural civilization around the Yellow River. A full regime then emerges roughly between 200 BCE and 200 CE when soy acquired several important roles in the intensive, high-yielding, cropping systems in North China. Soybeans provided soil improvement, survival insurance, food and sometimes feed for the great masses of mainly self-sufficient peasant households under the Han Dynasty. It was nevertheless also used for commerce, military rations and increasingly sophisticated food traditions. In the 3rd century CE, the Northern Chinese agricultural system disintegrated with the collapse of Han power, thus initiating a phase of rupture. This period, lasting until the 900s, is also characterized by the center of agricultural production and commercialization moving to the Southern Yangtze River valley. Chapter 2 covers this first cycle of change, organized along the lines of roots-regime-rupture.

The second soy cycle (1000–1850)

At the first phase of the second soy cycle (roots – between 1000 and 1600), soybeans expanded frontiers in the north of China and found a new geographical complementarity with the South's rice through trade. At the same time, slow innovation, with new soyfood processing practices, spread throughout medieval Southeast Asia. Moreover, Europeans went out into the world. The regime proper takes hold around 1600, lasting more or less until 1750. During this phase, soybeans were used for multiple purposes throughout Southeast Asia, such as savory flavoring, fertilizer enabling the simplification of landscapes and feed to increase pig meat production. The section also explores the burgeoning role of foreign soybean trade. During this period, Europeans took over an important part of the Asian intra-regional soy sauce trade and soon also integrated soy in the trans-oceanic trade, as soy sauce became an object of oriental exoticism in European upper-class cuisine. Further, European and American scientists and travelers took an interest in expanding knowledge about this "miracle bean" during the Enlightenment. This second cycle enters the rupture phase between 1750 and 1850, when the "Fall of the East" and the "Rise of the West" became reflected in soy history. Chapter 3 covers this second cycle of change, organized along the lines of roots-regime-rupture.

The third soy cycle (1860–today)

We argue that the roots to the current regime are to be found in the period between 1860 and the 1940s, when soy began to be massively traded across long distances by the end of this period. In the context of imperial rivalries, China gradually lost its grip on Manchuria. While Manchuria became the world's primary site for soy – in high demand as fertilizer and food in Japan and as a substitute for other oils and fats (for industrial purposes as well as for margarine) in Europe and the United States. Soy's cheapness and versatility were its most appreciated attributes in Europe. During this period, soybeans also became an increasingly established crop in the US Midwest, and powerful actors – such as the American Soybean Association (representing soybean farmers), Cargill, ADM and a few others (who were crushers/traders) and the USDA (representing the state) – became important soybean chain drivers. Interest in soy as feed also began during this period. This "root" period of our third regime cycle fits rather well with distinctive temporal shifts and configurations in McMichael's first food regime. By contrast, what we see as our current regime (1950 to today) – when soybean production became industrialized, monoculturalized, financialized and ultimately driven by increased meat consumption under the Great Acceleration – overlaps with both the second and third food regimes. We agree with McMichael that the period after the 1980s (his third food regime, then) is articulated through the increased corporate domination of the agrofood system, but we see it more as a phase shift than a shift in regimes. In our analysis, these tendencies (industrialization, financialization, etc.) were already evident in the 1950s. Accordingly, we see the whole period from 1950 until today as one regime – and one of the main characteristics of this regime is precisely the trend of acceleration. We have not, as of yet, seen the rupture of this regime, but we do see an increasing number of challenges to the present order and a rising degree of contestation.

While roots and regime belong to the same cycle of change, we have divided them into two separate chapters, with Chapter 4 covering the roots period, 1860–1950, and Chapter 5 offering a deep exploration into the third regime, covering the 1950s to the present. Since the soybean in this third cycle of change has evolved into a prominent part of the increasingly complex agrofood system with multiple central functions and roles all over the globe, we believe that both the roots and regime each need a chapter of their own. In addition, as the pace of change is faster and the amount of written records dealing with different aspects of soy is immense and growing, it would have been impossible to deal fully with this increasing complexity in only one chapter.

The long history of the soybean is thus understood to involve three distinct cycles of change – each including roots, regime and rupture (with the exception of the present regime, where it is too early to say whether it has entered a rupture phase). While, as we have explained above, we depart from the periodization of the food regime approach, we have still taken inspiration in its focus, concepts and comparative historical perspective.

While throughout this book we focus on dominant sets of dynamics, values and functions, it is important to remember that reality is messier – full of exceptions, alternatives and contestation. Besides the inspiration provided by the food regime analysis for the identification of regimes and focus areas, for the later periods – after 1860 – we contrast the specific configuration of soy with the picture provided by the food regime literature, explicitly addressing the question of when the soybean fits with wider agrofood configurations and when it does not.

In addition to these cycles of change – roots, regime and rupture – we also consider what the whole long world history of soy says about wider historical processes of uneven agrofood globalization. In this way, besides periodizing the long soy history into three "roots-regime-rupture" cycles, we have also considered the whole period from domestication until today, i.e. the *longue durée* of soy. Thus in (the concluding) Chapter 6, we piece together the long-term patterns of production, commercialization and consumption of the soybean (Braudel and Wallerstein 2009, 180, 191) and discuss their contemporary implications.

In this analytical part, we also reflect on the soybean and the wider systems it is embedded in using insights from our theoretical frameworks, addressing key aspects of political economy and resilience thinking. We probe the capacity of the soybean to deal with change throughout history and then further reflect on the capacity of the overall agrofood system to deal with (un)anticipated changes. Moreover, we reflect on challenges, opportunities and potential regime shifts in the coming agrofood system.

Our history of the soybean, then, is grounded in world systems traditions, such as GCC and food regimes and with insights from resilience thinking over the *longue durée*. This enabled us to undertake an in-depth understanding of ecological effects, while avoiding the pitfall of many resilience studies: the lack of any thorough analysis of power relations and deep history.

Methods and sources: critical reflections

There are innumerable challenges, risks and pitfalls involved in a research project that aims to grasp the shifting roles of the soybean throughout history – spanning thousands of years and the whole world. The task is not made easier by the fact that the theoretical traditions that we just presented pledge allegiance to some form of holism, some kind of "total history". Clearly, our ontological and epistemological assumptions begin from the premise that reality is complex, changing and interconnected – everything matters and everything is related in some way and on some particular timescale. What follows from this is the need to fully comprehend all the different parts of any system in order to grasp the whole. Within the limited time and resources given, however, we aim to at least start to weave together some key themes, well aware that experts in specific subfields or subperiods will certainly have important nuances to add to our accounts. The wide empirical scope of this study (both in terms of temporality and spatiality) and the variety of traditions, methodological approaches, types and uses of sources, etc., have made the source-critical task more difficult and at the same time more pressing. Here, we will say a few words about some of the main challenges we have faced and some of the main ways in which we have tackled them.

One obvious danger shared by all global history scholars is the prospect of reproducing sweeping generalizations which lack substance or present mere reflections of special interests, rather than findings from the most rigorous empirical research. "It is in the nature of global history to have to rely on secondary sources" (Widgren 2016, 85) and accordingly "the limits of critical world history are determined by the quality of the existing literature" (Myrdal 2016, 49). In addition, it is often hard for scholars of world history to critically scrutinize their available sources, since the wide scope of their research necessarily makes them move beyond their own area of expertise (McNeill 2016, 22). The broad scope of our study, for example, makes use of a wide range of literature dealing with different historical epochs and places, representing divergent perspectives and research traditions from many disciplines. Besides the many history subjects, we also draw on the disciplines of archeology, ecology, agronomy and anthropology. The high reliance on secondary sources from a wide range of fields thus increases the risk of unwillingly or uncritically reproducing some common, widespread and enduring simplifications, over-generalizations, biases, pre-conceptions and/or prejudices (Widgren 2016, 85; Graeber and Wengrow 2021, 18).

One way of minimizing this risk is to be aware of common myths and misconceptions, in order to be able to critically scrutinize them as they appear in the sources. For example, one recurrent narrative is that of empty or under-utilized land in areas that were, in fact, populated; this is a myth that is often used to legitimize different types of colonization and settler projects (Widgren 2016, 91–92). Such tropes have frequently appeared in both the source material and literature. Another recurrent narrative reflects history as a linear process, where societies move from simple to complex. In agrarian history, this view can sometimes appear as a kind of emplotment, by virtue of which agrarian systems move along a sequence from foraging and pastoralism, to cultivation and ever-intensive forms of agronomy. As Widgren argues, while human use of the environment in the long term has obviously progressed toward more intensive forms, the historical sequence must be used with caution, partly because there are many exceptions where intensive permanent agriculture has predated shifting cultivation, and also because foragers and pastoralists still exist and thrive – it is not an evolutionary ladder. Rather, different farming systems seem to reflect possibilities and constraints defined by the wider political economy (Widgren 2016, 94–95). The notion of history as linear – moving from simple to complex – is also often intimately intertwined with the idea of Europe representing the most advanced and superior stage in history. Eurocentrism (or any national variation on the theme of Western superiority) frames Europe as the "primary engine and architect of world history, the bearer of universal values and reason, and the pinnacle and therefore model of progress and development" (Sundberg 2009, 638).

While Eurocentrism and different kinds of European nationalisms have proliferated in Western literature for some time, scholars now point toward an emerging Sinocentrism, particularly within Chinese academia, preoccupied with the rewriting of more patriotic Chinese national history meant to elevate traditional identity in the 21st century (Perez-Garcia 2021, 43). Both Eurocentrist and Sinocentrist myths may serve particular interests, but the tendency to present either Europe or China as a bearer of specific and perhaps superior qualities is common also among scholars who do not see themselves as either Eurocentric or Sinocentric and who lack ideological motives or interests in reproducing these narratives. Of course, awareness is not enough. It is hard, not to say nigh impossible, to avoid methodological nationalism entirely. There is, for example, a tendency to analyze different aspects of the world from the standpoint of the nation-state as an analytical unit – e.g. the case of world production, trade and consumption of soybeans. This is partly due to the high reliance on written sources which create a strong bias toward the affairs of nations and states (McNeill 2016, 21). Both historical and contemporary statistics on agricultural production, trade and consumption are almost exclusively based on national accounts and national boundaries rather than in accordance with, say, biomes or transnational private ownership. Methodological nationalism is thus difficult to escape and it taints the stories that can be told.

In addition, most Western world history scholars have relied on texts written in the languages they master, which meant ignoring texts written in, for example, Afroasiatic or Sino-Tibetan languages and overly relying on texts from Anglo-Saxon speaking countries (Myrdal 2016, 73). In the research for this book, comprehensive bibliographies and sourcebooks on soy provided by the *Soyinfo Center* (2022) have been extremely helpful in reducing both the long-standing language barriers and the common (over)reliance on existing literature.[2] These resources are produced from *Soyinfo Center*'s computerized database, *SoyaScan*, which contains 94,400 records on soybeans and soyfoods from 1100 BCE to the present, mostly in published documents, but also across 7,800 unpublished archival sources. In this repository, each non-English source in the bibliography features an English summary of the main soybean-related content and often provides long excerpts on the passages that refer to soybeans or soy-related topics. The chronologically ordered bibliographies also include original interviews and comments on the incumbent source from soybean history scholars. Such vast material has allowed us to gain a fuller sense of the multiple uses and meanings of the soybean across the world since ancient times – and it has allowed us to come close to thousands of primary sources through the long, translated excerpts of the originals.

The fact that we have relied heavily on this archive, however, may have simultaneously created the risk of exaggerating the role of soybeans in production, trade and use (consumption) throughout world history. This is an obvious danger for all studies of any specific commodity in world history, and it is inherent to the method of mainly considering sources and literature that mention the object of study, while not regarding the texts that talk about production, trade and uses that, in fact, do not mention it at all. This risk may be even further exacerbated by the fact that the *Soyinfo* bibliographies exclusively cover texts that explicitly deal with soy and moreover they only reproduce the specific parts that mention soy. Being aware of this risk, we have adopted a wide range of strategies to make sure that we not only understand soybeans, but also the relative role and weight of soybeans in relation to other commodities in the system at any given period. We have tried to minimize this problem by triangulating sources and literature from different places and language areas, not least by involving a competent Chinese-speaking research assistant, Huai-Tse Yang. In this way, we were able to compare scholarly discussions about soy in different language contexts. In addition, we have benefitted from the recent digitalization of many archives, especially the USDA, the Dutch Royal Library (*Koninklijke Bibliotheek*), the Swedish royal library (*Kungliga biblioteket*) and the *Dutch East India Company* (VOC).

Of all the strategies we have used to tackle the abovementioned risks, including overestimating the role of the soybean, the most important one is probably our constant dialogue with our experienced and well-read project colleagues: Economic History Professor Ulf Jonsson and Agrarian History Professor Janken Myrdal. Their combined contribution has provided us with a solid sense of soy's role in its wider context. We have also discussed sources and interpretations with other scholars and experts in specific historical periods as well as thematic and geographical points of attention – not least with the soybean history researchers who participate in the international network *Soy in the Anthropocene*. We have thus managed the risks of over-generalizations, misunderstandings and unintentional biases by discussing our ideas and drafts with scholars covering a broad array of periods. Moreover, as we navigated the massive amount of primary, secondary and even tertiary (literature on literature) sources, we have systematically asked critical questions concerning implicit assumptions, hidden interests, myths and other types of potential biases. An important tool here is to use triangulation, proactively seeking to find other sources that can validate, problematize or nuance claims. We may still have missed, misunderstood or misinterpreted details of the long history of soy. These potential flaws are, of course, completely our own responsibility.

Notes

1 While the terms "Global Value Chain" (GVC) and "Global Commodity Chain" (GCC) are often used interchangeably, we use GCC in this book.

2 The *Soyinfo Center* was founded by William Shurtleff and Akiko Aoyagi in 1976, and since that time they have produced more than 50 books on the history of soybeans and a wide range of soy related processes, products and activities.

References

AMIS. 2022. "Market Database." Agricultural Market Information System. USDA. https://data.nal.usda.gov/dataset/amis-market-database (Accessed 2022-10-17).

Bair, Jennifer. 2005. "Global Capitalism and Commodity Chains: Looking Back, Going Forward." *Competition and Change* 9 (2): 153–80.

———. 2009. *Frontiers of Commodity Chain Research*, Stanford: Stanford University Press.

Bair, Jennifer, Mathew Mahutga, Marion Werner, and Liam Campling. 2021. "Capitalist Crisis in the 'Age of Global Value Chains" *Environment and Planning A: Economy and Space* 53 (6): 1253–72.

Baraibar Norberg, Matilda. 2020a. "Agrofood Globalization: The Global Soybean and Beef Commodity Chains." In *The Political Economy of Agrarian Change in Latin America*, edited by Baraibar Norberg, 117–63. Cham: Palgrave Macmillan, Springer Nature

———. 2020b. "Introduction to the Political Economy of Agrarian Change in Latin America." In *The Political Economy of Agrarian Change in Latin America*, edited by Baraibar Norberg, 1–56. Cham: Palgrave Macmillan, Springer Nature.

———. 2022. "Sojización as a New First Movement: A Polanyian Analysis of the South American Soybean 'Boom.'" In *The Age of the Soybean: An Environmental History of the Soyacene during the Great Acceleration*, edited by da Silva, and de Majo, 91–114. Winwick, Cambridgeshire: The White Horse Press.

Biggs, Reinette, Maja Schluter, and Michael L. Schoon. 2015. *Principles for Building Resilience: Sustaining Ecosystem Services in Social-Ecological Systems*. Cambridge: Cambridge University Press.

Braudel, Fernand, and Immanuel Wallerstein. 2009. "History and the Social Sciences: The Longue Durée." *Review (Fernand Braudel Center)* 32 (2): 171–203.

Carpenter, Steve, Brian Walker, J. Marty Anderies, and Nick Abel. 2001. "From Metaphor to Measurement: Resilience of What to What?" *Ecosystems* 4 (8): 765–81.

Carton, Wim. 2018. "On the Nature of the Countermovement: A Response to Stuart et al.'s 'Climate Change and the Polanyian Countermovement: Carbon Markets or Degrowth?'" *New Political Economy*, 1–6.

Crumley, Carole. 2006. "Historical Ecology: Integrated Thinking at Multiple Temporal and Spatial Scales." In *The World System and the Earth System: Global Socioenvironmental Change and Sustainability since the Neolithic*, edited by Hornborg and Crumley, 15–29. Walnut Creek, CA: Left Coast Press.

Drayton, Richard, and David Motadel. 2018. "Discussion: The Futures of Global History." *Journal of Global History* 13 (1): 1–21.

Du Bois, Christine M. 2018. *The Story of Soy*. London: Reaktion Books.

FAOSTAT 2022. "FAOSTAT Statistical Database." HD9000.4. Library of Congress Online Catalog (1,343,705). Rome: FAO.

Fernandez-Stark, Karina, and Gary Gereffi. 2019. "Global Value Chain Analysis: A Primer." In *Handbook on Global Value Chains*, edited by Ponte, Gereffi and Raj-Reichert, 54–76. Cheltenham: Edward Elgar Publishing.

Frank, Andre Gunder, and Barry K. Gills. 2000. "The Five Thousand Year World Systems in Theory and Praxis." In *World System History: The Social Science of Long-Term Change*, edited by Denemark, Friedman, Gills, and Modelski, 3–23. London and New York: Routledge.

Friedmann, Harriet, and Philip McMichael. 1989. "Agriculture and the State System: The Rise and Fall of National Agricultures, 1870 to the Present." *Sociologia Ruralis* 29 (2): 93–117.

Gereffi, Gary. 1994. "The Organization of Buyer-Driven Global Commodity Chains: How U.S. Retailers Shape Overseas Production Networks." In *Commodity Chains and Global Capitalism*, edited by Gereffi and Korzeniewicz, 95–123. Westport, Connecticut: Praeger.

———. 2014. "Global Value Chains in a Post-Washington Consensus World." *Review of International Political Economy* 21 (1): 9–37.

Gereffi, Gary, Miguel Korzeniewicz, and Roberto P. Korzeniewicz. 1994. "Introduction: Global Commodity Chains." In *Commodity Chains and Global Capitalism*, edited by Gereffi and Korzeniewicz, 1–15. Westport, Connecticut: Praeger.

Graeber, David, and David Wengrow. 2021. *The Dawn of Everything: A New History of Humanity*. London: Penguin/Allen Lane.

Gunderson, Lance H. 2001. *Panarchy: Understanding Transformations in Human and Natural Systems*. Washington, DC: Island Press.

Gutman, Graciela, Roberto Bisang, Pablo Lavarello, Mercedes Campi, and Veronica Robert. 2006. "Les Mutations Agricoles et Agroalimentaires Argentines de Années 90: Libéralisation, Changement Technologique, Firmes Multinacionales." *Région et Développement* 23: 215–46.

Hamilton, Gary G., and Gary Gereffi. 2009. "Global Commodity Chains, Market Makers, and the Rise of Demand-Responsive Economies." In *Frontiers of Commodity Chain Research*, edited by Bair, 136–62. Stanford: Stanford University Press.

Holling, C. S. 1973. "Resilience and Stability of Ecological Systems." *Annual Review of Ecology and Systematics* 4 (1): 1–23.

ISAAA. 2019. *Global Status of Commercialized Biotech/GM Crops in 2019: Biotech Crops Drive Socio-Economic Development and Sustainable Environment in the New Frontier. ISAAA Brief No.* 55. Ithaca, NY: Cornell University.

Jarrick, Arne, Janken Myrdal, and Maria Wallenberg Bondesson. 2016. "Globalization and World History: An Introduction to Studies of Methods." In *Methods in World History: A Critical Approach*, edited by Jarrick, Myrdal, and Wallenberg Bondesson, 7–18. Lund: Nordic Academic Press.

Jeff, Pratt. 2009. "Incorporation and Resistance: Analytical Issues in the Conventionalization Debate and Alternative Food Chains." *Journal of Agrarian Change* 9 (2): 155–74.

Kuhn, Thomas S. 1970. *The Structure of Scientific Revolutions*. Second Edition. Chicago: University of Chicago Press.

Levin, Simon A. 1999. *Fragile Dominion*. Cambridge, MA: Perseus Books Group.

Levin, Simon, Tasos Xepapadeas, Anne-Sophie Crépin, Jon Norberg, Aart De Zeeuw, Carl Folke, Terry Hughes, et al. 2013. "Social-Ecological Systems as Complex Adaptive Systems: Modeling and Policy Implications." *Environment and Development Economics* 18 (2): 111–32.

McMichael, Philip. 2009. "A Food Regime Genealogy." *The Journal of Peasant Studies* 36 (1): 139–69.

———. 2015. "Value Theory and Food Regime Temporalities." *Review* (Fernand Braudel Center), 38 (4), 307–28.

———. 2016. *Development and Social Change: A Global Perspective*. 6th ed. Thousand Oaks, CA: Sage Publications.

———. 2021. "Food Regimes." In *Handbook of Critical Agrarian Studies*, edited by Akram-Lodhi, Dietz, Engels and McKay, 218–32. Cheltenham: Edward Elgar Publishing.

McNeill, John Robert. 2001. *Something New under the Sun: An Environmental History of the Twentieth-Century World*. First published as a Norton paperback. New York: W. W. Norton.

McNeill, John Robert 2016. "Historians, Superhistory, and Climate Change." In *Methods in World History: A Critical Approach*, edited by Jarrick, Myrdal, and Wallenberg Bondesson, 19–44. Lund: Nordic Academic Press.

Mintz, Sidney W., Chee-Beng Tan, and Christine M. Du Bois. 2008. "Introduction: The Significance of Soy." In *The World of Soy*, edited by Du Bois, Tan, and Mintz, 1–23. Singapore: NUS Press.

Morgan, Kevin, Terry Marsden, and Jonathan Murdoch. 2006. *Worlds of Food: Place, Power and Provenance in the Food Chain*. Oxford Geographical and Environmental Studies Series. Oxford: Oxford University Press.

Myrdal, Janken. 2016. "On Source Criticism in World History." In *Methods in World History – A Critical Approach*, edited by Jarrick, Myrdal, and Wallenberg Bondesson, 45–84. Lund: Nordic Academic Press.

———. 2022. "A Theory Kit for World History." Science report 7. Agarhistoria. Uppsala: Department of Urban and Rural Development, Swedish University of Agricultural Sciences.

Nicolescu, Basarab. 2014. "Methodology of Transdisciplinarity." *World Futures* 70 (3–4): 186–99.

Oliver, Tom H., Emily Boyd, Kelvin Balcombe, Tim G. Benton, James M. Bullock, Deanna Donovan, Giuseppe Feola, et al. 2018. "Overcoming Undesirable Resilience in the Global Food System." *Global Sustainability* 1, E9.

Ortiz, Andrea Monica D., Charlotte L. Outhwaite, Carole Dalin, and Tim Newbold. 2021. "A Review of the Interactions between Biodiversity, Agriculture, Climate Change, and International Trade: Research and Policy Priorities." *One Earth* 4 (1): 88–101.

Peipoch, Marc, Mario Brauns, F. Richard Hauer, Markus Weitere, and H. Maurice Valett. 2015. "Ecological Simplification: Human Influences on Riverscape Complexity." *BioScience* 65 (11): 1057–65.

Perez-Garcia, Manuel. 2021. *Global History with Chinese Characteristics: Autocratic States along the Silk Road in the Decline of the Spanish and Qing Empires 1680–1796*. Singapore: Palgrave Macmillan.

Ponte, Stefano, Gary Gereffi, and Gale Raj-Reichert. 2019. "Introduction to the Handbook on Global Value Chains." In *Handbook on Global Value Chains*, edited by Ponte, Gereffi and Raj-Reichert, 1–27. Cheltenham, UK; Northampton, MA, USA. Edward Elgar Publishing.

Raikes, P., M. F. Jensen, and S. Ponte. 2000. "Global Commodity Chain Analysis and the French Filiere Approach: Comparison and Critique." *Economy and Society* 29 (3): 390–417.

Sarukhán, José, Anne Whyte, R. Hassan, R. Scholes, N. Ash, S. T. Carpenter, P. L. Pingali, et al. 2005. *Millennium Ecosystem Assessment: Ecosystems and Human Well-Being*. Washington, DC: Island Press.

Soyinfo Center. 2022. "Website." https://www.soyinfocenter.com/. (Accessed 2022-10-17)

Steffen, Will, Wendy Broadgate, Lisa Deutsch, Owen Gaffney, and Cornelia Ludwig. 2015. "The Trajectory of the Anthropocene: The Great Acceleration." *The Anthropocene Review* 2 (1): 81–98.

Sundberg, Juanita. 2009. "Eurocentrism." In *International Encyclopedia of Human Geography*, edited by Kitchin and Thrift, 638–43. Amsterdam: Elsevier Science.

Walker, Brian, Crawford S. Holling, Stephen R. Carpenter, and Ann Kinzig. 2004. "Resilience, Adaptability and Transformability in Social–Ecological Systems." *Ecology and Society* 9 (2): 5.

Wallerstein, Immanuel. 1974. "The Rise and Future Demise of the World Capitalist System: Concepts for Comparative Analysis." *Comparative Studies in Society and History* 16 (4): 387–415.

———. 2009. "Protection Networks and Commodity Chains in the Capitalist World-Economy." In *Frontiers of Commodity Chain Research*, edited by Bair, 83–90. Stanford: Stanford University Press.

Westley, Frances, and Philip S. Miller. 2013. *Experiments in Consilience: Integrating Social and Scientific Responses to Save Endangered Species*. Washington, DC: Island Press.

FAO, IFAD, UNICEF, WFP and WHO. 2021. *The State of Food Security and Nutrition in the World 2021: Transforming Food Systems for Food Security, Improved Nutrition and Affordable Healthy Diets for All*. Rome: FAO.

Widgren, Mats. 2016. "Four Myths in Global Agrarian History." In *Methods in World History: A Critical Approach*, edited by Jarrick, Myrdal, and Wallenberg Bondesson 85–106. Lund: Nordic Academic Press.

Wilkinson, John, and David Goodman. 2018. "Food Regime Analysis: A Reassessment." In *Ecology, Capitalism and the New Agricultural Economy*, edited by Gilles Allaire and Benoit Daviron, 142–162. London: Routledge London: Routledge.

Williams, James W. 2014. "Feeding Finance: A Critical Account of the Shifting Relationships between Finance, Food and Farming." *Economy and Society* 43 (3): 401–31.

2 The first soybean cycle (domestication to 900 CE)

The deep history of the soybean goes back several thousand years. We can still aspire to reconstruct some fragments – thanks to new archaeological discoveries and techniques – but most of it remains, as most human prehistory, irreparably lost to us. The earliest Chinese written sources on soybeans, as we shall see, only date back to the 9th century BCE, and they survived only because they were fortunately recovered and canonized during the Han Dynasty's (206 BCE–220 CE) adoption of Confucianist philosophy as its official doctrine.

The subject of the present chapter is the first soybean cycle, including its roots, regime and rupture. We will begin our study with a critical survey of the extant sources. This is followed by an exploration of the roots, where we map what we know about the earliest ancient written history of the soybean in North China. Then, we examine the different roles of soy during the first "soy regime" under the Han Dynasty. Finally, we cover the centuries-long process of regime disintegration and rupture which takes us almost to the 9th century CE. This chapter closes with a reflection on the shifts and continuities in the roles of the soybean under this first cycle.

A critical note on sources

Most of what we know about the roles of the soybean and the agrofood systems in which it was integrated during ancient times comes from a rich cache of early Chinese written historical accounts (Bray 1984, xxiv). These texts map a written tradition which appears unbroken since the early Zhou Dynasty (1046 BCE–246 BCE). Such a tradition testifies to a high degree of sophistication, enabling an advanced level of systematized knowledge, certainly unique by world historical standards and far superior to the ancient agricultural texts from the Indo-European world (Ibid, 52–53). Something similar can be said of agricultural treatises, which, as Myrdal argues, were read by few but likely reflect the high degree of sophistication of *de facto* agricultural traditions and practices (2020). These early works indicate that the soybean had been well known and cultivated since time immemorial along the Yellow River. But are these comparisons adequate? It needs to be remembered that all original manuscripts from the Chinese Bronze Age were

DOI: 10.4324/9780367822866-2

lost, and what we have access to are long, reprinted fragments and extensive quotations from early works which were often included and commented on in later texts (Bray 1984, xxiv).

One prominent example is the seminal *Shijing* or *Shih Ching* [*Book of Poetry*, sometimes referred to as the *Book of Odes*, or *Classic of Poetry*, or *Book of Songs*] from the 9th century BCE. The *Shijing* is a collection of over 300 poems of different genres, ranging from folk songs collected by Zhou Dynasty officers, to songs relating the life of the nobility, as well as poems in praise of the rulers and their life and hymns written for religious ceremonies (Giles 1901; Dobson 1964; Frankel 1978; Buckley 1993). Following an edict issued by the "tyrannical" emperor Shih Huang Ti in 213 BCE, the *Shijing* was destroyed (burned or buried) along with many other texts authored by Confucian scholars (Keng 1974, 391). Fortunately, however, all 305 poems of the book were recovered only decades later, most likely thanks to the oral tradition of scholars rather than by the actual salvaging of inscriptions upon tablets and silks (Ibid, 395). From this point on, the *Shijing* became one of the classics – as part of the Han Dynasty's official adoption of Confucianism as the guiding principle of Chinese society. In fact, many early texts on poetry, philosophy and agriculture were selected to become classics in this way by the bureaucratic state that emerged under the Ch'in (221 BCE–206 BCE) and the subsequent Han empire and became canonized through official education. The poems also constitute a small archive of crucially important historiographical sources (Hymowitz 1970, 415–17; Huang 2008, 45).

According to the Chinese history scholar Michael Loewe, the *classics* were closely associated with Confucius' teachings. In them, the "holy scripture" served as a source of ideological authority for the exercise of temporal rule. This narrative was upheld by loyal and educated statesmen, and it constituted the fundamental principles of governance for millennia to come. While these seminal works and classics provided important historical clues, it is also worth remembering that they represent the world through the eyes of the Han-Chinese officials and their particular values, interests and norms.

Besides canonizing earlier texts, loyal Han Dynasty officials were assigned to write huge state-commissioned compilations on both farming and "official history". These works drew extensively on earlier texts, incorporating long quotations from many far older manuscripts, but they simultaneously collected and wrote down songs, oral myths and traditions. Typically, these compilations also included chapters dealing with the present, built on contemporary observation. The most well known is the monumental historical text *Shiji* [*Records of the Grand Historian*] by Sima Qian from 90 BCE, which sets a standard for later government-sponsored historiography. Crucially, as we shall see, it contains important information on early soy trade. Another source is the well-known agricultural treatise *Fan Shengzhi shu* (or *Fan Sheng_shich Shu*) [*Fan Shengzhi's Manual*], written by the Han Dynasty official Fan Sheng (or Fan Shengzhi) during the 1st century BCE. This constitutes a central source for our analysis of the roles of soy in the ancient Chinese farming system.

Fan Shengzhi shu divides its coverage more or less equally between agriculture in much earlier times and detailed observation and information about contemporary (1st century BCE) farming in North China. One chapter, for instance, includes an account of the legendary prehistoric emperor Shen Nung (or Shennong), known as "the Divine Farmer". Shen Nung is described as the "Father of Agriculture", said to have lived around 2700 BCE and credited with having classified plants according to both their nutritional and medicinal value, placing soybeans among the five principal and sacred crops together with barley, wheat, millet and rice (Kiple and Ornelas 2000; Kiple 2007, 45; Shurtleff and Aoyagi 2021, 38–39, 74–78). While the myths of emperor Shen Nung are often cited in the soybean literature as factual, more recent research has cast doubt on the actual existence of this emperor, as well as on the perception of these five crops as sacred at that time (Cowan, Watson, and Benco 1992, 4; Kiple 2007, 45; Peruchi Moretto, Nodari, and Nodari 2022, 22). This clearly illustrates the problems involved when using a written text from one period as a reliable guide to a much earlier one. The temporal distances are dizzying: the Han Dynasty official Fan Shengzhi wrote the extensive agricultural treatise more than 2,000 years ago, incorporating stories and events that allegedly occurred more than 2,000 years before his own birth. Beyond these half-historical and half-mythical accounts, however, the *Fan Shengzhi shu* provides extensive information on farming as it was actually practiced in the 1st century BCE in the Guan Zhong region (an arid district of the Middle Yellow River). Fan Shengzhi included his own detailed field observations from among farmers in the area. Reading the voluminous treatise, it becomes clear that Fan Shengzhi aims for a high-yielding agricultural system, centered on many different crops and with a wide variety of seeds, where attention is given to the specific conditions of each area. Such explicit intentions reflect the policy of a state that thought of agriculture as the root of prosperity and saw history as a source of eternal good, wisdom and virtue. At the same time, of course, it is impossible to say with any real certainty whether the farming practices and techniques described in this text were representative, or even common, among the Han farmers in the area. The uncertainty is compounded by the complete absence of any description of non-Han Chinese (a feature which *Fan Shengzhi shu* shares with most other records from the Han Dynasty).

Quantitative information is provided only occasionally in the writings from this period, and there is a preponderance of information about political matters in the urban centers and comparatively little about events in the provinces (Loewe 1986, 5). In addition, like many of the classical works from the Han era, *Fan Shengzhi shu* was eventually lost during the 11th century CE, under the Song Dynasty. The detailed agricultural calendar *Simin yueling* (sometimes spelled *Ssu Min Yüeh Ling*) by Cui Shi (or Ts'ui Shih), which dates to 160 CE, faced the same fate under the Song Dynasty (Hsu 1980, 216–18). What we know from them thus comes in the form of preserved excerpts, fragments and commentaries in a wide range of other texts. The fact that the original manuscripts are lost adds yet another layer of distance and potential distortion of the historical records.

Despite all these limitations, which are not uncommon in the literature of early agrarian history, *Fan Shengzhi shu*, *Simin yueling* and other works from this time have a great deal to say about the uses and roles of the soybean.

The canonized historiography *Hou Hanshu* [*History of the Later Han*] compiled by Fan Ye (398–446 CE) is another important work that includes many stories and information on the different roles of soy in ancient times. *Hou Hanshu* was written in the 5th century, but most of the book talks of much earlier periods and draws extensively on earlier texts, such as the previously mentioned *Shiji* from 90 BCE. Fan Ye worked as a state official, but after a ruffle with the emperor's brother, his political career was brought to a sudden end in 432 CE; he compiled *Hou Hanshu* in exile (Shurtleff and Aoyagi 2021, 70–71). Fan Ye's compilation techniques are understood to involve, essentially, a process of rewriting which relies heavily on previously compiled historical works (L'Haridon and van Ess 2019), they have also been found to include some archaisms (Loewe 1986, 4).

One of the most important extant sources into lost early texts on Chinese agriculture is the famous *Qimin Yaoshu* (or *Chhi Min Yao Shu)* [*Essential Techniques/ Arts for the Common People*], completed by Jia Sixtie during the *Six Dynasties* period, somewhere between 533 and 544 CE. Many of the texts dealing with agriculture from the Han period – including the previously mentioned *Fan Shengzhi shu* and *Shiiji* – have survived thanks to long excerpts in this book, which contains a bibliography including many other specialist agricultural works, none of which have been preserved (Huang 2000, 123–29). *Qimin yaoshu* includes 92 chapters, divided into ten books. As Francesca Bray has put it, the "quotations in the *Qimin yaoshu* are our main or only source for several of these works" (Bray 2019, 359).

Apart from a unique compilation of earlier records, *Qimin yaoshu* provides important information about cultivation techniques, crop plants, animal husbandry and food processing in the Northern China of its time. Half of its content consists of original texts, while the other half is made up of quotations drawn from around 160 works that span the seven centuries preceding *Qimin yaoshu*'s composition (Bray 2019, 356). In his Preface, Jia Sixie states that the material in his book comes from four sources: (1) the Chinese classics; (2) contemporary books, proverbs and folk songs; (3) information gathered from experts and (4) his personal experience, or "original material" (Shurtleff and Aoyagi 2012, 23). Jia Sixtie was an estate owner, farmer and former official of the Northern Wei government and his extensive and erudite treatise covers subjects as diverse as bringing new lands under the plough, rotating crops, grafting fruit trees, using safflower to tint rouge, treating scabies in sheep and making different kinds of dishes and drinks. Very early on, the work became:

> [A]n obligatory point of reference for anyone writing on agriculture or attempting to raise local farming standards, whether in a private or an official capacity. As such, the *Qimin yaoshu* has come down to us almost intact, unlike all the earlier agricultural treatises to which it refers.
>
> (Bray 2019, 355)

Francesca Bray points out that the *Qimin yaoshu* set the pattern for later agronomic treatises and shaped the conceptualization of many farming issues (Ibid, 355–57). As for its technical advice, many later agronomists writing on farming in North China would claim that the sophistication and efficacy of the techniques recommended in the *Qimin yaoshu* had never been surpassed (Huang 2000, 123–29).

While invaluable on the different uses of soybeans, the same caveats apply to the *Qimin yaoshou* as to the sources already discussed above. We do not know whether the information about farming provided by Jia Sixtie is representative for the whole farming system in the North, but most scholars argue that the *Qimin youshou* was probably written with big commercial estates – such as the one owned by Jia Sixtie – in mind (Bray 2019, 368–69). The previously mentioned agricultural calendar *Simin yueling* was, in a similar way, written for a landowning class engaged in commercial farming activities – based on the work of their laborers, tenants or slaves – where soybeans were a cash crop. The practices of soy farming from these treaties may in fact differ substantially from the practices of small-scale peasants – the great majority of the population. In addition, it is impossible to say for sure whether the authors of these great works could have particular interests in distorting the reality about farming in some way or another.

Generally, no written sources are free of bias or isolated from the perspective of an author active at a specific time and place. It seems that all state-commissioned writings from the Han Dynasty onwards were submerged in Confucian ethical virtues. According to Loewe, this includes a notion that "the members of the community are seen as bound together in the service of their ruler, each one acting according to his own capacity and social station", thus reflecting the ideals of "a sophisticated, rank-conscious society, whose hierarchies rested on the distinctions laid down in Confucian lore" (Loewe 1986, 16). Social distinctions also responded to the needs of the imperial government to fill the ranks of an expanding civil service and to make membership of that service a matter of pride. In this way, together with the growth of the organs of administration, a professional class of officials also appeared, neatly differentiated by grade and salary.

Our access to the passages on soybeans in the classical sources has depended heavily on the *Soyinfo Database*, compiled by William Shurtleff and Akiko Aoyagi.[1] The *SoyInfo Center*, which, in turn, makes heavy use of the 1958 book *Dou-lei, or Doulei Shangpian [Varieties of Beans or Selected Sources on Legumes]* published by the Chinese Agricultural Series and compiled by CN Li (sometimes referred to as Li Cangnian, Li Ch'ang-Nien or Li Zhangian). This book contains reprinted texts from 69 original books about legumes, particularly soybeans, published in China from 1100 BCE to 1958. *Dou-lei* has never been translated into English, but the *SoyInfo Center* has translated the passages on soybeans, often with the help of soy-history scholar HT Huang. While many of the excerpts about soybeans from original books that we use in this chapter are translated passages from the CN Li compilation, we only refer to the original works and the *SoyInfo* bibliographies wherever the passages on soy are reprinted.

Besides the source problems we share with all historians working with early Chinese history, we face an additional challenge in that we have prioritized texts that explicitly mention the soybean, a natural consequence of our wide use of the *SoyInfo Database*. This approach to sources runs the risk of giving soybeans too central a role. However, we have tried to mitigate this risk by moving beyond the soybean-related excerpts. For each important "soybean source", we have looked into other texts which mention the same source. In this way, we have also drawn extensively on the rich field of Chinese agrarian history, which allows us to gauge the weight of the soybean more correctly. Thus, the early (re)written records – found through the *SoyInfo Database* – have been meticulously combined with secondary sources. Fortunately, the agrarian history of China is well covered in the literature thanks to the abundance of original texts and excellent scholarship. For the historical period we are concerned with here, the seminal researchers Joseph Needham, HT Huang and Francesca Bray have published impressive works based on ancient records on the history of agriculture, science and civilization in China. Many of these studies have moved beyond the focus on written sources of traditional history scholars and additionally build on archeological and forensic evidence.

Roots: a long prologue (origins–200 BCE)

The earliest history of the soybean remains shrouded in mist, but we know that humans probably domesticated the wild soybean (*Glycine soja*), which was naturally distributed throughout East Asia, through trial and error as early as 9,000 years ago. While it is likely that domestication occurred more or less simultaneously in different places across East Asia, whole-genome sequencing, pollen profiles and new archeological discoveries suggest that the earliest source of domestication of today's soybean (*Glycine max*) occurred in North China along the banks of the Yellow River (Peruchi Moretto, Nodari, and Nodari 2022, 22). Radiocarbon dating of crop remains from sediment sampling (floating) in North China has confirmed that soybeans existed in the area from at least 7000 BCE (Lee et al. 2007; Zhou et al. 2017). Moreover, soybean remains have been found at more than 30 sites in China, spread throughout a long period ranging from the Peiligang culture (7000 BCE) to the Han Dynasty (Liu and Chen 2012, 87–88). The earliest remains identified as soybeans in North China have been uncovered in Jiahu (7000–5500 BCE) and Bancun (ca. 5500 BCE), both in Henan province, and from Yuezhuang in Shandong province (ca. 6000 BCE). In Jiahu, located in the central plain near the Yellow River, as much as 14% of the total seed remains found at this Neolithic site were soybeans, indicating that they were commonly cultivated plants collected for food. Further, it is clear that the area of this plant's early dispersal during the Neolithic period was the entire middle and lower Yellow River region (Ibid, 87).

Despite many scientific advances, a number of controversies remain. Some researchers, for instance, claim that multiple and divergently successful domestication efforts took place across East Asia (Peruchi Moretto, Nodari, and Nodari 2022, 20–22). They have likewise noted that recent progress in whole-genome sequencing and archeological techniques seems to suggest that soybeans existed in places like Japan, Korea and South China in the beginning of the Bronze Age (Jeong et al. 2019), where they might have been used as a companion legume to rice (Ponting 2007). Meanwhile, others have argued that there was a "single origin" in China, but that cultivating techniques and soybean varieties soon spread into Japan and the Korean peninsula (Kiple 2007, 48–49). Relatedly, some argue that domestication first occurred in the Northeast of current day China, while others point to the region between the Yellow River and Huai River to the Yangtze basin in Southern China. Still others suggest that the domestication of soy happened in Japan (Wang, Li, and Liu 2012). In fact, it is difficult to establish the exact time(s) and place(s) for domestication with any confidence, because there is no simple way of distinguishing the wild from the domestic forms of soybeans – domesticated plants tend to be larger than their wild relatives, but this process is not always straightforward. In addition, the earliest domesticates played rather minor roles for several millennia; it took some time before they became the dominant staples or sources of protein in the subsistence system during the middle Neolithic (ca. 5000–3000 BCE) and even later periods (Liu and Chen 2012, 91). While uncertainty remains, the Yellow River basin in North China is a strong candidate as the world's original domestication sight. If it was not the first, it is in any case an area where the earliest known large-size soybeans date back to 7000 BCE, suggesting that they have been grown there for no less than nine millennia. In addition, early agrarian civilization in North China produced a strong tradition of writing, which is still with us today, passed down through reprinted fragments from early imperial times. Very little recorded history can inform us about the soybean's different roles and functions in other areas at this time.

The soybean was not the first domesticated plant in North China. Hunter-gatherers were already farming broomcorn millet (*Panicum miliaceum*) and foxtail millet (*Setaria italica*) at small scales there. This happened possibly 1,000 years before the domestication of soybeans, leading to the initial appearance of domesticated millet. Over time, a variety of other domesticated crops and animals appeared. It is thought that rice, millet, soybeans, pigs, dogs and perhaps chickens were domesticated indigenously; whereas wheat, barley, sheep, goats and horses were introduced, already domesticated, from elsewhere (Ibid). Soybeans were probably not highly valued crops in their early stages, and were likely only eaten if necessary (Lander and DuBois 2022, 30). Around 2000 BCE, however, farmers began cultivating soybeans that were larger and with higher oil content – the result of a long process of selection by farmers (Ibid, 32). Studies using stable isotope analysis of human and animal bone show that millet continued to be the central crop for millennia and that it heavily dominated consumption along the Yellow River around 1,000 years before the Common Era (Zhou et al. 2017).

Agriculture and human settlements inaugurated a new set of relations between humans and nature including a whole range of (mostly unintended) consequences. Agriculture inherently disrupts processes of natural succession and biodiversity. As peasants began clearing land, removing all the interdependent species and replacing them with one or a handful of new ones, a dramatic process of *ecological simplification* began (Friedman 2000; Kiple 2007, 64). But where the annuals were planted, the banished species tried their best to return, and the bare earth encouraged the rise of perennial weeds. Peasants had to put an impressive amount of labor into weeding, as well as chasing away birds and other animals that otherwise would eat the crops (Friedman 2000). While ecosystems became simplified, social organization became more complex. Notwithstanding the long and hard work demanded by agriculture, it also created a food surplus which allowed for a division of labor and concentration of power, which in China gave rise to a highly stratified early agricultural civilization (Kiple 2007, 62). In addition, from around 3000 BCE, the climate in the Yellow River valley became cooler and drier, which made it possible for previously flooded zones to be settled and exploited and also probably led to an intensification of agricultural practices. According to archeological findings on the plains of the Yellow River, by far the most important crop at this time was millet, but it was integrated in a complex, multi-crop system, which at this time also included rice, wheat, soybeans and hemp (Zhao 2011; Hu 2018).

Between 3100 BCE and 2300 BCE, the soybean started to spread over a larger territory – first to the South of China and then to Korea and Japan (Peruchi Moretto, Nodari, and Nodari 2022, 23). However, there is no place where the soybean has left as many traces in the historical record as in North China. The main reason is that from the times of the Zhou Dynasty an increasing number of written accounts started to emerge. In them, one of the oldest written signs is "crop", which derives from a character depicting millet, illustrating that millet was the chief staple. Four bronze inscriptions from the early Zhou Dynasty include the ancient character *shu*, technically meaning legume, but generally accepted to be the first written record referring to soy (Hymowitz 1970, 415–17; Bray 1984, 511–12). Soon a burgeoning written tradition of agricultural treatises, poems, biographies, philosophical dissertations and historical commentaries appeared, of which several mentioned the soybean.

No original texts have survived from the Zhou Dynasty, but we still know quite a lot about them through reprinted fragments. One prominent example is the abovementioned collection of poems *Shijing*, from around the 9th century BCE. The geographical area covered in *Shijing* spans over the lower Yellow River valley in the North and also today's Central China, where the soybean is described as one crop among many integrated in an intensive cropping system (Huang 2000). The book mentions over 130 different kinds of plant names and crops, among them the soybean (Keng 1974, 395). By contrast, only 12 poems mention domestic animals (Myrdal forthcoming). This may reflect that while this early agrarian system also had domesticated animals, animal protein was seldom part of ordinary people's diet, and there seems not to have been any consumption of dairy products among the Han-Chinese people (Zhou et al. 2017; Myrdal forthcoming).

One of the book's most interesting poems mentions the boiling of the beans, while another refers to the season when soybean leaves are eaten (Shurtleff and Aoyagi 2021, 35–37). The book uses two different verbs, *cai* and *huo*, in reference to the collecting of soybeans. It has been suggested that *cai* indicates gathering wild soybeans while *huo* instead refers to harvesting domesticated varieties of soy (Liu and Chen 2012, 86). This could indicate that wild soybean varieties were still common at this time. But it could also be that the different ways of collecting soybeans refer to earlier times, since *Shijing* also tells stories that refer to many years before. For example, one story mentioning soybeans provides something of a history of the Xia Dynasty (ca. 2100–1600 BCE) (Lee et al. 2007).

Other reprinted fragments from early ancient Chinese history corroborate that soybeans were quite common in the area. One illustration comes from *Fanzi jiran* (or *Fan Tzu Chi Jan* or *Chi Ni Tzu*) [*The Book of Master Chi Ni*], from about the 3rd century BCE:

> The five grains (*wugu*) are the root of the ten thousand people, the treasure of the realm. The grains that thrive in the east are wheat or barley (*mai*) and rice (*dao*), in the west hempseed (*ma*), in the north soybeans (*shu*), and in the center millet (*ho*).
>
> (Shurtleff and Aoyagi 2021, 45)

Still another important Zhou text that mentions the soybean is *Chunqiu Zuozhuan* [*Master Zuo's Commentary on the Spring and Autumn Annals*], from around 360 BCE. The *Zuozhuan* is a detailed commentary on the *Chunqiu* – the first Chinese chronological history whose object is the Spring and Autumn period (770–476 BCE), but while it provides a different account of the same events as the *Chunqiu*, it contains more extensive narrative accounts and background materials. One story involves a person who could not distinguish between soybeans (*shu*) and wheat (*mai*) during the rule of Chenggong (around 573 BCE). Another story relates how the soybeans were killed off by a heavy frost during the rule of Dinggong (509 BCE) (Shurtleff and Aoyagi 2021, 38–39). These early texts bear witness to the fact that soybeans were well known and important.

During the late Zhou Dynasty, society started to disintegrate in the wake of intra- and interstate struggles. The power of the old aristocracy collapsed and gave way to a territorial monarchical state (the short-lived Ch'in Dynasty) and, with it, the official recognition of private land ownership and the appearance of independent small farmers. The Ch'in state also made systematic use of its public land to attract new settlers (Hsu 1980, 10–13). This process was bolstered by the newly acquired ability to mine and smelt iron, as well as to produce bronze and to cast new farming tools and weapons. The use of fire, moreover, was much further advanced in China than in Europe, which explains the early use of iron in tools exposed to heavy wear, such as plows or harrows. Compounded, these developments allowed for the expansion of the agricultural frontier (Roberts 2004,

141–45). Iron tools also increased agricultural productivity, creating a surplus which could sustain the expanding ruling class, military troops and skilled craftsmen, and allowed for an increasing division of labor. The rise in agricultural production also led to a sharp population increase (Roberts 2004, 141–45; Zhang and Liu 2014).

Chinese civilization and agriculture (already relatively sophisticated and productive under the Zhou Dynasty) continued developing along the Yellow River in the North. The years between the 5th and 3rd century BCE were tumultuous, and it was not until after the unification of China that a more developed agrarian civilization emerged. While the hardy and drought-resistant millet remained the most important staple, soybeans also played several important roles in the agrofood regime which developed under the unified Han Dynasty and to which we now turn.

The first soy regime (200 BCE–200 CE)

According to archeological evidence, the Han Dynasty (206 BCE–220 CE) witnessed the greatest geographical expansion of soy in China (Liu and Chen 2012, 90). It also shows that it was mainly larger varieties of domesticated soybean – originating in the North – that dispersed throughout North, Central and South China. The diffusion was facilitated by the expansion of the Han Dynasty's administrative and economic power and by relatively significant demographic growth in the South due to migration (Liu and Chen, 2012, 91). During this period, soybeans exist wherever Han-Chinese people settled: still, this thematically organized section will focus on the roles and functions of the soybean in the North, where most of our surviving records originate.

While the historical records from the Han Dynasty are fragmented, it is still possible to see that from the vantage point of the soybean, the systems of farming, trading and food cuisine were sufficiently stable to constitute a proper agrofood regime in North China during this entire period.

Soybeans in cultivation: an integral part of intensive North-Chinese agriculture

> Calculate the acreage to be covered by soybeans for members of the whole family according to the rate of 5 mou (3,045 square yards or 0.65 acres; 1 mou = 609 square yards or 0.13 acres) per capita. This should be viewed as basic for farming. In the third month, when elm-trees are fruiting, plant soybeans on upland fields whenever it rains. Use 5 sheng (3.5 cups) of seeds per mou when the soil is mellow and not cloddy, but more seeds if the soil is not so. As late as twenty days after summer solstice soybeans may still be planted. [...] Soybean seedlings break ground as though with a helmet on top, so it is not necessary to plough very deep. Don't cover the seed with too much soil after planting. Too thick a cover renders the bean-stalk bent-necked; With proper care, the yield [of soybeans] from a good field may attain

> 10 shi per mou [equivalent to 2.7 metric tons/hectare] from poor land up to 5 shi (1 shi = 16.7 liters). To cultivate soybeans in shallow pits: Dig pits 10.5 inches or 26.4 cm (1 chi and 2 cun) apart. One ditch thus holds 9 plants, and one mou (609 square yards or 0.13 acres) contains 6,480 plants. 7.4.1 To plant soybeans in shallow pits: Make pits 5 inches or 11 cm (6 cun) across and deep, and 17 inches or 44 cm (2 chi) apart. Thus, one mou holds 1,680 pits. After making a pit, take 0.7 cup (1 sheng) of well ripened manure, mix well with earth from the pit, then return the mixture to the pit. Just before planting, pour 2.1 cups (3 sheng) of water into every pit, then place 3 beans therein. Cover with earth, but do not make the layer too thick. Press down with the palm of hand to ensure close contact of seeds with soil. Use 1.4 cups (2 sheng) of seeds and 74 gallons (16 shi and 8 tou) of manure per mou. Hoe when 5 or 6 leaves have appeared fully. Water if drought occurs; 3 sheng (2.1 cups) of water per pit. A full-grown man works 5 mou (0.65 acres). In the autumn, the yield may attain 16 shi per mou (= 67.7 bushels/acre) (= 67.7 bushels/acre).
>
> (Shurtleff and Aoyagi 2021, 55–56)[2]

The above text originally comes from the 1st century BCE agricultural treatise *Fan Shengzhi shu* and it is the first time the modern term for soybean (*ta tou)* appears. The author, Fan Shengzi, an agricultural development commissioner, based his observations mostly on the Guan Zhong region – an arid district at the midpoint of the Yellow River (Ibid, 55–56). His advice reflects deep knowledge of the specificities of the crop as well as about the roles of climate and soil conditions. Farmers were expected to not only follow simple universal guidelines – as no one solution fits all – but also to take into consideration the particular circumstances of each place for each crop. Fan Shengzi concludes that "with proper care" a good field can yield "10 *shi* per *mou*", or approximately 2.7 metric tons (MT) per hectare (MT/ha). Such expectations are perhaps a little on the optimistic side. Today, with all the technological innovations at our disposal, the five-year average yield of soybean production in China is 1.86 MT/ha. In Argentina, by contrast, the average yield of soybean is 3.89 MT/ha (USDA 2022; Langemeier 2021). Aside from considering the climate, soils and seeds, *Fan Shengzhi Shu* goes to great lengths to extol the importance of timing when planting soybeans and other crops (Shurtleff and Aoyagi 2021). In North China, where frost comes early and moisture evaporates rapidly in the spring, it was vital to time cultivation just right (Hsu 1980, 5–6). Similarly, the calendar *Simin yueling* – often translated to *Monthly Ordinances for the Four Classes of People* – written by Cui Shi in around 160 CE, includes detailed instructions on when to plant and harvest, when to prepare which food and medicine, when to take care of silkworms and weave silk, when to pray and offer sacrifice to ancestors and when to buy and when to sell what during the year (Hsu 1980, 216–18).[3] Likewise, the author of *Qimin Yaoshu*, Jia Sixtie, remarked on the importance of climate and soil conditions and to closely observe the plants in order to be able to harvest on time, as well as the specific requirements of different soybean varieties (Shih 1962, 44–45; Shurtleff and Aoyagi 2012, 24). In addition, the great landowner Jia Sixtie recommended reserving the best fields for millet, wheat and rice and using more marginal lands for soy (Lander and DuBois 2022, 33), clearly reflecting an early awareness that the soybean could provide a rather stable yield even in less than perfect soils. While Sixtie (among many) argued that the soybean was hardier and more robust than many other crops, he still included careful instructions for soybean cultivation:

> Use a wheat field as a base, and plant 3 sheng of seeds per mou. Broadcast the seeds and use a plow to form a narrow and shallow channel; level it plane. If the weather is dry, the stems will be coarse and sturdy, and the leaves sparse. If there are too few seeds, the seedlings will not grow tall; if the seeds are placed too deep, the seedlings will not be able to emerge from the soil. If the soil is too damp, first plow deeply, then broadcast the seeds away from the plowed furrows and level the soil. Do not do this if the soil is not too damp. In the 9th month, if you see leaves close to the ground turning yellow and about to fall, immediately harvest the crop. Even if the leaves do not turn yellow, they can easily start to rot. If you do not harvest, the wind will quickly strip the leaves, and the rain will rot the stem. The crop will be ruined.
>
> (Shurtleff and Aoyagi 2012, 23–24)

As illustrated in the sections about soybean production from *Fan Shengzhi shu*, *Simin yueling* and *Qimin Yaoshu*, Chinese farming ideals included knowledge and labor-intensive techniques. These were by no means exclusive to growing soybeans. For instance, Sixtie listed no less than 86 varieties of spiked millet and 18 of other kinds of millet, recording their specific qualities, such as resistance to drought and insect pests. *Simin yueling*, with its impressive number of crops and varieties, again provided far more space to millet and wheat than to soybeans (Myrdal forthcoming).

Besides crop-specific advice, these treatises also include generic advice, where the observance of time and weather and soil conditions was emphasized as a central pillar for successful farming. These methods were reiterated and developed in several other texts (Hsu 1980). One example is the 4th century CE *Guanzi*, or *Kuan Tzu* [*Book of Master Kuan* (or Master Guan)], about which Needham wrote: "[I]t must be one of the oldest writings on geo-botany in any civilization, bearing every evidence of compilation following actual surveys of territory, farmland, neighboring wilds, hill and mountain" (Needham, Gwei-djen, and Huang 1986, 6:48). *Guanzi* includes, for example, a categorization of arable land in relation to features such as the depth of the water table or altitude, and he additionally identifies 15 different types of hilly areas based on the same criteria. He also describes the plants and trees best suited for each type and gives a full list of trees and plants which one may expect to find at different elevations. It also gives an ecological gradient – where plants were arranged in order from those in lake water to those in dry areas. A long section also divides the soils of the Nine Provinces into three separate classes, according to their productivity, and each of these classes, in turn, is subdivided into six further types, with details on the expected agricultural yields and a wide selection of trees and plants which do best there (Needham, Gwei-djen, and Huang 1986, 6:48–51). In synthesis, the book argues that each plant and soil need to be carefully combined with regard to their respective requirements (Sterckx 2019, 283), and it provides a guide to manage for optimal agricultural results. This knowledge-intense and place-based way of farming resembles what today is referred to as "precision agriculture". It is indeed mind-boggling that this radically site-specific and adaptive management already existed in China 2,000 years ago - without today's digital technology.

The high level of sophistication and knowledge expressed in these works reflects the fact that Confucianism saw agriculture as the root of state prosperity and power. Nearly every emperor of the Han Dynasty proclaimed that farming and farmers were "the root of all under Heaven" (Sterckx 2019, 311). Philosophers and farmers were described as the brains and sweat of a moral and productive society, which ultimately rested on a directly taxable, self-sufficient and stable peasantry (Ibid, 312).[4] The state was involved in the development of agricultural production, investing in irrigation, infrastructure and agricultural extension, as well as in the compilations of the agricultural textbooks and encyclopedias we have discussed (Loewe 1986, 16–17). The state was also involved in frontier land expansion and in building up public granaries as famine relief. Moreover, Han officials wanted to sustain a large population to enable a large army. Agriculture proved mostly equal to this task and managed to sustain population growth, except for occasional famines or external shocks (Bray 1984, 52–53). Indeed, when the first official census was conducted, in the 1st century CE, it reported an astonishing total of 60 million people. Even if such a number seems implausibly high, it is clear that China was large and densely populated with the majority living in the North (Bray 1984, 7). A low estimate is that around 40 million people inhabited this area, which was roughly about the size of France (McNeill 2001, 38). But not everyone agrees: according to Loewe, for example, population statistics from the 2nd century CE were probably based on a real count and were therefore likely to be quite accurate, barring textual errors (1986, 5). In any case, there is consensus that the population grew considerably during the 1st century of Han rule, to at least twice its original size. It is more difficult, however, to know if improvements in the living conditions of ordinary people kept pace with the rates of demographic growth. As Loewe argues:

> It remains open to question to what extent the unified empires of Ch'in and Han maintained easier conditions of living or imposed harsher burdens on the population than the localized kingdoms of China that preceded or followed them. Nor can any answer be given to the question whether the enlarged and sophisticated civil administration of Han provided the people of China with a more secure and prosperous life, or made its principal impact as an instrument of oppression.
>
> (Ibid, 16)

However, as several surviving texts from the Han period indicate, a standard landholding of one typical Han farming household was a little under two hectares of land (Hsu 1980, 9, 16–17, 23).[5] In *The Book of Master Xun*, written by the Confucian philosopher Xun Kuang around 240 BCE, it was instead assumed that the standard landholding for one farming household was 100 *mu*, which would be approximately 6.6 ha (Loewe 1993, 178–88; Shurtleff and Aoyagi 2014, 46).[6] Perhaps Xun Kuang was wrong, or perhaps the average area declined steeply during the Han period. In any case, according to Hsu's estimates, the standard landholding made a 28.5% production surplus in a normal year, which was converted to cash for taxes as well as payment for clothing and other goods not produced on the farm (1980, 235). During the recurrent droughts or floods, however, this surplus all but disappeared, and there were recurrent famines. Several texts also mentioned that the farmers were often burdened with taxes (various land, *corvée* and conscript taxes), as well as miscellaneous exploitation by local authorities (Ibid, 23–24). While taxes on land were lowered under the period and some texts indicate that in difficult years farmers were given the option of paying duties with millet and beans, many farmers still ended up losing their land to richer people, becoming tenants on their former land and forced to pay 50% of their harvest for rent (Ibid, 16–17, 23). However, as Hsu remarks, rulers occasionally attempted to curb increasing land concentration by putting limits on farm size and the amount of slaves and at other times redistributing land (Hsu 1980, 22–28, 241). At the same time, after pressure from the aristocracy, the government abandoned the ancient land distribution system: the well-field system was removed around 11 CE. Many rulers, trained in Confucian texts, proposed land reform in attempts to return to the ancient well-field system; in these endeavors they sometimes made alliances with the peasantry against the landed elite (Bray 1984, 423). However, Loewe underscores that by the end of the Han, social distinctions that rested on wealth and landed property had been sharpened (Loewe 1986, 16). Nevertheless, compared to ancient Europe, it seems clear that land ownership in China was more evenly distributed, and farms worked by a single family were the norm for most of China's history (Bray 1984, 7–8).

Notwithstanding uncertainty, shifts and unevenness, it seems that the majority of the people were farmers and that they owned a small plot of land of around two hectares on average. This, in turn, was expected to feed eighth mouths – the couple, the husband's elderly parents and four children – and provide silk and hemp for domestic use.[7] While animals were not at the center of this agricultural system, most peasants still had some draught animals, and some pigs and chicken for meat and as a source of manure (Hsu 1980, 9; Bray 2018; 2019, 361–67). Some farmers also kept a few sheep for wool. These animals, while relatively few, still needed to be fed, and here the soybean fulfilled yet another function (Bray 2018; 2019, 361). Black soybeans, in particular, are mentioned in several texts as good animal feed (Bray 1984, 4), and, as we shall see in more detail soon, black soybeans were also considered a bulwark against famine. In this way, they served the dual purpose of providing good animal feed during the good years and famine relief during the bad (Shurtleff and Aoyagi 2021, 124).

However, despite the incorporation of new frontier land and the issuance of land grants, population growth led to a significant decrease in the average size of peasant land holdings (Isett and Miller 2016, 28–29). As the average peasant farm size was quite small already at the beginning of the Han Dynasty, and since it shrank significantly during the period, farmers were pressed to take out more harvest from less land. Accordingly, an increased amount of labor was deployed to prepare the land, and continuous manual weeding and soil preparation were carried out (Isett and Miller 2016, 30–31). These activities were facilitated by the popularization of iron tools (Hsu 1980, 4), in addition to techniques such as deep plowing and irrigation. While iron tools had been invented prior to the Han Dynasty, the increasingly powerful state took a much more proactive role in further developing and diffusing the technology (Hsu 1980, 4). In addition to putting more labor on the land, farmers gradually abandoned the tradition of allowing it to fallow as pressure for land-use intensification rose (Bray 1984, 7–8; Isett and Miller 2016, 28–29). However, while all agricultural activities reduce the nutrient supply of the soils, nutrients – especially nitrogen and phosphorus – are depleted much more rapidly if land is used intensively and if the soil is not allowed to rest (McNeill 2001, 22–23). This, in turn, limits plant growth, which was readily observed and discussed in the sources. Shih, for instance, pointed out that if a field gave poor harvests it could be fallowed for one year, while in yet another text, he argued that medium or inferior land had to be fallowed every other year (Shih 1959, 8; 1962, 44). The pressure on the land was intensifying, however, with more and more people needing food. Fallowing became a last resort (Bray 1984, 7, 162, 429–30). Thus, in order to sustain the high productivity, compensatory techniques had to be adopted – and the soybean played an important role in this respect.

One such compensatory technique was the use of a large variety of crops, including legumes, in diversified crop rotation systems (Myrdal forthcoming). While there were several legumes around, as well as several varieties of soybean, yellow soybeans seem to have been the most important in China, though also "lesser beans" (as they are labeled in the treatises) were common. Fan Shengzh's agricultural treatise was particularly concerned with the benefits of cultivating these legumes in the quest to restore field fertility (Hsu 1980, 12–13). As he argued, soybeans could be rotated with winter wheat and millet in order to improve the soil and increase output, getting a third harvest over a period of two years (Bray 1984, 432–33; Isett and Miller 2016, 30–31). Other alternative schemes were longer and more complex – as much as five years – and included millet, hemp, soybeans, mung beans, wheat and sesame (Bray 1984, 56–59, 431).

Soybean plants have a symbiotic relationship with a specific bacterium that fix atmospheric nitrogen in the soil in a form that roots can assimilate (Lander and DuBois 2022, 30). While this nitrogen-fixing capacity was not properly understood until much later, the Chinese knew from experience that legumes were good for the soil, even when it was being used more intensively (Hsu 1980, 12, 85; Lander and DuBois 2022, 32). The relation between land rotation and the preservation of soil fertility would be further explored and systematized in Jia Sixtie's *Qimin Yaoshu*. As he remarked, the soybean could be planted ahead of cereals – as a "forerunning crop" – in order to maintain the land; moreover, soybeans, he argued, were ideal in rotation with millet, since they could be sown at any time between early spring and early summer (Shih 1962, 44–45; Shurtleff and Aoyagi 2012, 24). Jia Sixtie also pointed out that if no rotations were made for a long time, the soil would become so exhausted as to need a fallow period (Bray 2019, 363). Crop rotations with soybeans and other legumes thus made up for some of the shortcomings of intensive continuous cropping and seem to have been an important driver behind the expansion of soybean cultivation (Hsu 1980, 12–13).

Besides rotations, the utilization of fertilizers was also important. Animal manure and human "night soil" had been used for a long time in China, possibly as early as the Shang Dynasty (1600 BCE–1046 BCE). However, with the intensification of farming, complementary fertilizers became necessary. Moreover, before the Han Dynasty, cattle rearing had been relatively common, but as land became more scarce, pastoral land use was considered too inefficient (Bray 1984, 3–5). What this meant, then, was that manure from animals probably became less abundant as animals disappeared from the farm. Moreover, the Chinese plow was light and could easily be managed by a single beast, so there were few draught animals (Hsu 1980, 7; Myrdal forthcoming). Thus, with the decrease of animal husbandry, green manure became increasingly important as a fertilizer. Several passages of *Qimin Yaoshu* described how "organic fertilizers" – mostly different kinds of beans – were used as green manure, probably by plowing the entire plant under (Bray 2019). In this way, soybeans could improve the quality of the soils not only as a nitrogen-fixing rotation crop. While mung beans were described as the best in enriching the soils, soybean plants were also mentioned as suitable, albeit less desirable (Loewe 1986, 16–17; Zhou et al. 2017; Bray 2019). In short, in order to maintain high productivity levels in intensively farmed soils, farmers adopted practices of crop rotations with legumes, fertilizing and continuous weeding. Soybeans clearly played a significant role in this system.

Notwithstanding the fact that the state-commissioned book *Fan Shengzhi shu* and the other official texts on agriculture might tend to exaggerate the level of sophistication of Chinese agriculture, it seems clear that a high-yielding and labor-intensive, crop-centered system had emerged along the Yellow River during the Han Dynasty. This had not been the simple result of a particularly benign climate, and neither was it the product of remarkable soil conditions. Quite the contrary, the plains of the Yellow River suffered from erosive soils and high rain variability which caused droughts and floods, as well as salinization of the soil. The driest zones along the Yellow River were and still are in its upper reaches, where farmers resorted to irrigation (Bray 1984, 109). These canal systems tended to silt up, a problem exacerbated by deforestation. At the same time, the river emptied further down onto a flat plain which made irrigation difficult. To capture and retain the water from the monsoon season, in late summer, village wells became the norm; but irrigation accelerated salinization of the soil. In the lowest reaches of the river system by contrast, drainage was the main problem (Ibid, 101; 121). Rather than optimal biophysical conditions, then, three factors tend to do a better job at explaining how the Han Dynasty could sustain such a high rise in population: first, very high levels of knowledge and skill; second, the process of state-led frontier expansion and finally, a marked increase in peasant (tenant and slave) labor. As Myrdal has observed, on a regional level, agrarian technology is clustered into technological complexes (2022, 16). Soybeans under the Han Dynasty integrated such a technological complex characterized by crop-centrism, high variety of crops, complex rotations (and sometimes intercropping), fertilization and iron tools. This technological complex, in turn, rested on a deep understanding of soils and the optimal combination of crops and varieties in different environments and conditions.

Soybeans in trade: relatively predictable and cheap

For ancient Chinese thinkers (as for their much later Physiocratic counterparts, perhaps), the idea of economic value outside of agriculture was inconceivable. Accordingly, while agricultural production was seen as life sustaining, trade was seen in a comparatively unfavorable light, essentially looked upon as mere speculation with the single aim of making profits at all costs (Sterckx 2019, 312). Grain was understood to stand in an opposite relation to gold, and, in the official discourse, peasants ranked second only to the "gentleman-scholar" who served the monarch. The peasant stood above the artisan and the merchant (Isett and Miller 2016, 26). This contrasted with the idealized rural life expressed in Greco-Roman literature; Chinese texts stressed that the emperor himself would plow millet and that the empress would tend the silkworms (Ibid). The emperor oftentimes struck alliances with peasants in his attempts to diminish grandee power and oppose hereditary rank and wealth. While Confucian ideals were not too concerned with economic matters and tended to concentrate on more noble pursuits, such as warfare, sacrifice or farming activities (Milburn 2007, 37), many ancient texts nevertheless mention trade and commerce in ways that indicate that they were still important and prevalent: this included the trade of soybeans.[8]

> The first type of foodstuffs was fine millet, and was the highest class of goods, and cost seventy cash; the second type of foodstuffs was panic millet, and was the middle class of goods and cost sixty cash a picul; the third type of goods was red beans, which was the lowest class of goods and cost fifty cash a picul. The fourth type of foodstuffs was rice, and he ordered that this be the upper class of seed, and a picul cost forty cash; the fifth type of foodstuffs was barley, and it was the middle class of seed, and cost thirty cash per picul, and the sixth type of foodstuffs was soya beans, and was the lowest class of seed, and cost twenty cash a picul. The seventh type of foodstuffs was kuang millet, which were classed as food that grew wild, and therefore had no price, and the eighth type of foodstuffs was fruit, which also grew wild, and had no price. There were no ninth or tenth types.
>
> (Milburn 2007, 23)

The above quote comes from *Jinizi [Book of the Young Master of Accountancy]*, written during the last years of the Warring States period (280 BCE–221 BCE).

It offers an early reminder of the wide extent to which soybeans were consumed and traded in North China, as well as their relative cheapness compared to other crops. These two characteristics (widespread availability and relatively low value) are, as it happens, key to understanding the roles of soybeans in the commercial system which began to emerge along the banks of the Yellow River *(Huang He)* approximately 2,000 years ago. This book also contrasts to the Han rulers' apparent disregard for economic activity. The original book was lost, but long fragments were subsequently circulated across the Chinese world. According to the previously mentioned *Shiji* from 90 BCE, *Jinizi* was written by an individual named Ji Ran, although in other texts it has been argued that Ji Ran was a fictional figure (Milburn 2007, 24–26, 38). At any rate, its basic structure is straightforward: the sage Jini advises the King of Yue on a range of moral and economic issues. Jini, for instance, defends state involvement in the nation's economic affairs, with the aim of increasing wealth and, in turn, achieving a better position to undertake prolonged wars of conquest (Ibid, 37). Moreover, according to Jini, the good ruler should consider the cyclical nature of the economy (expressed in terms of ebbs and flows, as well as of yin and yang) and act in a counterbalancing way – an early echo of Keynesian countercyclical policies, perhaps. In times of droughts or floods, for instance, the good ruler should use their access to food from other climate zones where the harvest had not been lost.

> If there is a time when people are selling, then wise men go against. [...] If the ruler is able to thoroughly understand ebb and flow, and employ the wise and use the able, then goods will come to him from a thousand li away. If he is not well-versed in these matters, then even things found within a hundred li cannot arrive. The price of things sought by rulers increases ten times, [the laws] that they choose are priceless. A ruler can take advantage of this ebb and flow, but he ought not do it personally. He should see where the people do not have enough and where they have too much, then he should make commands about this in order to benefit [his people].
>
> (Ibid, 36)

The King of Yue is persuaded by Jini and claims that his advice would now guide the government and be followed for generations. Subsequent Han emperors seemed to have taken the advice to heart and the rulers made large-scale purchases in years of good production and sold in years of poor production, buffering the effects of famine (Wright 1979, 93–94). Considering the robustness of soybeans as a crop, and the fact that it was easy to store, it is not surprising that China's first granaries, developed during the Han Dynasty, included soybeans for use in times of famine or bad harvest (Huang 2008, 49; Shurtleff and Aoyagi 2021, 70–71).

Another influential suggestion from Jini was that the good ruler should regulate prices and see that they do not fall too low and harm the farmers, but also see that they do not rise too much and injure the tradesmen:

> If farmers are harmed then grass and trees are not cleared, if tradesmen are injured then goods are not put on the market. Therefore if [the price of] rice does not go higher than eighty nor lower than thirty, farmers and tradesmen both benefit from it. Therefore, those who ruled states in the past made this the basis of their economic policy, and official markets were opened.
>
> (Milburn 2007, 33–35)

This idea was also reflected in the Han granary system, which worked as a grain price stabilization mechanism as well as a buffer against famines (Wright 1979, 93–94).

As the fragment from *Jinizi* which opens this chapter indicates, soybeans were sold and bought, but their price was set significantly lower relative to other cultivars. This likely reflects the fact that the soybean was less valued than millet, rice, barley and red beans, but also that the robustness of the soybean rendered a stable supply and also that it was a cash crop (Shurtleff and Aoyagi 2014, 70–71). This interpretation is further strengthened by Fan Ye's historiography *Hou Hanshu*, in which the tale of *Chimei Zai* takes place against the backdrop of a famine in which all food was so expensive that one catty of gold would only buy 5 *dou* of soybeans; many in the end, resorted to cannibalism.[9] As another of the stories in *Hou Hanshou* makes clear, when food was so scarce that even soybeans were expensive, there was nothing left to hope for (Shurtleff and Aoyagi 2021, 67).[10]

Texts such as *Jinizi* and *Hou Hanshu* indicate that trade was common and that soybeans had a relatively predictable price (lower than that of many other food crops, but rose when harvests were poor and during years of famine). Other sources from this period also show that Han rulers disapproved when farmers left agriculture and moved into such "unproductive" activities as trade, on the grounds that this contributed to labor shortages in farming, and ultimately of food (Hsu 1980, 24–39). For example, the court of emperor Wen (179–167 BCE) concluded that there were too few producers for too large a number of consumers and urged the "idling and parasitic" population to return to the farms. During the reign of emperor Wu (141–87 BCE), some private warehouses and storage facilities used in commerce were even destroyed and the state seized the lucrative iron and salt trades (Hsu 1980, 39–40). Nevertheless, during the Han Dynasty, business activities flourished, and there was a spectacular wave of urbanization, in spite of the official discourse extolling the virtues of working the land. Goods from the countryside or abroad flowed in and out of the cities via waterways and sometimes roads. In this context, soybeans and soy products were commercialized and traded by merchants in the Yellow River valley since at least as early as the 3rd century BCE (Lander and DuBois 2022, 37).

Our sources from this period speak of millions of *shih* of grain having to be imported into the capital district to feed its several-million strong population.[11] This was also a time of increasing regional specialization in specific staples (Hsu 1980, 37). In addition, while most independent farmers managed to effectively use their small parcels of land to become self-sufficient in food, they typically had to buy other goods, as well as paying government levies and shrine fees, from the sale of their surplus harvest (Ibid, 235). Trade was thus well developed, and the state, all things considered, seems to have been more supportive of commerce than its pronouncements suggested. Even if they were only graced with "low formal status", merchants gained influence and power during this period. Meanwhile the state ultimately acted to facilitate commerce, in particular, by investing in transport projects and other infrastructure.

A key development under Han rule is the appearance and spread of new soy preparation techniques, perhaps particularly, *shi,* the fermented paste of black soybeans, which became an important commercial food product. The previously mentioned canonized historiography *Hou Hanshu* (398–446 CE), for example, includes a story about a 1000 earthenware vessels of mold-fermented cereal grains and fermented black soybeans (*shi*). According to its author, Fan Ye, this account was already included in *Shiji*, or *Shih Chi* [*Records of the Grand Historian*], written by Sima Qian in 90 BCE (Huang 2000, 336; 2008, 49, 51). Whatever the case, what is important for our purposes here is that the aforementioned vessels of *shi* were, we are told, articles of commerce. As new food processing techniques spread, then, trade with soy became increasingly important. By the 1st century CE, it was even capable of generating fortunes. According to another story in *Qian Hanshu,* or *Chhien Han Shu* [*History of the Earlier Han Dynasty*] from around this time, two of the seven most wealthy merchants in its narrative had built their wealth on the *shi* trade (Huang 2000, 337; 2008, 49).

It is clear that trade in soybeans began to expand especially midway into Han rule; what is much more difficult is to estimate the relative importance of this domestic trade. At least officially, the merchant profession was despised and so it was considered unworthy of attention by the ruling elite (Milburn 2007, 19). Thus, historians typically know less about Chinese domestic trade than about the long-distance exchanges along the Silk and Spice Routes. The commercial links connecting China with the Mediterranean and the Middle East generated extensive and recurrent exchanges of goods, ideas, technologies and people (Frank and Gills 2000, 6). However, they were typically restricted to goods that could travel for long periods by pack animals without being spoilt, and for which value exceeded the transport costs (in other words, goods like silk, porcelain, pottery, metals, timber, textiles, animals, slaves, wine, rice, dates and spices). Where the better records of trade exist, soybeans are generally absent.

Soy for human consumption

Soy as food appears in many ancient texts even if it was, as yet, far from a cornerstone of Chinese cuisine. The *Dongguan Hanji* written by Bang Gu, Liu Zhen and others in 270 CE, is a very relevant source here. It includes a biography of the emperor Guangwu (the first emperor of the Late Han Dynasty, from 5 BCE to 57 CE) and makes the claim that hemp and soybeans were abundant during the early years of his long reign. Crucially, it includes a biography of a local official called Liu Ping, who was appointed minister of the imperial clan in 60 CE due to his virtue, loyalty and fearlessness (de Crespigny 2006, 537). In one passage, he appears starving and scrounging for food, when he is suddenly assaulted by a group of also-very-hungry bandits who are planning to have him for supper. Liu explains that he is trying to find something to eat for his elderly mother. If they let him fulfill this filial duty, the tearful Liu pleads, he will come back to them and surrender. The robbers take pity on him and set him free, expecting to never set eyes on him again. But Liu is nothing if not a man of his word and returns to the bandits. They are surprised to see him, and moreover, totally stunned by his sense of nobility. "We have never seen anyone who is as virtuous and faithful as you". Once more they let him go, except this time, they also gave him three *sheng* (about 6 dl) of soybeans (Shurtleff and Aoyagi 2021, 67–68).

These and other stories bear witness to the fact that the soybean was a fairly common source of food. While soy was part of the diet, the most common meal in North China at the time came from grains and other starches (Bray 2019, 361). Millet and wheat were the most important food crops in the intensive cropping system of the North, but rice was produced in growing volumes thanks to, in great part, the Han Dynasty's investments in irrigation. Thus, rice began appearing as an import from the humid and rain abundant Southern provinces.

The most common staple crops consumed during the Han Dynasty were wheat, barley, rice, foxtail or broomcorn millet and beans. The standard diet would have contained rather few vegetables compared to later periods (Hsu 1980, 89), and meat consumption began diminishing as population increased and ever smaller areas of land were available for animals (Lander and DuBois 2022, 32). Dairy products were usually not consumed at all, mostly because North Chinese agriculture was intensive and crop-centered, with few animals (Zhou et al. 2017, 217). Han governors praised cultivation but considered pastoral practices as inherently destructive and wasteful (Loewe 1986, 16–17). Pastoral activities were also seen as "barbaric" and associated with the nomadic, non-Han peoples (Kiple 2007, 68). Thus, cattle herds shrank and almost disappeared as pastureland was almost completely eliminated from China during this period (Bray 1984, 432; Kiple and Ornelas 2000, 422; Isett and Miller 2016, 30). Other legumes and fish provided most of the supplemental proteins in a diet composed mostly of grains (Bray 1984, 4–5). But soy had specific qualities as a food stuff. For instance, it was able to thrive in a region like the Yellow River plains, characterized by high climatic variability and often challenging soil conditions, which made it an important insurance crop. Additionally, soybeans could be stored up to a full year and, so, be kept in preparation for times of famine (Shurtleff and Aoyagi 2014, 38–39, 65–66, 74–78). Thus, even if soy would still take centuries before becoming a central element of the Chinese diet, it already played a vital function as a robust crop which could feed a growing population, especially when the harvests failed and famines ensued.

Soybeans as starvation food

Soy's ability to yield a harvest even when all other crops failed also made it important in feeding the poor (Huang 2000, 293). This is how it appears in a folk song from the 1st century CE recounting the failure of irrigation works in Ju-naan (modern Anhwei), after which nothing was left to eat but soybeans and yams (Bray 1984, 514). The aforementioned agricultural treatise *Fan Shengzhi Shu* also insists that good soybean yields were easily obtained even in the worst years (Lander and DuBois 2022, 33). According to the information that other sources offer on average acreage (2 ha) and family structure (households of eight people), such targets appear unlikely for most farming households. If Fan Shengzhi's advice were to be followed, each family would have to use 1.619 ha for soybeans, leaving only 0.3 ha of the standard landholding for other uses. All the same, the sources clearly treat soy as a reliable crop.

Now, the fact that soy was valued as a buffer against hunger did not mean that it was appreciated on culinary grounds: far from it. The same chemistry that protects the soybean against pests makes it partially indigestible for humans and conducive to flatulence and stomach pain (Huang 2000, 293; 2008, 46–48; Mintz 2008, 59). The most efficient way known at the time to break down its indigestible carbohydrates was boiling it into a porridge or gruel (*tou fan*), or alternatively, into a very soft, thin liquid bean congee (*tou chu*, *douzhou*, or *choushu*). The *Monthly Ordinances for the Four Classes of People* includes detailed instructions on when best to fry and boil soybeans, showing that these were fairly common activities. Boiled soybeans were probably the most common for human consumption, particularly in the form of congee. Eating them was thus often referred to as "sucking soybeans" (Huang 2000, 294, 314), which was a sign of severe hardship. *The book of Master Xun* discussed earlier illustrates this well:

> If the 10,000 things lose their balance, events change their order. Above, the heavens miss their timeliness. Below, the earth misses its bounty. In the middle, the people lose their harmony. The world is in a state of drought, as if it is being burned. The sage Mozi may have his gown and belt, but he will only be able to suck soybean congee, and drink water. What worth is left to life?
>
> (Shurtleff and Aoyagi 2021, 46–47)

As seen in this quote, soybean congee was ultimate hardship food – the only food that might be available even when "the world is in a state of drought". While prolonged boiling of soybeans was common as a means to overcome the indigestible elements and turn soybeans edible, it didn't do much to improve their taste (Huang 2000, 294, 314; 2008, 47–48).

This sentiment is echoed in a fascinating story, *Tongye* [*Contract Between an Indentured Servant and his Master*], written by Wang Pao in 59 BCE. In it a man named Wang Tzu-yüan is staying at the house of the widow Yang Hui. One day, he asks her slave, Pien Liao, to get him some wine. The slave replies that he has only been instructed to guard the tomb, not to buy wine. Enraged by what he considers his insolence, Wang, buys Pien Lao from the widow and writes a very long and detailed contract stating the latter's duties and including the requirement that he only eats cooked soybeans (*fandou*) and drinks water. When the list of his new duties is read, Pien Liao breaks down, sobbing:

> [I]f things will be as Master Wang has said, I'd rather return to the yellow earth [the world of the dead], and let the earthworms penetrate my forehead. If I had known earlier what would happen, I would have bought wine for Master Wang and dared not to offend indeed.
>
> (Reprinted in Hsu 1980, 231–34; cf. Huang 2000, 294; 2008, 48; Shurtleff and Aoyagi 2021, 3).

Of course, the bottom line is that slaves should obey their masters (and masters' friends); but the story shows beyond any doubt that boiled soybeans were shorthand for an utterly joyless diet and possibly, existence.

Unpalatable or not, the poor probably ate more boiled soybeans than they would have liked; a dismal culinary fate they shared with the military. In the *Dongguan Hanji*, mentioned above, we have the story of a general who is preparing the infamous soybean congee (*douzhou*) for the exhausted emperor and his troops during a grueling military operation. On the second day, the emperor tells his generals that the soybean congee had helped lessen their hunger (Shurtleff and Aoyagi 2021, 68). The background to the tale is very real. The Han Dynasty felt constantly threatened by neighboring peoples from the North, as well as by factional struggles, banditry and uprisings. As the territory of the Han empire had expanded considerably, new administrative districts were established in the Northwest and Northeast (Loewe 1986, 17). One particularly serious security threat came from the powerful confederation of the Hsiung-nu tribes, and, in response, the state increased the size of the army substantially. But raising an army required food production, transport and storage (Milburn 2007, 31). And, as we have already seen, the soybean was not only nutritious, but also easy to store for long periods (Kiple and Ornelas 2000, 422–23; Mintz 2008, 59). The *Guanzi*, for example, openly states that the ultimate purpose of increased agricultural production is to be able to expand and sustain a strong military force: "[I]f grain is abundant then the state is rich, if the state is rich then the army is strong, if the army is strong then the battles are won, if battles are won then territory will be expanded" (Milburn 2007, 37). It is worth pointing out here that Chinese sources often counted soybeans as grain.

Besides different ways of cooking soybeans (congee, gruel, boiled), the introduction and spread of the rotary mill during the Han Dynasty allowed for grains and beans to be ground into flakes, which in turn could be compressed into cakes. The introduction of the mill significantly improved the status of wheat – which together with millet now became the preferred crop. Ground wheat allowed for such highly appreciated food dishes as noodles (Bray 1984, 459; Huang 2008, 51). Isotope data analysis from the period shows that there was a substantial rise in wheat consumption (Zhou et al. 2017, 217). However, milling technology did not have a similar revolutionary impact on the status of soybeans. While grinding soybeans into cakes became a nutritious and practical food for soldiers and travelers on the move, they were still far from being considered tasty and remained difficult to digest. Wheat, then, gained precedence over beans during this period (Kiple and Ornelas 2000, 423).

Early processed soy foods and ritual and medicinal uses

Eventually, more sophisticated preparation and processing techniques began to enhance culinary appreciation of soy, and none more so than fermentation. While it is likely that fermentation was developed in the context of alcohol production – and probably discovered by accident – it soon turned out to be useful in a wide variety of other food processing techniques, for both preservation and taste enhancement (Huang 2000, 593). By far the most common such fermentation procedures in processing soyfoods were based on molds rather than yeasts. This was in stark contrast to the West, where mold was more often met with suspicion and a sense that the food had been spoilt. In any case, thanks to fermentation, soy became more digestible and, at the same time, its flavor, texture and aroma were significantly improved. What is more, the required cooking time was shortened while storage and shelf life were prolonged. In the long run, this process turned soybeans into a more desirable and popular product. However, the full story of fermented soy is long and incremental and does not completely unfold until the Ming Dynasty. Since the historical origins of this unfolding lie in this first cycle, we will confine ourselves to overviewing them briefly here.

Jiang and *shi* are mentioned in fragments that have survived from the 1st centuries of the CE. One example is the 2nd-century CE text *Chi Chiu Phien* [*The Handy Primer*] by Yen-Shih Ku, which talks about *jiang*, a fermented paste or sauce made from soybeans or chopped soybean (or groats; see Huang 2000, 346–47). According to HT Huang, *shi*, (fermented black soybeans) however, was already regarded as a culinary necessity of daily life around 100 CE (Huang 2000, 336; Shurtleff and Aoyagi 2014, 65–66). *Jiang* and *shi* are not the only fermented soybean products in the sources. In the *Monthly Ordinances for the Four Classes of People*, some texts refer to *modu* (or mo-tu paste), described as a fermented paste made from chopped soybeans. The book includes careful instructions on how and when to prepare *modu*, and – when the product is finished months later – how to use it as a base for fish and meat sauce, as a basic soy sauce, or for pickling melons (Hsu 1980, 217). The soy sauce mentioned here seems to refer to a simple liquid soy condiment which preceded the invention of modern soy sauce (Shurtleff and Aoyagi 2012, 22). While both *jiang* and *shi* are certainly mentioned in several Han texts, none of them explain all the steps involved in their preparation. The first known document which includes a full description is the *Qimin Yaoshu*, discussed above, which came several centuries after the Han (Huang 2000, 337; Shurtleff and Aoyagi 2012, 21). Soybean fermentation required a number of imperative conditions and involved a chain of elaborate and labor-intensive steps over 10–15 days (Liu 2008, 443; Liu and Chen 2012, 90; Shurtleff and Aoyagi 2012, 25); only then could the actual fermentation (needing another month) begin, before the procedure was complete. Another important processed soybean product (in this case, usually not fermented) that is believed to have emerged during the late Han Dynasty was soy curd (*doufu*) or tofu. While the detailed evolutionary refinements that had to be in place in order to develop tofu are unclear, there are texts and archaeological evidence (for instance, a mural in a tomb from the Later Han period) which show that tofu existed and suggest that its preparation was similar to methods used today.

The earliest archaeological finds of fermented soyfoods support the written sources and throw light on yet another important function of processed soybeans. In particular, the *Ma-wang-tui* site, found in 1972 and dating to about 160 BCE, is located in the central Yangtze valley region and consists of two small saddle-shaped hills. The finds have unlocked relevant information about diets and early use of fermented soy (Buck 1975). In what was later named Han Tomb No. 1, a fermented soybean paste mixed with wheat flour, *jiang*, and fermented and boiled black soybeans, *shi*, were found in pottery jars together with a 1,000 other objects around and above the well-preserved remains of a woman, later identified as the wife of Li Ts'ang, first Marquis of Tai, who reigned from 193 BCE to 186 BCE (see Huang 2000, 346). Another Ma-wang-tui feature (Han Tomb No. 3) contained the corpse of a young man in his thirties (believed to be Li Ts'ang's son); with it were grains, vegetables and meat. In both tombs, the grains included rice, wheat, oats, millet and hemp seeds; also present were fruits and vegetables, meats and most importantly for our story, soybeans (Buck 1975, 39). The tombs at Ma-wang-tui give important insights to what foods the upper classes appreciated and had access to, at a time when the newly emerging culture of the Western Han was established in North China (Ibid, 73). The tomb also contained bamboo slips on which were written several food characters, including those for *shi*, *jiang* and a grain-based mold ferment called *qu* (Huang 2000, 25, 346). These and other archaeological finds then indicate that soy was a regular component of offerings used in funeral rituals, which means that soy also had a ritual value, at least in its processed forms.

Besides food, soybeans were also used medicinally. Different soybean varieties and forms were understood to have different properties. The 200 BCE text, *The Yellow Emperor's Classic of Internal Medicine: Questions and Answers* by Huangdi Neijing Suwen, discusses the ways in which different varieties and forms of the soybean affect the body. The fragments that have survived from the Classical pharmacopoeia *Shennong Bencaojin* [*Heavenly Husbandman*] (sometimes *Shennong bencao jing* or *Shm Nung P'en Tshao Ching*) constitute a compilation from the 2nd century CE and explore, among other things, the health benefits of boiled yellow soybeans, including for swelling (when used externally), or using soybeans curls (dried strips) to treat numbness in the joints and muscles (Needham 1971; Shurtleff and Aoyagi 2014, 65–66). The same book warns about prolonged ingestion of black soybeans, which will make the internal organs feel heavy (Shurtleff and Aoyagi 2014, 143). Another important work about the therapeutic value of soybeans comes from *Shiliao bencao* (*Compendium of Diet Therapy* [*Materia Dietetica*]), which was written by Meng Shen in 670 CE. According to this work, boiled into a liquid form, soybeans can eradicate poisons from the system and cure gastric fever, paralysis, pains, difficulty in passage of urine and other bladder troubles, improve circulation of the blood, improper heart, liver, kidneys and stomach function and even remedy the chills. The yellow bean, according to Meng Shen, can be used to increase lung power, make the body plump and beautify the complexion (Shurtleff and Aoyagi 2014, 89–90).

These cultural and medicinal functions of soy, as well as the early culinary role of processed soy products, are obviously interesting aspects of the role it played during this period, but they are not themselves integral to what we have been describing as the regime. Once more, it would still take centuries before tofu, soy sauce and other soyfoods would become widespread and common (Huang 2000, 358–60). That development is the object of coming chapters; let us now turn our attention to the rupture of this first soy cycle.

Rupture: disintegration of the Han Dynasty and population decline in the North (200 CE– 900 CE)

The Han Dynasty was marked by a remarkable increase in both food production and population. By the period of the later Han (25-220 CE), however, floods and droughts caused severe food shortages. Combined with turbulent civil unrest, the population started to decline. Moreover, the imperial impulse for agrarian frontier expansion ran up against a desire for rural stability, which depended on hard-working peasants feeding themselves and generating annual grain surpluses. The cost of expanding and defending the empire fell squarely on the shoulders of the peasantry. As Isett and Miller have put it:

> In a population of fifty-four million, one million served in the army, another thirteen million performed a month of corvée labor every year, and every adult male paid the annual poll tax. Freeholders paid an additional fixed land tax, which was not especially onerous, taking about one-fifteenth of the harvest.
>
> (2016, 28–30)

In addition, droughts brought severe famines, which caused yet more fighting and unrest. While the emperor sided with the peasants rather than the great land-owners, the latter accumulated extensive landholdings allowing them to keep their own standing armies, destabilizing and threatening the power of the state. Over the long term, the big landowners were stronger than they had ever been since the beginning of the dynasty. In addition, civil unrest caused many peasants to abandon their land, seeking shelter as dependents or serfs on fortified farming estates and, "even in times of peace, most peasant families lived on a knife edge between sufficiency and hunger" (Bray 2019, 371). The growing independence of families belonging to the landed elite was affecting political cohesion and the stability of imperial authority in a way that presaged the breakup of the Han Dynasty in 220 CE with the abdication of the last Han emperor, Hsien-ti.

Besides the power struggles and upheavals, it is possible that the fall of the Han Dynasty was also spurred by the effects of environmental changes. For example, the Yellow River was changing its course, which disrupted Eastern China and caused massive destruction and loss of life (Loewe 1986, 16). In addition, the increasingly intensive agriculture and widespread use of iron tools caused severe erosion and soil degradation in some areas. Thus, notwithstanding all the efforts to maintain the soil in healthy conditions, the Chinese agricultural system began to feel its limits. Land impoverishment and degradation set in as a recurrent threat to productivity. Sometimes soil loss was so severe that the land had to be abandoned. In addition, iron tools made forest clearings easier, expanding the agrarian frontier, and in the longer term, resulting in faster erosion washing away fertile lands into reservoirs and lakes (McNeill 2001, 35–38). While order and "rules of nature" were explicit ideals for agricultural production in the official discourse, forests and water resources were degraded as agriculture expanded over vast areas (Bray 1984, 1; Huang McBeath and McBeath 2010, 15–16). Indeed, agricultural production was influenced by thought systems emphasizing the protection of nature and especially natural resources. One ideal from the pre-Ch'in into the Han Dynasties was that everything should be done in strict accordance to the aforementioned "rules of nature". This early environmental ethic was combined with a social ethic, identifying the rule of nature with the social order. But for all its agroecologically sound wisdom, some scholars have in fact cast doubt whether there ever was much meaningful consideration toward, much less deference to, the biophysical limits of expansion (McNeill 1998; Smil 2004).

After the fall of the Han Dynasty, North China was plagued by wars, political turbulence, frequent upheavals and invasions from nomadic peoples (Loewe 1986, 15–17). In time, it was the Northwesterners who swept into the cities of Ch'ang-an and Lo-yang, driving the dynasty to found a new capital in the South in 317 CE. China disintegrated into fragmented congeries of kingdoms, referred to as the *Six Dynasties*, and characterized by a more "decentralized polity, a hybrid ruling elite, the appearance of a manorial type of economy, the emergence of organized religion, and a heavy reliance on close patron–client ties between upper-class men" (Dien and Knapp 2019). While this period was politically unstable and fragmented, it was, nevertheless, a time of progress in terms of technology, medicine, astronomy, mathematics and cartography. Adherents of Buddhism, as well as Daoist [Taoist] communities became increasingly influential during the *Six Dynasties* (Dien and Knapp 2019). The flood basin of the Yellow River remained the agricultural, economic and demographic center, but the pastoral people on the steppe successively entered into the farming regions of Northern China.

Under Northern Wei (386–534 CE), the government established a sort of land reform, based on the restoration of the ancient equal-field system and aimed at increasing the viability of peasant farms. This system allocated land to peasants at the notional rate of 80 *mu* of land for grain and 20 *mu* for hemp per person (one *mu* was approximately 666.5 square metres, or 0.066 hectares), enough to provide for basic subsistence and the payment of taxes in grain and cloth. As well as securing a free peasantry as the tax and labor base for the state, the reforms were also designed to reduce the power of big landowners (Bray 2019, 369–71). The equal-field system seemed to, at least, slow the pace of accumulation of lands by wealthy families, even if a significant amount of farming activities remained in large estates (Isett and Miller, 2016, 28–29). Besides reform, the pressure on land lessened as population growth halted and even started to fall. According to Myrdal, the Chinese population around the year 600 CE reached a nadir of around 40 million. Of course, these figures need to be taken with caution, as they probably only include the Han Chinese: what's more, in difficult times, the fall in population tends to be overestimated due to tax evasion strategies and/or de-urbanization (Myrdal, forthcoming). In any case, the North – which by the end of the Han period had harbored three fourths of the population (Ibid) – started to lose some of its population to the more humid (and wet-rice dominated) South. Accordingly, the intensity of agriculture in the North decreased, opening up the possibility to have more land under pasture and with that expanding the influence of nomadic people who did not practice intensive agriculture.

This shift is clearly illustrated by the previously mentioned seminal work *Qimin Yaoshu,* which Jia Sixie was writing at a time when the Tuoba Wei had ruled Northern China for over a century. The influence of food and agricultural traditions from pastoral peoples – centered in dairy and animal husbandry – is particularly prominent in the contemporary (544 CE) section of the book, in stark contrast to other segments drawing on previous agricultural texts (Bray 2019, 356, 367). Jia Sixtie also considered how best to produce animal feed for sheep and goats and mentioned how female ducks were fed with soybeans and millet to produce many eggs (Shurtleff and Aoyagi 2021, 75–79). Additionally, he advocated sowing one *qing* (roughly 6 ha) of soybeans for every thousand head of sheep, which is indicative of the scale that he envisaged as manageable (Bray 2019, 368). Plainly, animal husbandry increased in the wake of an inflow of pastoral nomadic groups and a drop in population. This released some pressure on lands and brought a short-lived "pastoral turn" (Bray 2018). It was thus the heyday of the manorial enterprises which typically raised large numbers of livestock for both profit and pleasure, but which would end up disappearing from the Chinese diet under later dynasties, as we shall see in the coming chapter.

Qimin Yaoshu is not only illustrative of the "pastoral turn" of the North but also of the fact that many big landholdings managed to escape any division of land linked to the previously mentioned equal-field system. These estates, such as the one owned by Jia Sixtie himself, typically used large amounts of human labor (from slaves to wage workers), land, equipment and draft animals, and had a strong commercial orientation with access to several local markets (Bray 2019, 371).

After the renewed unification, with the final conquest of South China by the Sui Dynasty in 589 CE (established in 581 CE in North China), the Northern Chinese agricultural system disintegrated as the center of agricultural production and commercialization. These gradually moved to the Southern Yangtze valley, alongside the central nodes of governance and administration. Under Sui rule, the borders of the empire expanded into Central Asia, but the dynasty was short-lived and overthrown again by a series of rebellions starting in 613. The general shift from the North to the South continued under the T'ang Dynasty

(618–884), and, although several centuries had passed since Han rule, the T'ang state administration included many features developed under the Han. For example, Confucianism continued to be central and the administrative cadres were thoroughly schooled in Confucian thinking, which gave primacy to order and tradition. One important element here was the notion of the *Mandate of Heaven*, symbolized in the emperor as the superior being governing his subjects and life of the kingdom (Perez-Garcia 2021). Also among the continuities was the reliance of the bureaucracy on members of a highly educated class grounded in Confucian writings and other *classics*, as well as the notion of China as an agrarian state *par excellence* (Loewe 1986, 15).

The T'ang rulers further developed the state by selecting the most talented candidates for bureaucratic service through a system of entry examinations (Perez-Garcia 2021, 81). The move southward was also spurred by increased investments in infrastructure, e.g. the Grand Canal, as well as by a new elite class of landed gentry that promoted agriculture in the South (Kiple 2007, 44). As the Chinese economic center gradually moved south, wet rice took over millet as the dominant crop (Bray 1984, 9, 427). This shift began to consolidate at the turn of the first millennium; we discuss these new dynamics fully in Chapter 3.

Reflections on the soybean and its functions

While agricultural change always involves social-ecological alterations, not all agricultural systems are equally simple and/or extractive. Illustrated by the different practices of, and roles played by, the soybean (and many other crops, such as rice, millet and wheat), we saw that the agrarian system of imperial China was characterized as, first, land- and labor-intensive, second, a pattern of plot ownership probably much more evenly distributed than in Europe and, third, a sophisticated system of crop rotations involving a relatively high degree of crop variety and diversity. As we have seen, many ancient texts testify to the wide array of cultivars grown and show ancient farmers tending their small fields not only with care, but with skill (Bray 1984, 6: 7–8). According to the *Qimin Yaoshu*, soybeans typically entered this system of rotation in different combinations with millet, hemp, mung bean and sesame, and in ambitious rotation schemes which lasted around five years (Bray 1984, 6: 56–59). Besides rotations, the Chinese often followed the practice of intercropping, or mixing in different types of crops (Ibid, 6: 432–33), all aimed to ensure both high yields and long-term soil sustainability. The high relative intensity of Ancient Chinese agriculture was also made possible by different methods which assisted in restoring the stock of nutrients, such as night soil and animal manure along with a wide array of other fertilizers (Federico 2010, 8–9). In this way, the nitrogen fixation capacity of soybean plants (while not fully understood in those terms) made them popular as forage crops that could enrich soils and plants were also frequently ploughed under as green manure, without being harvested. In short, while beans were certainly eaten, their most important value seems to have been as a green manure for improvement of the soil (Hsu 1980, 86).

All this doubtless reflects an impressive knowledge of soils, but it must also be underlined that technology and population growth eventually unleashed fundamental changes within ecological structures and functions, both at local and even regional scale. Forests and other natural vegetation were subjected to the plow as agriculture expanded and intensified. Environmental historians have shown that the first significant pressures on forests resulting from expanding agriculture probably emerged as early as 100 BCE, just as Han rulers were promoting clearings and pushing into the agricultural frontier. Simultaneously a number of animal species – e.g. elephants and tigers – started to disappear from the North China Plain (Sterckx 2019, 268). The extensive deforestation was one of the causes of the recurrent flooding of the Yellow River (Ponting 2007, 74). In addition, other destructive agricultural practices, like draining wetlands, channeling rivers, razing forests and, particularly, forested mountainsides, resulted in erosion which not only removed precious topsoils, but drenched valleys with silt. Critically, big cities emerged during the Han Dynasty. As urbanization took off, an increasing amount of food had to be produced further away from those consuming it. Consequently, rural agricultural land started for the first time to systematically "export" nutrients to the cities. This uneven relationship in terms of ecological resources between town and countryside only deepened as the decades rolled on. Soy had emerged as a cash crop for the first time in its history during the Han and the *Six Dynasties*. Most food was consumed by the same hands that produced it – the exception being the surplus consumed by the military and the state administration, or used in exchange with milk and meat-producing nomadic groups.

One of the most striking features of this highly productive agricultural system was the way in which it centered on crops and gave little importance to livestock (in stark contrast to the traditional mixed-farming systems of Europe). The crop-centrism of imperial Chinese agriculture was probably the result of multiple reinforcing elements, such as the Confucian emphasis on tradition, "laws of nature" and contentment (in one formulation: eating more soybeans and less meat). However, perhaps the most important reason was the effort to sustain a large population (including a large army), combined with an early awareness of the large amount of resources (particularly land) that a meat-based diet required. Thus, for the emperors who sought to sustain a growing population (and to avoid popular revolts), vegetarianism was perhaps a necessary ideal in order to make the scarce resource (land) produce enough food. Pastures and forests were converted into croplands as new areas were incorporated under the Han culture and urbanization.

Within a value system so centered on sustaining a large population with the least amount of resources possible, and given the availability of soybeans – whose many qualities were, as we have seen, well known – one might expect soybeans to have become the main Chinese staple. Indeed, while the soybean played many functions, its key role was to feed humans especially during bad times (Lander and DuBois 2022, 29).

But, as we have also shown, soy was far from being anybody's favorite food: it needed years of development in food processing methods before it was appreciated for its flavor. Food choices are of course related to availability. As Sidney Mintz reminded us in his seminal *Sweetness and Power*: "human beings never eat every edible and available food in their environment" (1985, 3). But despite its taste, as we saw, the soybean's role as provider of proteins was nevertheless stressed as vital, especially in hard times. The advances which would take soybeans from subsistence food to a cornerstone of Chinese cuisine were taking place at the end of this period (Huang 2000, 16). Popular habits shifted slowly and gradually. Only after the Northern food regime began to unravel, and only after the agricultural, economic and demographic center moved south under T'ang and Song rule, did soyfoods begin to be appreciated by "everyone" in China.

In short, the soybean was by no means among the most important crops in the knowledge-intensive and highly productive agricultural system that developed along the Yellow River, where the majority of croplands were grains – particularly millet and wheat. It seems, nevertheless, to have served several important functions, including improving the soils; acting as fertilizer; and feeding animals, soldiers and the poor and, step by step, being processed – through ever more intricate food-preparation techniques – into a wider range of more popular food products. Due to China's hegemonic position in East Asia, processed soy became the central food and flavor throughout the region, as we shall see in the next chapter.

Notes

1 We have used the extensively annotated bibliography and sourcebook *Early History of Soybeans and Soyfoods Worldwide (1024 BCE to 1899)* from 2021 compiled by William Shurtleff and Akiko Aoyagi, providing keys and information about sources considering the early, pre-European, history of soybeans. This book (1,283 information-dense pages) is the most comprehensive study ever published about the early history of soybeans and soyfoods. It lists, in chronological order, detailed information about all known documents and commercial products containing soybean in all languages (in all, it covers 3,531 published documents).

2 This *SoyInfo Database* source is based on the English translation with facing Chinese text: Shih, Shêng-han. 1959, *On "Fan Sheng-chih shu": An Agriculturistic Book of China Written by Fan Shêng-chih in the First Century* B.C. Peking, China: Science Press. (Kexue Press). 68 p.

3 Of course, the calendar goes well beyond soy in its scope: it also includes suggestions of when to repair houses and irrigation canals, as well as when "one can marry a wife", when a husband and wife should sleep separatelyor when to carry out funerals for poor dead people (who have been dead for a long time). Besides time, instructions often consider weather conditions and the forces of *yang* and *yin* (Hsu 1980, 217–28: Shurtleff and Aoyagi 2012, 22).

4 However, it is also clear that while agriculture is deemed essential, a good Confucian prefers to become a state official rather than to till the land himself (Sterckx 2019, 313).

5 Hsu here is mainly building on reprints from the previously mentioned *Qian Hanshu*, or *Chhien Han Shu* [*History of the Earlier Han Dynasty*].

6 Xun Kuang is sometimes spelled Xun Qin. He lived 313 BCE–238 BCE.

7 At this time, the well-field system was still very common. This was a type of land redistribution where the aristocrats could remain as the owners, but land was divided into nine identically sized sections; in which the eight outer sections were privately cultivated by peasants and the center section was communally cultivated on behalf of the landowning aristocrat. The produce of the privately cultivated land was entirely for the farmers only. Produce from the communal fields, worked on by all eight families, went to the aristocrats, which, in turn, could go to the king as tribute (Hsu 1980, 9).

8 See *Shiji* [*Records of the Grand Historian*] by Sima Qian (or Ssu-ma Ch'ien) from 90 BCE, which listed many important Chinese urban centers, and their trade of basic goods with their vast "hinterlands" (Hsu 1980, 36–37).

9 A *dou* is equal to about 1/10 of a bushel in the late Han Dynasty.

10 Like many of our sources, Hou Hanshou includes much re-writing of earlier texts (Loewe 1986, 4; L'Haridon and van Ess 2019, 1–6).

11 The *shih* was the standard measure of grain capacity during the Han Dynasty - 1 *shih* was roughly equivalent to 29.3 kilograms (Loewe 1961; Hsu 1980, 103).

References

Bray, Francesca. 1984. "Science and Civilisation in China. Volume 6, Biology and Biological Technology. Part 2, Agriculture", In *Science and Civilisation in China*, edited by Needham. Cambridge: Cambridge University Press.

———. 2018. "Where Did the Animals Go?" In *Animals through Chinese History: Earliest Times to 1911*, edited by Sterckx, Siebert, and Schäfer, 118–38. Cambridge: Cambridge University Press.

———. 2019. "Agriculture." In *The Cambridge History of China: Volume 2: The Six Dynasties, 220–589*, edited by Dien and Knapp, 2:355–73. The Cambridge History of China. Cambridge: Cambridge University Press.

Buck, David. 1975. "Three Han Dynasty Tombs at Ma-wang-tui." *World Archaeology* 7 (1): 30–45.

Buckley Ebrey, Patricia. 1993. *Chinese Civilization: A Sourcebook*. 2nd ed. New York: The Free Press.

Cowan, C. Wesley, Patty Jo Watson, and Nancy L. Benco, eds. 1992. *The Origins of Agriculture: An International Perspective*. Washington, DC: Smithsonian Institution Press.

de Crespigny, R. 2006. *A Biographical Dictionary of Later Han to the Three Kingdoms (23–220 AD)*. Leiden, The Netherlands: Brill.

Dien, Albert E., and Keith N. Knapp, eds. 2019. "Introduction." In *The Cambridge History of China: Volume 2: The Six Dynasties*, edited by Dien and Knapp, 2:1–24. The Cambridge History of China. Cambridge: Cambridge University Press.

Dobson, W. A. C. H. 1964. "Linguistic Evidence and the Dating of the Book of Songes." *T'oung Pao* 51 (4/5): 322–34.

Federico, Giovanni. 2010. *Feeding the World: An Economic History of Agriculture, 1800-2000*. Princeton: Princeton University Press.

Frank, Andre Gunder, and Barry K. Gills. 2000. "The Five Thousand Year World Systems in Theory and Praxis." In *World System History: The Social Science of Long-Term Change*, edited by. Denemark, Friedman, Gills, and Modelski, 3–23. London and New York: Routledge.

Frankel, Hans H. 1978. *The Flowering Plum and the Palace Lady: Interpretations of Chinese Poetry*. New Haven, CT and London: Yale University Press.

Friedman, Harriet. 2000. "What on Earth Is the Modern World-System? Foodgetting and Territory in the Modern Era and Beyond." *Journal of World-Systems Research* 6 (2): 480–515.

Giles, Herbert. 1901. *A History of Chinese Literature*. New York: Appleton-Century.

Hsu, Cho-Yun. 1980. *Han Agriculture: The Formation of Early Chinese Agrarian Economy (206 BC–AD 220)*. J.L. Seattle and London: University of Washington Press.

Hu, Yaowu. 2018. "Thirty-Four Years of Stable Isotopic Analyses of Ancient Skeletons in China: An Overview, Progress and Prospects." *Archaeometry* 60 (1): 144–56.

Huang, Hsing-Tsung 2000. "Science and Civilisation in China. Volume 6. Biology and Biological Technology. Part 5. Fermentations and Food Science". In *Science and Civilisation in China*, edited by Needham. Cambridge: Cambridge University Press.

———. 2008. "Early Uses of Soybean in Chinese History." In *The World of Soy*, edited by Du Bois, Tan, and. Mintz, 45–55. Singapore: NUS Press.

Huang McBeath, Jenifer, and Jerry McBeath. 2010. *Environmental Change and Food Security in China*. Advances in Global Change Research. Dordrecht: Springer.

Hymowitz, Theodore. 1970. "On the Domestication of the Soybean." *Economic Botany* 24 (4): 408–21.

Isett, Christopher, and Stephen Miller. 2016. *The Social History of Agriculture: From the Origins to the Current Crisis*. Lanham, MD: Rowman & Littlefield.

Jeong, Namhee, Ki-Seung Kim, Seongmun Jeong, Jae-Yoon Kim, Soo-Kwon Park, Ju Seok Lee, Soon-Chun Jeong, et al. 2019. "Korean Soybean Core Collection: Genotypic and Phenotypic Diversity Population Structure and Genome-Wide Association Study." *PLoS One* 14 (10): e0224074.

Keng, Hsuan. 1974. "Economic Plants of Ancient North China as Mentioned in 'Shih Ching' (Book of Poetry)." *Economic Botany* 28 (4): 391–410.

Kiple, Kenneth F. 2007. *A Movable Feast: Ten Millennia of Food Globalization*. Cambridge: Cambridge University Press.

Kiple, Kenneth F., and Kriemhild Coneè Ornelas, eds. 2000. *The Cambridge World History of Food*. Cambridge: Cambridge University Press.

Lander, Brian, and Thomas David DuBois. 2022. "A History of Soy in China." In *The Age of the Soybean:* An Environmental History of the Soyacene during the Great Acceleration, edited by da Silva and de Majo, Chapter 2, 29–47. Winwick, Cambridgeshire: The White Horse Press.

Langemeier, M. 2021. "International Benchmarks for Soybean Production." June 30, 2021, Purdue University, Center for Commercial Agriculture. https://ag.purdue.edu/commercialag/home/resource/2021/06/international-benchmarks-for-soybean-production-2021/.

Lee, Gyoung-Ah, Gary W. Crawford, Li Liu, and Xingcan Chen. 2007. "Plants and People from the Early Neolithic to Shang Periods in North China." *Proceedings of the National Academy of Sciences* 104 (3): 1087–92.

L'Haridon, Béatrice, and Hans van Ess. 2019. "Fan Ye and the Hou Hanshu." *Monumenta Serica* 67 (1): 1–6.

Liu, KeShun. 2008. "Food Use of Whole Soybeans." In *Soybeans*, edited by Johnson, White, and Galloway, Chapter 14, 441–81. Grain Chemistry & Utilization Laboratory, USDA-ARS, Aberdeen: AOCS Press.

Liu, L., and Chen, X. 2012. *The Archaeology of China: From the Late Paleolithic to the Early Bronze Age*. New York: Cambridge University Press.

Loewe, Michael. 1986. "Introduction." In *The Cambridge History of China: Volume 1: The Ch'in and Han Empires, 221 BC–AD 220*, edited by Twitchett and Loewe, 1:1–19. The Cambridge History of China. Cambridge: Cambridge University Press.

———. 1993. *Early Chinese Texts: A Bibliographical Guide*. Berkeley, CA: Society for the Study of Early China.

McNeill. John R. 2001. *Something New under the Sun: An Environmental History of the Twentieth-Century World*. First published as a Norton paperback. New York: W. W. Norton.

McNeill, John R. 1998. "China's Environmental History in World Perspective." In *Sediments of Time: Environment and Society in Chinese History*, edited by Mark Elvin and Ts'ui-jung Liu, 31–49. Cambridge: Cambridge University Press.

Milburn, Olivia. 2007. "The Book of the Young Master of Accountancy: An Ancient Chinese Economics Text." *Journal of the Economic and Social History of the Orient* 50 (1): 19–40.

Mintz, Sidney W. 1985. *Sweetness and Power. The Place of Sugar in Modern History*. New York: Viking.

———. 2008. "Fermented Beans and Western Taste." In *The World of Soy*, edited by Du Bois, Tan, and Mintz, 56–73. Singapore: NUS Press.

Myrdal, Janken. 2020. "Agricultural Treatises in Eurasia before ca. 1500: The Iceberg Method, a Survey, and a Grand Hypothesis." *The Medieval Globe* 6 (2): 1–47.

———. 2022. "A Theory Kit for World History." Science report 7. Agarhistoria. Uppsala: Department of Urban and Rural Development, Swedish University of Agricultural Sciences.

———. forthcoming. *The Agricultural History of East and Southeast Asia*. Global Agricultural Systems. Uppsala: Department of Urban and Rural Development, Swedish University of Agricultural Sciences.

Needham, Joseph. 1971. "Science and Civilisation in China. Volume 4. Physics and Physical Technology, Part 3. Civil Engineering and Nautics". In *Science and Civilisation in China*, edited by Needham, Cambridge: Cambridge University Press.

Needham, Joseph, Lu Gwei-djen, and Hsing-Tsung Huang. 1986. "Science and Civilisation in China: Volume 6, Biology and Biological Technology, Part 1, Botany". In *Science and Civilisation in China*, edited by Needham, Cambridge: Cambridge University Press.

Perez-Garcia, Manuel. 2021. "Global History with Chinese Characteristics: Autocratic States along the Silk Road in the Decline of the Spanish and Qing Empires 1680–1796." Singapore: Springer Singapore.

Peruchi Moretto, Samira, Eunice Nodari, and Rubens Nodari. 2022. "Soybean as a Critical Resource: Domestication, Dissemination and Introductions (Glycine Max)." In *The Age of the Soybean: An Environmental History of the Soyacene during the Great Acceleration*, edited by da Silva and de Majo, Chapter 1:19–28. Winwick, Cambridgeshire: White Horse Press.

Ponting, Clive. 2007. *A New Green History of the World: The Environment and the Collapse of Great Civilizations*. New York: Penguin Books.

Roberts, John M. 2004. *The New Penguin History of the World*. 4th ed. London: Penguin Books.

Shih, Shêng-han. 1959. *On "Fan Sheng-Chih Shu": An Agriculturistic Book of China Written by Fan Shêngchih in the First Century* B.C. Peking: Science Press (Kexue Press).

———. 1962. *A Preliminary Survey of the Book Ch'i Min Yao Shu: An Agricultural Encyclopaedia of the 6th Century*. 2nd ed. Peking: Science Press (Kexue Press).

Shurtleff, William, and Akiko Aoyagi. 2012. *History of Soy Sauce (160 CE To 2012): Extensively Annotated Bibliography and Sourcebook*. Lafayette, CA: Soyinfo Center.

———. 2014. *Early History of Soybeans and Soyfoods Worldwide (1024 BCE to 1899): Extensively Annotated Bibliography and Sourcebook*. Lafayette, CA: Soyinfo Center.

———. 2021. *History of International Trade in Soybeans, Soy Oil and Soybean Meal, and Trade Policy (1859–2021): Extensively Annotated Bibliography and Sourcebook*. Lafayette, CA: Soyinfo Center.

Smil, Vaclav. 2004. *China's Past, China's Future: Energy, Food, Environment*. New York: Routledge Curzon.

Sterckx, Roel. 2019. *Chinese Thought: From Confucius to Cook Ding*. UK: Pelican Books, Penguin Books.

USDA. 2022. Foreign Agricultural Service, Crop Explorer World Agricultural Production https://ipad.fas.usda.gov/cropexplorer/pecad_stories.aspx?regionid=che&ftype=prodbriefs (Accessed 2022-11-12)

Wang, Ke-Jing, Xiang-Hua Li, and Yang Liu. 2012. "Fine-Scale Phylogenetic Structure and Major Events in the History of the Current Wild Soybean (Glycine Soja) and Taxonomic Assignment of Semi-Wild Type (Glycine Gracilis Skvortz.) within the Chinese Subgenus Soja." *Journal of Heredity* 103 (1): 13–27.

Wright, Arthur F. 1979. "The Sui Dynasty (581–617)." In *The Cambridge History of China: Volume 3: Sui and T'ang China, 589–906 AD, Part One*, edited by Twitchett, 3:48–149. The Cambridge History of China. Cambridge: Cambridge University Press.

Zhang, Qing, and Lifeng Liu. 2014. "Perspective on the Trend of Soybean Production and Trade in China." In *Proceedings of Selected Articles of 2013 World Agricultural Outlook Conference*, edited by Shiwei Xu, 19–28. Berlin Heidelberg: Springer.

Zhao, Zhijun. 2011. "New Archaeobotanic Data for the Study of the Origins of Agriculture in China." *Current Anthropology* 52 (S4): 295–306.

Zhou, Chunhui, Gregory Hodgins, Todd Lange, Kazuko Saruwatari, Nicholas Sturman, Lore Kiefert, and Klaus Schollenbruch. 2017. "Saltwater Pearls from the Pre- to Early Columbian Era: A Gemological and Radiocarbon Dating Study." *Gems & Gemology* 53 (3): 286–295.

3 The second soybean cycle (1000–1860)

This chapter covers the historical dynamics behind the roots, regime and rupture of what we call the second soy cycle. After this introduction, we first critically reflect on empirical sources before we delve into the four main chapter sections: roots, regime, rupture and reflection. While the roots and rupture sections are chronological, the regime section is structured around the main functions of the soybean in the agrofood system: cultivation, trade, cuisine, fertilizer and animal feed. The roots of the second cycle began to take hold with the long shift from China's North to the South. The origins of this gradual move, as we saw in our previous chapter, are already detectable in the processes which led to the unraveling of the first soy cycle, and which accompanied the fall of Han rule. It is under the Song Dynasty (960–1279), however, that this geographical reconfiguration is consolidated and – despite the relatively short Mongol interlude under the Yuan Dynasty (1271–1368) – completed during the Ming Dynasty (1368–1644). The development of soy continued roughly along the same patterns initiated under the Song: frontier expansion, slow innovation and the spread of new processing practices. A key distinctive element of the second soy cycle, however, is the fact that soybeans began to play an important role outside China, in East and Southeast Asia.

Beyond its roots, the regime phase of the cycle is discernible roughly between 1650 and 1790, under the Qing Dynasty (1644–1912), when soybeans acquired distinctive functions in production, trade and end uses, *inter alia*, as a typical component in traditional farming, as a commercial fertilizer on far-away lands – allowing landscape simplification abroad – or as animal feed – particularly enabling large-scale pork production. Last, but not least, it is under this regime phase that soy finally became a quintessential ingredient in Chinese and Asian cuisine, even though, as we shall see, it would continue to fulfill the purpose of feeding the military and those in need during famines. We also explore the burgeoning role of the soybean trade. Europeans had been exploring Asia and benefitting from the region's knowledge, technologies, resources and innovations for many centuries. After the conquest of the Americas and the rise of the Columbian Exchange, however, the focus shifted more aggressively toward export trade. Transoceanic commerce grew exponentially and soy sauce (referred to as "catch-up") emerged as an "exotic" high-cuisine ingredient in the West, where it began to be shipped as

DOI: 10.4324/9780367822866-3

transport costs fell. Growing interest in soy among Europeans was fueled by the popular appeal of travel stories and recipe books. Even if soy sauce never became one of the largest imports during this period, by 1680 import volumes were significant enough to be mentioned in Parliamentary debate as a threat to England's balance of trade. Simultaneously, the soybean became an object of scientific curiosity in the West. As mentioned, soybeans also grew in importance as a component in fertilizers, sustaining intensive rice and cotton production. Soy also became key as feed in an ever more specialized pork production. At the same time, soybeans remained central for the peasant households that produced them as both food and fodder. In short, during this regime, soy acted as a savory ingredient, fertilizer, feed and object of "oriental exoticism".

By the late 18th century, the Sinocentric soybean regime entered its rupture phase. As with any long historical wave, the scope of the second soy cycle involves an overlap of discontinuity and relative stability; but as we argue in our concluding reflections, the overall shifts in the production, commercialization, consumption and governance structures related to soy paved the way for an eventual third cycle (Chapters 4 and 5).

A critical note on sources

While China has a long and unbroken tradition of written records, the spread of printing technologies dramatically lowered the cost of publishing after the 10th century, and the number of books in circulation increased significantly. Indeed, under the Song Dynasty, advances in movable-type printing greatly accelerated the publication of texts. In addition to books, new items such as paper money and advertisements appeared at this time. In general, the Chinese state was able to disseminate both old and newly commissioned agricultural works and authorized new editions of the Confucian classics, works of history and philosophy and poetry, fostering a new readership and a new wave of scholarship. The number of published texts mentioning the soybean rose exponentially, leading to thousands of new documentary sources on soy, as well as the eventual appearance of new genres, such as cookbooks. All this is clearly reflected in the comprehensive annotated bibliographies by the *Soyinfo Center.* Many of the original Chinese books that the bibliographies refer to have been reprinted later. An important instance of this is the seminal *Doulei* [*Selected Historical Sources on Legumes*] compiled by Li Changnian, which was reissued by Nanjing's Agricultural University in 1958 and features excerpts on soy from 69 books written between 10 BCE and 1927. Many of the passages on soy were translated into English for the *Soyinfo Center* by the Chinese food historian HT Huang and are now available free online in the database's extensive bibliographies. The quick availability of these sources, however, provides no shortcut through the difficult, but necessary, exercise of trying to situate and interpret them in their correct historical context. As Chinese agrarian historians have pointed out, among the books dealing with agriculture in this period, one can find at least three distinct genres: (a) farming manuals, financed by the state and providing direct technical guidance for common people under the mentoring hand of state officials (the main readers of these treatises); (b) publicly sponsored agricultural books which might include technical information, but whose main purpose were the eulogization of state managers; and (c) guidebooks on how to run big farms, often authored by landowners for other landowners (Bray and Métailié 2001, 324). This triple categorization has helped us in contextualizing and interpreting the multitude of "soy stories" that have survived from this period. To offer but one example, state-sponsored texts differ from those by landowners in that they tend to focus on peasants and public officials, portraying proprietors as a disruptive element in an essentially dual contract between the state and the people.

As part of the effort to contextualize these written sources, we have also traced information about the general role of trade, population and production. In line with most economic history scholars, however, we see that there is a general lack of reliable statistical evidence allowing for precise quantifications (Deng and O'Brien 2015, 240). We have nevertheless tried to explore in some depth the use and trade with soybean cakes as fertilizer in 17th-century China, especially through the lens of different *gazetteers (difangzhi)*. Gazetteers were compiled by central and local officials between the 10th century all the way up to the late Qing Dynasty for all administrative areas of the Chinese empire. There are more than 3,000 volumes available on topics as varied as regional boundaries, topography, official buildings, water, population, agricultural production, taxes and trade. Even though they constitute an almost inexhaustible resource, gazetteers are nevertheless problematic, containing many inaccuracies, biases and errors. Their household, tax registration and production data are particularly unreliable. For one thing, taxes tended to be based on a headcount of working family members; households thus had an incentive to underreport their numbers (Perez-Garcia 2021, 101). In addition, local officials could have had good reasons to exaggerate or understate agricultural production, depending on shifting national policies (Shi 2017, 12–13). Accordingly, we have not used the gazetteers to make specific quantitative claims; but we do use them to locate provinces where soybean trade took place and to complement other sources and literature. The historical record, though, remains relatively incomplete, and thus, even when some broad strokes and general trends are apparent, we must allow for a relative lack of precision in making solid claims about the geographical location, degree of marketization and overall size of soybean production. The same goes for soy production and trade beyond China, in the Southeast-Asian region, where new and more numerous publications referring to soybeans emerged, especially in the Early-Modern period. As with the gazetteers, these sources are also often unreliable from a quantitative point of view, although much information can still be gleaned from them. For example, more than 900 Japanese culinary tracts have survived from this time, a significant increase from the half dozen or so which were written in the Early Middle Ages (Rath 2010). Of course, cookbooks are very all-encompassing, but critically, they do not always speak about the food that ordinary people eat. According to Eric Rath, for instance, these texts often discuss banned ingredients:

> In the Edo period, culinary books made it possible for hundreds, if not thousands of times more people to read about certain forms of dining and about dishes they could not legally consume or afford to assemble.
>
> (Rath 2010, 99–100)

Eventually, printing also cheapened drastically in Europe from the late 16th century, leading to a comparable proliferation of literature on soy and, predictably, new challenges to source criticism. As with other periods, we have complemented the bibliographies published by the *SoyInfo Center* with alternative sources and literature, in order to understand the relative role and weight of soybeans in relation to other commodities. In particular, the archival records of the *Dutch East India Company* (VOC) include a large number of lists, registers and handwritten letters amply documenting the incipient trade in *shoyu* (soy sauce). This archive also includes VOC's papers from its Japanese factories – in Hirado (1609–41) and Deshima (1641–1869) – which the National Archives of the Netherlands have recently digitized in their entirety (Roessingh 2022). Moreover, the *Soyinfo Center* has started to publish some transcriptions and scans from VOC records that mention soy, while the *Koninklijke Bibliotheek* (KB) (National Library of the Netherlands) has created a digital database, *Delpher*, which holds millions of Dutch texts coming from various collections, libraries, museums and heritage institutions (KB 2022). This database offers, among many other things, scanned original texts from Dutch newspapers from the early 17th-century listing and sometimes selling the soy which had just entered Holland through different ports. We have also used the digitized archives of the *Kungliga biblioteket*, in Sweden, and found lists of imported goods from China, including soy sauce, brought by the *Ostindiska kompaniet* (Swedish East India Company, 1750–1812), compiled in 1883 (Nyström 1883).

Roots (1000–1650)

Slow shifts and a new geographical complementarity in China

As we have seen in the previous chapter, China suffered a strong demographic contraction between 750 and 1250 (corresponding roughly to the second half of the T'ang and almost the entire Song Dynasties). Overall, the population in North China fell by around 30–45%, while the population in the fertile and humid South doubled (Myrdal forthcoming). Crucially, by the end of this critical phase, the demographic and economic center of China would always be in the South, on the lands along the Yangtze River and beyond (Bray 1984, 427). This geographical shift involved a slow but enormous change in the orientation of agricultural activity. Briefly put, intensive wet-rice production surpassed millet as the most important crop and rice became the staple of choice in Chinese food habits (Ibid, 48). The practice was spurred by the introduction of the *Champa* rice variety, which had been brought by Song agricultural officials from the Mekong Delta and handed out freely to farmers. *Champa* rice is drought resistant and ripens early: it thus helped support the new agricultural model and sustain average yearly population growth rates of around 1.2% (Banister 1992, 51; Lander and DuBois 2022). Tax revenues also increased in turn and were partly reinvested into improved methods of water control and irrigation, drainage projects and public granaries (Kiple 2007, 44; Barbier 2011, 22; Myrdal forthcoming).

While money, people and rice accumulated in the South, the epicenter of soybean production remained in the North. A more or less constant border around the 35th latitude line, would continue to separate the predominantly rice-growing areas south of the Yellow River from the grain and bean-oriented North (Perkins 1969, 42–43). The South also saw the rise of new urban centers, and while this region was capable of growing rice and, to some extent, beans, it was still not self-sufficient. Conversely, the old garrison centers in the North now found themselves needing to acquire surplus food from beyond their area. Consequently, extensive trade links were needed to fully exploit the agricultural complementarities between the North and the South and to feed and clothe their combined populations. Additionally, the government improved the canal system which dramatically decreased transport costs and enabled a unique degree of marketization over China's vast territory (Needham 1971, 310; Barbier 2011, 22). All this allowed further urbanization and played a pivotal role in the increasing exchange of goods.

China then entered a new phase of expanding economic activity and commercial profit (Fairbank and Goldman 2006), and indeed, if Needham is right, living standards during the Song Dynasty surpassed those of medieval Europe (2004). The agricultural complementarities between the North and the South thus meant that soy became a cash crop for trade; but while the soybeans came mainly from the North, all great technical and theoretical innovations in Chinese agriculture came from the more intensive South (Bray 1984, 6:6, 59). In any case, even if the agrarian frontier-based development model under the Song Dynasty had allowed for a complex and thriving economy, it was still plagued by internal and, especially, external threats. In 1127, the Song lost control of the traditional "birthplace of Chinese civilization" (the regions along the Yellow River) to the Jurchen tribe, which had rebelled against Song rule several decades before and had been making inroads across the Northern territories. The (Southern) Song empire held fast for at least another century, until it was finally crushed in 1271, not by the Jurchen, but by another threat: the Yuan army. China would be reunified, though not under the leadership of Han Chinese, but under nomadic Mongols led by Kublai Khan – one of the grandsons of Gengis Khan (Ponting 2001, 496; Verschuer and Cobcroft 2016, 299–300). From the world historical perspective of soy, however, the shift to Mongol rule was not that significant, at least if we follow the many texts written under the Yuan Dynasty. These include long excerpts from earlier documents – sometimes going back to antiquity – which the Mongols continued to treat as expressions of "eternal truths". For instance, the special handbook *Nung Shan Chi Yao* or *Nongsang jiyao* [*Fundamentals of Agriculture and Sericulture*], written in 1273, refers to many earlier texts and claims that the three soybean varieties grown in several Northern provinces since antiquity – white, black and yellow – all had different uses, but were all beneficial (Shurtleff and Aoyagi 2021a, 121–22). The yellow bean is said to be easy to prepare (Ibid). Another example is *Nung Shu*, written in 1313 by Wang Zhen, which redelivers information from several earlier works and talks about the agricultural system of North China, confirming that soybeans and other legumes were customarily grown before cereal grains in order to increase their yield (Ibid, 123–24). Wang Zhen found that the farming practices in the North were more labor saving than those in the South and sought to promote them (Bray and Métailié 2001, 326). The point is that texts like this, published under Yuan rule, helped to preserve the record of all previous roles of the soybean: keeping the soil healthy, safeguarding against famine, providing food for troops and the poor, appearing in all types of soyfoods and sauces, serving as fodder and so on (Shurtleff, Huang, and Aoyagi 2014, 116–58). And this was not the only continuity between Song and Yuan: much of the state apparatus remained intact. Despite their conquest, the Mongols still found themselves dependent on a Han-Chinese administrative elite in many respects. Most important of all, perhaps, is that the agrarian frontier-based development mode also persisted (Roberts 2004, 449; Barbier 2011, 22; Perez-Garcia 2021, 81).

As we have seen, many sources mentioning soybeans are little more than reiterations of earlier texts, but some novelties and shifts still appear. For example, texts like *Hanshi zhishuo* [*Master Han's Straight Talk*] from 1273, *Nongsang yishi cuoyao*, or *Nung Sang I Shih Ts'o Yao* [*Selected Essentials of Agriculture, Sericulture, Clothing and Food*], from 1314, emphasize the role of soybeans as feed for cattle and horses (Shurtleff and Aoyagi 2021a, 119–24). This can be pointing to the fact that that husbandry became clearly more important under the Mongols. The Han-Chinese agrarian system, as we saw in our previous chapter, was characterized by only a small number of larger animals (dairy products were frowned upon as "barbaric"). The Mongols, by contrast, brought a "pastoral turn" in the North, making fodder more important than it had been previously (Myrdal 2022, 17). The lower demographic pressure in the North also allowed more space for husbandry. Nonetheless, the point should not be overstated: most agricultural treatises from this period focus on crops whilst animal husbandry is not mentioned as often (Myrdal forthcoming).

From the late 1340s onwards, people in the countryside suffered famines resulting from droughts and floods. Adding to this vulnerable situation, the final 20 years of the Yuan Dynasty were marked by internal political struggles and social unrest. In many areas, the particular situation of the peasantry was desperate and the general conditions for the successful continuity of Yuan rule began to disappear (Barbier 2011, 22). In 1368, the Yuan military was defeated by the armies of the Ming Dynasty and retreated to the Mongolian steppe, which remained under the rule of Northern Yuan, together with Manchuria, until 1635; China returned to Han rule. As with previous political shifts, however, there were many continuities in the organizing principles and practices of the state administration. The Ming bureaucracy, for instance, kept the Yuan model of organization in three state commissions: civil, military and surveillance. Additionally, many Mongols continued serving as public officials until the late 16th century. While South China remained the economic, cultural and administrative center, the Ming initiated a revival of the North. As the northernmost borders were now secured against the "barbarians" by a strong state and army, the intensive agricultural system described in Chapter 2 – with crop rotation and the use of green manure, etc. – returned to the North (Bray 2019, 270). The North's brief pastoral turn under Yuan rule met its end, and, consequently, a sharp decline in meat and dairy production and consumption followed (Myrdal forthcoming). The land was managed in small parcels by peasants who, by and large, held direct access to the land they worked. While landholdings were, in general, small, land was still plentiful in comparison with the more populous South.

The evolution of soyfoods throughout East Asia

Nourishment and cuisine have been among the soybean's key functions since ancient times. We saw, however, that early soyfoods – often just beans cooked into granules or congee – were viewed as little more than a coarse sustenance for poor people and soldiers. As cultivating techniques, tools and different varieties of soy spread to Japan and the Korean peninsula from China, soy congee probably acquired similar roles there (Kiple 2007, 48–49). After all, soybeans had been moving back and forth between Korea and Japan since at least the 3rd century CE (Du Bois 2018, 26–27). But while congee remained a nutritional supplement (Huang 2000, 352), soybeans gradually went from being eaten mostly boiled or steamed, to being prepared in ever more sophisticated ways. The spread of soybean cultivation also reached the Indian subcontinent around the year 1000, introduced from China through the "silk route" (Hymowitz and Shurtleff 2005). While soybeans did not become an important crop in India at this time, the hardy black-seeded soybeans had some success in the hilly areas of Assam, Bengal, Manipur, and the hills of Khasi and Naga (Werner and Newton 2005, 4: 45). Thus, soy grew both in reach and sophistication.

The process leading to the improvement of soybean's flavor and digestibility, as mentioned in Chapter 2, was kickstarted with the development of new fermentation techniques, most importantly, mold (Huang 2000, 278). The method had been known since Han times, but it would take many centuries until fermentation was used in a wide range of soyfoods and sauces throughout China (Ibid, 595). For instance, fermented black soybean paste, *jiang*, which we encountered in the previous chapter, reached Japan before the year 1000. One popular theory is that first *jiang*, and later tofu, were brought over from China by a delegation of Buddhist monks (Huang 2000, 317–19). Accurate or not, it seems clear that Chinese Buddhist monks played an important part in disseminating soyfoods, tea and deep notions that killing animals was taboo (Kiple 2007, 49).

The development of food-processing techniques for soy is reflected clearly in the growing field of culinary treatises, which we have mentioned above. One example from the late Song Dynasty is *Wushi zhongguilu*, or *Zhonggui lu* [*Madam Wu's Recipe Book*], from 1200. It includes one of the earliest recipes for soy sauce and reflects the evolving trends in Eastern regional cuisine. Interestingly, this is also the first known cookbook written by a woman, even though all that is known about her is that she lived in Pujiang in Zhejiang province (Shurtleff and Aoyagi 2021a, 114–15). What this and other sources speak of is the appearance and spread of soy sauce even beyond China; in Japan it is thought to have appeared in 1228. In one account, the Buddhist monk Kakushin, who had returned home from Song Dynasty China, was fermenting soy soup (*miso*) when he discovered that the liquid which gathered on the bottom of the vats could be used as a tasty seasoning. He called it *tamari*, which is widely seen as the predecessor of soy sauce proper (Ibid, 116). While Buddhist monks clearly used both miso and soy sauce in Japan, it was not until the 14th century (under the Muromachi period) that soybean cultivation became common among farmers, who began to increasingly make their own miso.

Along with processing techniques and the spread of cookbooks, the rise of soy sauce was spurred by the appearance of big and growing cities and the accompanying affluent urban population which engaged in wage labor outside of agriculture and demanded different types of soybean foods. These new soyfoods within, for instance, the incipient city life in Hangzhou – the capital of Southern Song – are captured in the 1275 text *Mengliang Lu* [*Dreams of the Former Capital*] by Wu Zimu (Huang 2000, 396; Shurtleff and Aoyagi 2021a, 120–22, 6). One of its sections speaks of the goods sold on the city's many markets, among them, of course, soybeans as bean-flavored water (*douershui*), sweet soybean soup (*gandoutang*), young soybean congee (*douzizhou*), brownish-green soybeans (*geqingdou*), salted young soybeans or maybe green vegetable soybeans (*yandouer*), sugared yellow young soybeans (*douer huangtang*), fermented black soybeans (*shi*), fermented soybean paste (*jiang*), fermented black soybean sauce (*shizhi*), cooked soybeans ground to make a cake (*doutuan*) and pan-fried tofu (*doufu*) (Huang 2000, 329, 436). According to Wu, urban dwellers commonly purchased prepared or semi-prepared foods, at least in Hangzhou, and soybeans were often included in inexpensive meals such as soups and noodles. *Mengliang Lu* states that there are "seven necessities of life" which consist of "firewood, rice, oil, salt, soybean paste (*jiang*), vinegar and tea" (Huang 2000, 329, 436).

Numerous written sources from the 14th century cover the different types of tofu (Hymowitz and Newell 1981, 159–63). In the spread of these standardized ways of preparing tofu, the Buddhist monks – again – seem central in the spread of these techniques, as illustrated in this quote from the Chinese poem *Dougu*, written by Su Ping in 1500:

> The best is King Wainan's skill, you see the beauty when you peel. Ground in mortar and milk flows. Boil in water and it turns to snow. Soak in the jar and white curds show. Cut apart with a knife yet the jade is sound. Who knows the delicacy of the curd? Only the Buddhist and Taoist.
>
> (Shurtleff and Aoyagi 2021a, 136)

One of many varieties of tofu that emerged under this period was *sufu*, a highly flavored curd made in two steps: first, fungal solid-state fermentation and second, aging the resulting material in brine containing salt and alcohol (Han, Rombouts, and Nout 2001; Shurtleff and Aoyagi 2021a, 123). *Sufu* was a very versatile product, varying in flavor and color (red, white or grey), depending on method of fermentation used: mold-based, natural, bacterial or enzymatically ripened and dressing mixture (Han, Rombouts, and Nout 2001, 2–4). These soy curds were extremely popular perhaps especially on account of their high protein content. Meat consumption was low and probably falling. As we have just seen, dairy products had been widely consumed under the Yuan Dynasty, but had then been pushed out as a consequence of the cultural backlash against Mongol culture (Myrdal 2022, 17). The Song Dynasty used religious injunctions to consolidate an explicit aversion to beef and demographic growth left less room for pasturing livestock in the main arable regions (Chang 1977, 74, 201). Apart from being rich in key nutrients like calcium and protein, soy products had the important advantage of needing less land than that required by meat and dairy farming. Finally, draught animals were not seen as edible, but as part of the household (Bray 2018). In general, Chinese rulers placed a higher premium on the animal's contributions to agriculture than on the animal's status as a source of food (Shurtleff and Aoyagi 2014a, 100). This was particularly true for densely populated South China, where cattle were kept only as draught animals. In sum, given the predominance of small family farms, almost no land was used exclusively for pastures (Bray 1984, 6:4; Banister 1992, 51; Bray 2018).

As with soy sauce, tofu also spread beyond China, and again, Buddhist monks on the move seem to have been the conduit. In Vietnam, for instance, an old song suggested that "if you want to enjoy tofu with traditional soy sauce, sharpen your knife and scissors, shave your head and become a monk" (Du Bois 2018, 30). It was not only religious links, however, but critical commercial ones that facilitated the dissemination of different soy products such as tofu. Chinese merchants, for instance, were also instrumental in spreading their culinary culture throughout East Asia, including soyfoods. These traders had long-established links with many other realms in this region, going back at least to Han times, but the intensity of these links increased significantly during the 12th and 13th centuries. Chinese traders began making regular round trips to Southeast Asia and supplanted the Arabs as the foremost trading partners throughout the region, so much so that the Song Dynasty established a permanent navy in 1132 to protect and support Chinese commercial interests from the East China and Yellow Seas all the way to the Arabian Peninsula and the Red Sea (Needham 2004). During Mongol rule, Chinese merchants and government-sponsored navigation were present across the Indian Ocean and dominated the maritime trade in the Southeast (Roberts 2004, 459–60; Pomeranz and Topik 2006, 32–33; Gordon 2008, 185). China's seafaring and navigation capacities were unmatched in both distance and size, able to take three times more tonnage than any other navy (Reid 1993, 2:12). Unsurprisingly, China's dominant position in late-medieval East Asia led to a very broad dissemination of Chinese cooking methods and customs, particularly, to Japan, Korea and other nearby lands (Liu 2008, 442). Chinese traders spread across Southeast Asia, with many settling down in different trade hubs (Reid 1993, 2:15). Not least in Manila (the Philippines), settled Chinese traders, the so-called Sangleyes, became a major influence in, and probably shared their soyfood culture with, Filipino society long before Spanish colonization. However, transport prices were still high, so trade was dominated by high-value, non-bulky goods. Even in the absence of quantitative data, we know that soy products were not among the most important exchanged goods in the intensified intra-Asian commerce (Reid 1993, 2:12). We also know that Chinese soy-based culinary arts were spreading throughout the region, whether or not commercial exchanges were involved. In this way, *jiang* (soybean paste) and soy sauce grew in importance in Japan, Korea, Indonesia and India (Huang 2008, 50; Mintz 2008, 60); fermented black soybeans (*kuki* or *shi*), in turn, also began to spread in many East Asian countries (Shurtleff and Aoyagi 2021a, 4–5). A local variant of fermented soyfood even emerged in Japan – *natto*, made of soybeans processed with a bacterium called *bacillus subtilis*. A 12th-century legend tells of the struggle between the forces of Lord Natto and Lord Salmon – evidently, Natto's forces prevailed, though the point is that the tale can be interpreted as an allegory on the superiority of vegetarian over animal-based foods (Du Bois 2018, 33–34; Shurtleff and Aoyagi 2021a, 134). *Natto* is but one example: many other soybean foods began emerging in records from outside of China during this period, with their attendant origin stories and myths (Bray 1984, 6, 423, 514).

Alongside the different fermented and processed forms of soybean foods, whole pods of young soybeans, green soybeans and soybean sprouts were also popular from the late Song period onwards (Huang 2008, 51). One surviving text from the Song Dynasty, *Shan Chia Chhing Kung* [*Basic Needs for Rustic Living*], mentions a savory dish of sprouted black soybeans. To prepare it, one first had to soak the beans in water, grill them with oil, salt, vinegar and spices and finally roll them in a sesame pancake (Huang 2000, 296). During this period, sprouts apparently became popular and were sold in the markets of many cities. Sprouts represented a simple and convenient way of turning the rather indigestive soybean into a tasty and nutritious vegetable. Yet again, Buddhist monks seem to have been important in the spread of *edamamé* (green vegetable soybeans or green soybeans in pods). According to *Nichiren Shônin Gosho Zenshu* [*The Collected Writings of Saint Nichiren*], the well-known Buddhist Saint, Nichiren Shônin, left a note thanking a parishioner for the *edamamé* he left at the temple (Shurtleff and Aoyagi 2014a, 100, 118). However, the fact that sources such as the *Nichiren Shônin Gosho Zenshu* were produced by the monks themselves (who were among the key bearers of a written tradition) might give us pause to question if the traveling Buddhist monk was indeed so central in the dissemination of soyfoods.

Irrespectively of the role of monks, the most important innovation in processing soybeans was, in fact, crushing. The exact origin is unknown, but soy oil is mentioned already in the 10th century, both as an ingredient for caulking boats as well as for frying tofu. Soy oil is further praised for its nutritional and therapeutic uses in *Wulei xiangan zhi* [*Treatise on the Mutual Responses of Things According to Their Categories*] by Lu Zanning, written in 980 (Shurtleff and Aoyagi 2016b, 29). We know that crushing was happening from the turn of the first millennium, but the oil (or wedge) press itself would not be described until 1313 in

Wangzhen nongshu [*Wang Zhen's Agricultural Treatise*] (Shurtleff and Aoyagi 2021a, 124). About a fifth of a soybean's weight is oil, but extracting it is in fact very difficult and quite sophisticated oil presses are needed to produce the oil and soybean cake (the pasty residue left after crushing). No wonder, then, that soy oil makes a relatively late appearance on a scene which already featured rapeseed, almond and perilla seed oil since the 7th century (Chang 1977; Shurtleff and Aoyagi 2016b, 29). While soy oil could be used for many different purposes, edible oil seems by far to have been the most important; eventually reaching a slot among the proverbial "seven necessities of life" (Shurtleff and Aoyagi 2021a, 100, 113–14). Soybean cake, however, is less frequently mentioned in early sources. As we shall see, it would eventually gain prominence both as fertilizer and feed under the regime phase during the Qing Dynasty.

To conclude, once China began to master the art of processing the soybean into products like different types of bean curds, soy milk and particularly the fermented flavoring of soy sauce, the soybean was celebrated on gastronomical grounds for the first time (Mintz 2008, 64). While grinding different soybeans varieties to mix with flour to make noodles and pasta, or preparing soy-based sauces and other fermented soyfoods had already been practiced since ancient times, it took many more centuries for the techniques to be fully developed and popularized. Indeed, soyfoods did not spread completely throughout Chinese territory until the mid-Ming Dynasty (Ibid) – in the regime phase. Soyfoods also spread throughout feudal Asia under Chinese influence, but were soon appropriated and given a local, traditional inflection (Huang 2000, 317–19). Finally, the spread of crushing technology opened up key new roles for soy.

The Asian inward turn and the European outward thrust

In several ways, the Ming seizure of power in 1368 constituted an important political departure in the history of China, noticeably in terms of state ideology. Old Confucian ideals of harmony, self-reliance and tradition continued to hold sway throughout the long and fractured series of imperial shifts since Han rule and the Ming Dynasty was no exception. Its Confucian revival meant extolling the virtues of public service and agriculture, while decrying the decadence and lack of ethics of commerce (Barbier 2011, 90). Ming rule redoubled its Confucian credentials and dissociated itself sharply from Mongol openness to trade, expressing much more disdain and distrust for foreigners and merchants (Nakajima 2018) and turning its focus more closely on the agrarian economy (Barbier 2011, 175–80). As for trade, Ming rulers looked inward, first abandoning commercial routes which had collapsed with the bubonic plague and the fall of the Mongols and then severely restricting private foreign land and sea exchanges. In time, all foreign private commercial ventures were banned (Pomeranz and Topik 2006, 47; Colla 2008, 124–26; Nakajima 2018). All remaining international trade was instead restricted to the so-called "tribute system" – a highly ceremonial arrangement in which all foreign rulers had to send tributary missions to the Chinese emperor to prove their loyalty and submission to him (Colla 2008, 124–26), often by "kowtowing", that is, kneeling and hitting the floor with one's head. The Chinese then responded with gifts, typically books (particularly Confucian classics), musical instruments, silk, porcelain, paper money and finally, the coveted right of (heavily conditional) "mutual trade". Those "barbarians" who did not send tributary missions were only allowed very limited trading rights, i.e. exclusively in specified ports and during specific periods. Otherwise, their only remaining chance was to trade indirectly by having their goods included in the tribute offered by others (Pomeranz and Topik 2006, 12). As Gakusho Nakajima puts it, this was a centripetal system, whose overriding principle "was the unity of tribute and trade, which meant that there could be no trade without tribute" (2018, 142). Enforcement was weak though, and smuggling was widespread (Reid 1993, 2:13)

After the Black Death, European demand for Southeast-Asian goods – cloves, nutmeg, silk, spices, tea and porcelain – grew suddenly and rapidly. The old overland routes, as we saw, had become riskier and the Silk Road links had been severely disrupted, but major advances in European shipbuilding and navigation technology enabled new naval expeditions across the Atlantic, Indian and Pacific Oceans (Pomeranz and Topik 2006, 42; Barbier 2011, 180). As China turned inward, Europeans began pushing outwards.

Following the conquest of Malacca in 1517, Portuguese merchants arrived at Guangzhou Bay and established themselves as the dominant European maritime power of the 16th century. After much diplomatic tussling, the Portuguese had to settle for renting a permanent base in the island of Macau (just outside Guangzhou). From Macau, however, the Portuguese opened a direct European link (Colla 2008, 123), which became strategic when Ming emperor Jiajing banished all overland trade between China and the West in 1524 (Li 2019). Portugal managed to monopolize commerce between Japan and China in 1543. In 1554, they signed the Luso-Chinese Trade Agreement in which Ming rulers sanctioned "mutual trade" in Guangzhou Bay on condition that it was brokered exclusively by Chinese merchants and that the Portuguese pay taxes (Pomeranz and Topik 2006, 12; Nakajima 2018). In satisfying the largely symbolic conditions of the Chinese, however, Portugal thus acquired control of the vital Macao-Canton (Guangdong) axis, which opened up to further links into the Pacific (Perez-Garcia 2021, 15).

At this point, Europe was still a backwater in the periphery of the lucrative Southeast-Asian trade and China was by far the world's most powerful and advanced economy. The balance, however, was beginning to change. Ming rulers decided to replace depreciated paper and copper currency with silver and enormous amounts of Japanese silver flowed from Nagasaki into Chinese markets via Macao in exchange for Chinese products (Colla 2008, 124). Soon the region was awash with silver and not only from Asia.

In 1565, Andrés de Urdaneta discovered a safe route connecting Manila directly with Acapulco. From the Spanish Empire's perspective, the Acapulco – Manila line was part of a larger (24,000 mile) link in which goods could be sent across the Pacific, re-loaded on the "*Flota de Indias*" in Mexico City and shipped, via the Caribbean eventually to Seville. Manila – founded by the Spanish in

1571 – became the central node for Chinese trade with the Americas: silver from Acapulco would now arrive at South-Chinese ports and would be exchanged for silk, spices, cotton cloth and porcelain (Kiple 2007, 145; Barbier 2011, 225–26; Nakajima 2018). Exchanges between all the continents soon took off as never before (Gordon and Morales 2017, 3), and Europe's expanding territorial and commercial presence became mutually reinforcing (Barbier 2011, 225–26). Of course, commercial relations went well beyond flows of silver and other goods: seeds from a variety of New World crops (peanuts, pepper, maize, mulberry, sweet potato, other potatoes and tobacco) also made their entry into Asia via Manila (Kiple 2007, 145). This great transformation is what historians usually call the Columbian Exchange, and, as Alfred Crosby argued back in 1972 (not perhaps without some hyperbole), it constitutes the greatest ecological restructuring ever to take place, changing the international geography of food production forever.

Though backed by their states, who actively pushed the expansion of trade and conquest, 16th-century European merchants had to double as generals, courtiers and even suitors in order to be successful: many even took local wives to gain better access to Asian markets and societies (Pomeranz and Topik 2006, 36). Many Southeast-Asian women were merchants long before the arrival of the Europeans. To secure a woman's contacts and skills often required intermarriage (Pomeranz and Topik 2006, 28–29). All the same, when neither contracting, diplomacy or networking sufficed to procure goods or gain market access, direct use of force was generally the default (Ponting 2001, 525). The ill-treatment of locals was not a European monopoly, however, but shared by many indigenous elites (Pomeranz and Topik 2006, 61). But the difference is one of degree. As Andre Gunder Frank and Barry K. Gills argued, Europeans "specialized in exploiting global differences in resources, production and prices to maximize their profits as middlemen, and where convenient they used force to assure their own participation in this exchange" (2000, 7). Thus, while China was still the world's strongest and wealthiest country, the consolidation of the regime phase in our second soy cycle took root within a rapidly changing system of international relations and power. Moreover, during the six centuries long "roots" phase, soy became a central culinary ingredient throughout East and Southeast Asia. As we saw above, moreover, the production of tofu and new soy oil extraction techniques opened up new roles and functions for the soybean, for instance, as commercial fertilizer. The intensive agricultural systems of rice and other nutrient-demanding crops like corn which rose to prominence thanks to the Columbian Exchange led to an ever-growing need for soy.

The soybean regime (1650–1780)

During the interval between 1650 and 1780, the changes and developments mapped out thus far came to full fruition in the regime phase of the second soy cycle. Due to China's hegemonic power in the region, its technology, farming practices and culinary traditions spread to neighboring countries (Bray 2004, 10–15);

the many roles and functions of soy in this regime were no longer exclusively Chinese, but shared all over East and Southeast Asia. As with other sections of this book, our study of the regime explores the main functions of the soybean in the agrofood system: cultivation, trade, cuisine, fertilizer and animal feed.

Soybeans in cultivation: traditional roles alongside expansion

China at this time was mainly concerned with domestic and continental matters, with emphasis on stability and agriculture (Barbier 2011, 27–30, 235). The newly installed Qing Dynasty knew well that pressure on the peasantry had played its part in all dynastic changes in China (as it had in 1644, when the Manchus themselves seized power). Accordingly, previous efforts to foster social stability through the exploitation of frontier areas were intensified (Barbier 2011, 271). Clearly, official records of cultivated land areas from this period are not reliable (Shi 2017, 16), but many scholars have contributed with estimations. According to the environmental historian John Richards, the total amount of land cultivated in China expanded by 50–60 Mha between 1400 and 1800. China's population, meanwhile, went from 75 to 320 million (Richards 2003, 112–47). Expanding internal frontiers of settlement was the main factor behind economic growth and social change in early modern China, but the spiraling population growth pushed the limits of this frontier-agriculture model, with land-per-capita ratios actually falling – aggravated by the traditional practice of bestowing land equally among all sons (Isett 2007, 293). Qing rulers responded by intensifying frontier expansion through the building of dams and artificial reservoirs for the large-scale supply of regulated irrigation, which made possible the conversion of lowlands into rice paddy systems (Barbier 2011, 98–99). The rulers also intensified efforts to monitor harvests, creating and disseminating information and investing in ambitious infrastructure projects (McNeill 1998, 33–34; Pomeranz and Topik 2006, 57; Isett 2007, 7, 9, 241). These policies generated agricultural surpluses capable of supporting large urban-based populations of elites, soldiers, priests, artisans and intellectuals (Barbier 2011, 98–99).

Whereas rice and wheat were considered superior crops in China, soybeans were also important. North China remained the most important soy-producing hub.[1] This area was characterized by a (re)intensified crop-centered system, specialized in soybeans, cotton, sorghum, millet and wheat (Li 1998). The different varieties of the yellow soybean were the most commonly cultivated (Huang 2000, 345). Now as before, soy was used to maintain the fertility of the soil, ploughed under as green manure, or in crop rotations, providing nitrogen to future harvests. According to Kenneth Pomeranz, a typical North China farm would receive three crops of nitrogen-fixing soybeans over an average six-year rotation (2021, 227). Soybeans thus allowed for continuous cultivation, avoiding the need to fallow (Isett and Miller 2016, 31), a practice which became as vital again during the Ming Dynasty as it had once been under Han rule. In both cases, fallowing almost completely disappeared and crop rotation with soybeans became essential to sustain the North's demographic growth. Meanwhile, soy's role as cash crop also gained significance (Bray 1984, 6:514). As before, original texts from this period reproduce information from earlier works and offer advice on when to plant, weed and harvest different varieties of soybeans across the various climate zones and soils (Kuo 2013, 83–84). One example, beautifully covering the varieties and uses of soybeans, is the 1637 *Tiangong kaiwu* [*Exploitation of the Works of Nature*], by Song Yingxing:

> There is one variety called the Ta tou, or Large bean, which grows in two colors, black and yellow. Those black and yellow beans must be planted about the time of the ch'ing ming season. The yellow bean includes three kinds, as follows: Wu yueh huang tou, or Fifth month yellow bean, Liu yueh pao, or Sixth month bean, and Tung huang tou, or Winter yellow bean. The Wu yueh hung tou, or Fifth month yellow bean, yields few grains in its pods, while the Tung huang tou, or Winter yellow bean, yields at least double those of the preceding. The Hei, or Black variety, is harvested in the eighth month. It is customary when taking a long journey north of the Huai river, to feed the horse with black beans in order to make it strong. The abundance or scarcity of the Ta tou crop depends, first, upon the fertility of the soil, second, upon diligent weeding and cultivation, and third, upon sufficient rain and dew for moisture.
>
> (Shurtleff and Aoyagi 2014a, 160)

In the fertile and densely populated South, rice was still the most important crop by far. Paddy-rice production was intensive, requiring ample water reserves and fertile soil (Perkins 1969; Barbier 2011, 99, 170–71). Additionally, farmers in the South also increased the production of cotton (the next most common crop), mulberry for silkworms, tea, sugarcane and citrus fruits. While soybeans were not a very central crop in the South, they were nevertheless grown in the Southern provinces of Zhejiang, Hubei, Hunan, Jiangsu, Anhui, Jiangxi and Sichuan (Shi 2017, 101–02).[2] Soybeans and other dry crops could also enter crop rotations in the rice paddies. The practice of triple-cropping wheat with early rice and autumn soybean (or buckwheat or potato) was common in Zhejiang and Sichuan (Shi 2017, 75). In mulberry groves, intercropping was also common and soybeans were sometimes grown beneath the trees to raise soil fertility (Li 1998, 15, 51). As we will explore in greater depth in the section below on the role of soybean cakes as fertilizer, soybeans not only enriched soils when included in cultivation, but were also a potent and advanced commercial organic fertilizer for the intensive farming systems of the South. Bean cake from soy is estimated to have boosted agricultural output significantly where applied (Li 1998; Barbier 2011, 98–99).

Moreover, as we saw, the European conquest of the Americas not only gave way to a substantive flow of capital into China, via Manila, but also brought with it a voluminous transfer of plants and animals to Eurasia. China was fastest in the large-scale adoption of these new crops; peanuts, sweet potatoes, potatoes, maize, tobacco and yams began changing agricultural landscapes throughout the 17th century (Perkins 1969, 47; Crosby 1972, 150). Maize and sweet potatoes, were quickly adopted and cultivated on hilly uplands without irrigation, escalating problems of erosion and river siltation. China's population was able to grow once more (Kiple 2007, 145), but it did so based on a new wave of frontier expansion. Thus, as American crops spread across the Chinese Highlands, so did social and ecological risks (Mann 2011, 180).

Once more, despite all the revolutionary changes happening in this period, agricultural productivity did not undergo similar great changes, but remained generally stagnant (Perkins 1969, 56). Apart from the appearance of new crops and the use of bean cake as fertilizer (see below for more detail), increases in agricultural output mainly depended on continuous expansion of the agrarian frontier into hitherto unproductive hillside topographies. Millions of Han-Chinese migrants, in response to the inducements of internal settlement frontiers, moved long distances and provided a constant human flow into the West and its sparsely populated mountain areas (Richards 2003, 112–47; Krech, McNeill and Merchant 2004, 216). Pomeranz estimates that "long-distance migrations to underdeveloped parts of China during the late seventeenth and eighteenth centuries alone easily surpassed 10,000,000, with most of the colonists establishing freehold farms" (2021, 84). The hardy soybean also spread with migration. As overcrowded peasants sought new horizons in areas such as the Shandong peninsula and its highlands, more ecologically fragile and less productive soils were also incorporated. These marginal soils were unsuitable for grain production, but served for cash crops such as soybeans and peanuts. Their sale in turn raised cash to buy grain (Isett 2007, 244). Soybeans were also important in the still sparsely populated lands on the empire's periphery (at least during the early days of the new Qing Dynasty), e.g. Shandong, Anhui and Taiwan, where more millet, wheat, soybeans and timber were produced than consumed locally (Isett 2007, 13–14; Shao 2017). With time, however, these areas ceased to yield food surpluses. While total output was on the rise virtually everywhere, output per capita was not. Thus, as acute demographic pressures arose, surpluses tended to vanish. In the long run, food production could not keep up with population growth.

The fact that soy had already reached other Asian regions allows us the benefit of a comparative perspective, especially if we consider the case of Japan, which contrasts sharply with developments in China. The feudal, military Tokugawa shogunate (also known as the Edo shogunate or *Edo bakufu*) had ruled Japan since 1603. The Tokugawa government took several measures to increase food production, including soybeans among other crops (Isett and Miller 2016, 85–87). An edict passed in 1639 actually sought to force all peasants to plant soy and adzuki beans between their rice fields and farms. This significantly increased soy acreage (Shurtleff and Aoyagi 2014a, 184). As the Japanese Neo-Confucianist philosopher and botanist Kaibara Ekken (1630–1714) noted in his seminal *Yamato honzō* [*Medicinal Herbs of Japan*], soybeans became the second most widely produced crop after rice (Shurtleff and Aoyagi 2021a, 207). Whether empirically accurate or not, it is clear that almost all Japanese peasants at this time grew soybeans and made their own miso and soy output rose.

Agricultural production in Japan doubled between 1600 and 1850; its population, however, only rose by half (Roberts 2004, 475). Additionally, as we shall soon see, the use of soybean cake as fertilizer was much more extensive here than in other parts of Asia. In short, Tokugawa Japan managed to escape the Malthusian contours of the crisis afflicting China, which periodically struggled to match agricultural output with demographic growth. Meantime, the consolidation of soybean production in Japan would condition relations between both countries in the long term, when both powers would clash over Manchuria. This Northeast Chinese region (constituted of the provinces of Heilongjiang, Fengtian/Liaoning and Jilin) was the home of the ruling Manchus and had up to then avoided high population density. It was thus still capable of yielding an exportable surplus of soybeans and grains. The Manchu Qing rulers were keen to avoid antagonizing the more numerous Han, but were nevertheless ambivalent about the exploitation of their home region, which had long been demarcated as separate from China Proper (Isett 2007, 2–7). Indeed, a formal prohibition had banned Han-Chinese peasants from settling there (Deng and O'Brien 2015, 257; Shi 2017, 426). However, mounting population pressures became a strong push factor for Han-peasant migration into the region and such inflows continued through illegal purchases and squatting. While rulers tried to control these illicit settlements and sometimes returned colonists to China proper, many more managed to remain (Isett 2007, 7–9). Thus, after only a few decades of Qing rule, Han immigration is thought to have reached over 1 million (Pomeranz and Topik 2006, 56), and Manchuria quickly became one of the regions with the fastest rise in the production of soybeans and traditional grains. Additionally, maize had now entered into the North and was often intercropped with soybeans (Perkins 1969, 47–48). In the regime phase of this soy cycle, then, Manchuria gradually became one of the very few areas in China capable of generating soybean and grain surpluses, with ominous consequences to be discussed in the following section.

To conclude, soybeans under the regime phase still played many of its traditional cultivation roles – improving soils as a rotational crop, or green manure, while providing food and feed to peasant households, particularly in North China. At the same time, the soybean became an increasingly important cash crop and, along with the general agrarian frontier development model, soy expanded to new, and often more fragile, soils. Meanwhile a contrasting situation played out in Japan: there, agricultural output per capita remained stable (as population did not grow). Although some areas in China still yielded surplus production of grains and beans, Manchuria was the most important and, as such, it played a central role in the heavily regulated, yet expanding, soy trade, to which we now turn.

Soybeans in trade: expansion alongside restrictions

The commercial role of soybeans during the regime phase of the second soy cycle took shape in a context of some ideological ambivalence regarding the social standing of trade. While the Qing held fast to the neo-Confucian derision of profit making and speculation (extolling instead the virtues of public service and agriculture), there was considerable investment in trade infrastructure and numerous exceptions within an overall protectionist trade system. The commercial role of soy then developed and expanded alongside an imperfectly restrictive set of policies. Importantly, the commercial role of soy at this time acquired different characteristics depending on the geographical scale at which it took place. So our examination will proceed accordingly: we will first look at the soy trade inside China, then across Asia and finally, at the emergent global soy market.

Soy trade in China

From the mid-Ming Dynasty to the late-Qing Dynasty, the amount of agricultural output that reached the market increased significantly, reflecting a unique degree of internal marketization by world standards (Perez-Garcia 2021). Cash crops planted on a large scale included cotton, hemp, mulberry, tea, sugar cane, tobacco and soy. This level of domestic commercialization naturally tended to reinforce geographical specialization (Katō 1953; Shi 2017). Moreover, as contact with the outer world intensified, growing overseas demand for raw cotton, silk and tea created an incentive for many peasant farmers to shift production to non-food commercial crops, while buying food from other peasants (Marks 1998, 163). In addition, as observed by Pomeranz:

> [T]he Qing strongly encouraged rural women to spin and weave, both to bolster the economic stability of tax-paying peasant households and because the example of a mother at her loom was considered good for the moral education of her children. Officials distributed cotton seeds, printed instructional pamphlets, encouraged the teaching of relevant skills, and promoted the "man plows, woman weaves" division of labor as the basis of strong families.
>
> (Pomeranz 2021, 87)

The process advanced quickly and by the end of the 18th century, for instance, the agricultural economy of Lingnan had become thoroughly commercialized, with even peasant farmers in westernmost Guangxi affected by market demand centered on Guangzhou and the Pearl River Delta. Geographically, the North specialized in selling millet, soybeans, soybean cakes, sorghum and timber in exchange for the South's silk, cloth, paper, tea and cotton (Ponting 2001, 377; Pomeranz and Topik 2006, 42, 62; Huang McBeath and McBeath 2010, 16). Manchuria also emerged as an additional supplier of soy to the South during this period. Soybean and soy products became the most important ones in the southward exchange along the Grand Canal (Bray 1981, 83, 628–29). The main soy buyer was the populous Yangtze Delta, which by the mid-18th century had a population density over 2,590 people per square kilometer (Li 1998, 113; Pomeranz 2021, 85). Prices of grains, beans and cotton cloth in different regions began to move in parallel, confirming market integration (Isett 2007, 257). However, while the harmonization of commodity prices between some places was important, the market system was still fragmented at many levels and probably many commodity prices on local markets diverged from the more important and connected markets (Shi 2017, 198). Moreover, notwithstanding that the majority of the trade items were agricultural products, on average only about 10–15% of total harvest was surplus. Thus, in relative terms, most agricultural output never entered the market (McNeill 2001, 38–40; Isett 2007).

According to Chen Ciyu's estimates, cited in Li Bozhong (1998, 113): "5.2 million *shi* of beans and bean cakes from Manchuria annually passed the Huai'an customs astride the Grand Canal in the Qianlong period (1736–95)". Another study calculated that Manchurian soy oscillated between five and ten million *shi* annually, with almost everything going to the prosperous region of Jiangnan – encompassing Shanghai, part of Jiangsu, Anhui, Jiangxi and Zhejiang provinces (Li 1998, 210; Fan 2007). However, still another study, based on the tax revenues from the soy trade, indicates that while volumes were expanding rapidly (tax revenues from soy trade doubled in less than a decade), the total volume of exported soybean and bean cake from Manchuria in the second half of the 18th century was approximately 1.2 million *shi* (Katō 1953: 601–05; Isett 2007, 229). While the figures discussed are empirically uncertain (Shi 2017, 98), there is less controversy surrounding the fact that demand for soybeans, soy oil and bean cake greatly increased since the early days of Qing rule, leading to soaring prices (Katō 1953; Li 1998; Isett 2007, 230–45). The relative prices of cloth, silk and sugar deteriorated in favor of grains and beans. The rising value of soy reflected its high demand – as commercial fertilizer in the South, as everyday food and flavor in Chinese cuisine and as feed for swine. These roles will be explored in coming sections, but something important here is that while prices of soybeans were relatively high and increasing – one *shi* of soy was exchanged for more *shi* of cloth – soybean output did not fully respond to rising prices. The reasons for this were manifold, but two stand out among the most important: population growth coupled with land scarcity (which we have discussed above) and the roles played by public policies (including trade regulations), peasants and traders, to which we turn now.

Ambivalent trade regulations, peasants avoiding markets and greedy merchants

Public policies both enabled and restricted trade. A key galvanizer of internal market integration was the infrastructural improvement of waterways, connecting the Yellow River valley in the North and Northeast (the most important soybean surplus areas) with the Yangtze valley in the South, via the Liao River and its tributaries. The centerpiece of this fluvial network was the Grand Canal, which linked Hangchow further South, and also connected these channels to the Beijing and Tianjin areas. The Qing Dynasty spent 10% or more of their imperial budget on expanding on this project, the longest artificial waterway in the world at the time (Mann 2011, 184). All this greatly facilitated fast and cheap long-distance trade, coupling dense networks of urban centers. Another trade-promoting policy shift was that Qing rule now demanded citizens to pay at least part of their taxes in money, which meant that farmers (the great majority of the population) had to sell produce to acquire currency for their taxes. Commercialization was thus facilitated by the monetization of taxes, as it was by the inflow of silver from Japan and the New World (Li 1998).

This combination of public investment in trade infrastructure and monetization created ideal conditions for a "soy boom" in Manchuria, where the land was extensive, fertile, sparsely populated and temperate – perfect for soybean cultivation. But the boom would have to wait for over a century. One key element holding it back was the Manchus' ambivalent trade policies. On the one hand, the central government wanted to redistribute the grain and beans produced in Manchuria, Jilin, Heilongjiang and other peripheral regions with a favorable ratio of land to labor, to more densely populated areas. On the other hand, provincial governors, by contrast, created different obstacles to hinder further exploitation and trade (Isett 2007, 259). Besides restricting Han-Chinese peasants immigration, the Manchus also imposed a formal ban on exports of grains and soybeans. Manchu local officials argued that shipments from the Northeast would empty their grain stores and raise local prices to unbearable levels, thereby threatening regional political stability and food security (Ibid, 245). The central government attempted to remove these hurdles, arguing that output in Manchuria was sufficient and rising. When that failed, the state took advantage of the fact that the ban on food exports had not been extended to state procurements and licensed merchants to ship grain and beans from Manchuria to other public granaries in areas experiencing shortage (Ibid, 248–49). At this time, the Qing Dynasty oversaw a huge network of both state-run and private granaries that stockpiled grain in good years and sold grain below market price in bad years for price stability and famine relief (Pomeranz and Topik 2006, 49–53). At its peak, the Chinese public granary system was a comprehensive empire-wide system, ultimately reaching into each of the 1,300 counties and holding some 45 million bushels of grain (Spence 1993; Marks 1998, 229). However, the system required huge economic and organizational resources to work well. By contrast, it was periodically short on grain and plagued by corruption and smuggling activities (Marks 1998; Isett 2007, 237). In light of these limitations and the mounting demand for grain and beans across the empire, Qing rulers finally withdrew from the direct management of food supplies, ending public procurement and, instead, bought grains and beans exclusively from private merchants (Marks 1998).

While, the formal ban on exports of grains and beans from Manchuria, was removed, a quota limiting Manchurian exports remained: only vessels carrying maximum 100 *shi* of soybeans were allowed to trade (Isett 2007, 246–52). The state also reserved the right to intervene in the grain trade in order to secure social stability or to discipline merchants whenever their interests were contrary to the aim of the state. As before, the rationale behind keeping these residual restrictions was that Manchurian people depended on grains and soybeans for food, but it would not be able to compete on the free market with the prices offered in urban areas. In this way, purchasing power was considered an unjust and illegitimate mechanism for resource allocation (Katō 1953, 600–02). The rules and laws of commerce reflected a difficult act of balancing the interests of three very different groups: first, urban dwellers and fertilizer-craving farmers in the South wanted more trade; second, leaders and officials in the state apparatus (mostly Han) were generally also in favor of increasing commerce for stability and finally, the local administrators and some Manchu officials generally wanted to keep grains and beans where they were produced to keep local prices down (Isett 2007, 242). Liberalization was successful in the long run, with increasingly strong merchant guilds (Ibid, 248). By the late 18th century, output was deemed high enough that exports would not trigger local shortages: and only then were all restrictions on exports from Manchuria finally relaxed (Ibid, 228, 250).

Trade regulations and restrictions were not the only factor limiting the division of labor and urbanization inside and between regions. While protectionist trade policies clearly played a role in restricting soy exports from Manchuria in spite of rising relative prices, low productivity increases in agriculture also played a role (Isett 2007, 236–37). As shown earlier, the initial abundance of land disappeared (per capita acreage steadily declined) which gradually shifted activity from fallow farming to permanent cultivation (Ibid, 301). With less land and cheap labor, peasants spent more time collecting, mixing and spreading fertilizers, weeding and irrigating their fields, instead of adopting new technologies that would increase labor productivity (Perkins 1969, 57; Isett 2007, 293–94, 297–300). For example, most farmers did not own draft animals that could allow an increase in productivity (Perkins 1969, 57–58, 92). At the same time, wealthier peasants used cheap labor rather than cutting production costs by improving tools and reducing labor inputs. Thus, bigger commercial farmers were no more successful in terms of labor productivity (Isett and Miller 2016, 92–113).

Another factor contributing to the general lack of response to price signals was that peasants tended to avoid market integration and cash-crop specialization (Isett 2007, 296–97). For the majority of Manchurian peasants, access to land was obtained through long-established customary practices and legal possession rather than through market relations. This was also true for labor (non-paid family work), tools (inherited or self-made), expertise (experience and communal exchange) and exchange (through local institutions). In this way, Manchuria stood in contrast to the farmer in the South, where tenancy was high and wage labor common and where peasant attachment to the land was extremely high (Perkins 1969, 8, 91–92). As a result, the Manchu farmer did not have to maximize monetary returns and could sidestep the competitive pressures of market integration (Isett 2007, 12–13): since land and labor were not commodified, peasants were not forced to cut costs by all means to maximize profit.

Still, the lack of commercial response to the rising prices of soy cannot be entirely understood without the role of powerful merchants. Merchant houses controlled the bulk of trade in soybeans, *shi*, tofu, bean cake, soy sauce and other soy products (Lander and DuBois 2022, 37). These big merchants – organized in powerful *liangzhans* – used their superior market information to reduce competition, squeeze out profit, suppress peasant earnings and increase their own political power. One classic merchant strategy, for example, was hoarding. The 17th century *Yueshi bian* [*Viewing the World*] by Ye Mengzhu, describes in detail how money could be made by hoarding soybeans (Shurtleff and Aoyagi 2016b, 40–41; 2021a, 170–77). Merchants bought from the peasants when the price was low (typically, just after harvest), stored the purchased produce and waited until prices had risen again, pocketing the difference (Isett 2007, 269). Other than hoarding, merchants were able to gain and consolidate economic and political influence through guilds. The most important soybean and bean cake trader at the time of the Jiaqing emperor (1796–1820) was the Shanghai Soybean Guild (*douhang*) and within the *douhang*, the *Dongqi* merchants controlled the wholesale distribution of both locally produced and imported soybeans (Hsu 1998, 365). The *Dongqi* worked in alliance with hinterland merchants operating in the Northeast, forcing petty merchants to market the bulk of their yields through them. In this way, the biggest merchant houses could make substantial profits on regional disparities in agricultural production and consumption. Gains were nevertheless not reinvested in production, but rather used for enhanced political influence through the establishment of merchant guilds and monopolies (Isett 2007, 237–49, 267).

While the Qing rulers tried to regulate the activities of merchants and keep their influence in check, guilds and merchant associations grew increasingly powerful and, from the 18th century on, were able to resist state oversight (Isett 2007, 254; Perez-Garcia 2021, 97). Moreover, the imperial government's withdrawal from the procurement of food, mentioned above, opened a space of even greater influence for merchants. Additionally, the state also deputized some traders for the collection and remittance of commercial taxes, granting them monopoly rights in mediating commercial transactions and collecting a broker's fee (Isett 2007, 265). However, while land taxes accounted for around 80% of tax revenues (Sng 2014, 110), the power of guilds increased and expanded where an important part of commercial activities were controlled through informal networks, fostering a culture of non-compliance which bypassed state supervision and even commanding their own militia. All in all, trader organizations gradually emerged as a state within the state. The constant fear of uprisings limited the Qing emperors' will and ability to confront the powerful Han-Chinese merchants (Perez-Garcia 2021, 120).

In short, soybean trade expanded, but soybean output did not respond to rising demand, an anomaly which can be explained by a combination of factors: land constraints, lack of productivity increases, trade restrictions and the hoarding and smuggling of powerful merchants. The Qing Dynasty's contradictory approach to soy trade, moreover, was not confined to Manchuria, but also marked their foreign economic relations.

The European take over of intra-Asiatic trade

The Qing Dynasty shared the Ming's disinterest and suspicious attitude to foreign trade. Ideally, international commerce was kept exclusively within the "tribute system" (Nakajima 2018). However, by the early 17th century, Chinese private merchants and state trading missions dominated the maritime trade of East and Southeast Asia, from the Bay of Bengal to the developing ports in southern China (Reid 1993; Ponting 2001, 367; Pomeranz and Topik 2006, 32–33). And, as American silver flowed in and the tax system changed from *corvée* to silver, the Qing emperor's regulations could be more frequently bypassed by corrupting local government officials and Chinese merchants in the areas of the Guangdong and Fujian provinces (Perez-Garcia 2021, 81). Moreover, despite the Qing's aversion to trade in general and their reservations toward foreign trade in particular, its tribute system and commercial isolation were full of loopholes and a myriad of exceptions, including whole periods when the ban on Chinese private trade was lifted (Reid 1993, 2:18; Isett 2007, 7; Nakajima 2018; Perez-Garcia 2021, 40–44). Even if China's foreign trade was increasingly characterized by European, and later American, clippers arriving to China's ports (loading up with tea, silk, sugar and porcelain bound for their home markets), the part played by Chinese ships was far from negligible. Their numbers must have been impressive and, as the provincial governor Li Shizheng noted, "in any given year, a thousand ships come and go [from Guangdong]" (Marks, 1998). Whether Governor Li had statistics on the numbers of ships passing through the various ports or whether he was just guessing, he conveys the sense of a fairly large fleet of Chinese-owned and -manned junks traversing the region in the years after the coast was opened in 1685.

China's "imperfect" commercial isolation during the 17th and 18th centuries was maintained against a background of a deep historical transformation: the rise of maritime commerce on a world scale and the passage from the "Age of discovery" to the "Age of European imperialism". The West's naval superiority quickly enabled it to dominate the key trading routes of the world economy and establish "ocean empires" (Barbier 2011, 238). Intra-Asiatic trade networks came increasingly under the hold of European companies, expanding rapidly from their hubs in Macau, Guangzhou (Canton), Formosa (Taiwan) and Manila and enabling the exchange of spices, silk, tea, grain, rice, sugarcane and soy sauce (Frank and Gills 2000, 8).

The Iberian Union, the dynastic merger between the Kingdom of Portugal and the Crown of Spain (from 1580 to 1640) was the first to hold a near-monopoly position in Far Eastern trade for some decades, with Macau at the center. The Iberians, via Jesuit missionaries, also expanded their presence in Japan (Shurtleff and Aoyagi 2021a, 137). While many written records have been lost, there are documents talking about Iberians exporting wheat flour, biscuit, soybeans, oil (probably soy oil) and pig's trotters from Nagasaki to Manila (Tremml-Werner 2015, 249).[3] The Iberian hold was soon lost, however. Spain and Portugal's reach into East Asia would soon implode due to a lack of manpower, a shaky financial base and the aggressive strategies of the Dutch East India Company (*Vereenigde Oostindische Compagnie*; VOC). While VOC had not been able to expel the Portuguese from Macau, it had sent expeditions against them and succeeded in establishing indirect trade links with inland China (Pomeranz and Topik 2006, 12). Moreover, from 1603, Japan was ruled by the increasingly nationalist and anti-Catholic Tokugawa (or Bakufu) regime. When the Dutch arrived in Japan in 1608, they were granted access to all ports by the Shogun, as well as the privilege to hold a yearly audience with the emperor. In 1613, the English also established a factory on the island and attempted to capture richly-laden Portuguese and Chinese vessels sailing to and from Nagasaki (Roessingh 2022, 7–11). Japanese rulers, however, adopted increasingly isolationist policies. After 1640, foreign trade was seriously restricted by the state, leaving little except for some traffic in silver and silk (Pomeranz 2021, 242). In the end, all European merchants were banned from trading with and within Japan (Kiple 2007, 147). The single notable exception was Dutch merchants, who had managed to convince the Japanese emperor that they were only interested in trade and not in religion. The VOC thus managed to become the only European company allowed to trade with Japan during this period of isolation, opening a factory on the islet of Deshima in Nagasaki, from where it controlled all of the country's European trade until 1854 (Roessingh 2022, 8).

Dutch (as well as British and to a lesser degree French) merchants began to take over many key Portuguese trading posts in the lucrative spice and slave trade (Perez-Garcia 2021, 16). Alongside VOC, the British East India Company (EIC) was also successful in obtaining the control of profitable trade lines in key natural resources and products (Barbier 2011, 238). Moreover, both VOC and EIC were private companies, chartered by their home states to conduct trade, make war, establish colonies, conduct diplomacy and even strike their own coins (Chaudhuri 1978; Pomeranz and Topik 2006, 142). While their goals were in themselves simple (buying cheap in the East and selling dear in the West), VOC and EIC are considered the first instances of the modern joint-stock company, effecting a formal separation between ownership (anonymous and private) and control, while simultaneously relying on a strong support from the state. But achieving dominance required large investments and the formation of a trading-capital fund and here the Mercantilist logic of states and merchants colluding to wrest control over trade from rival powers played a key role. Pomeranz and Topik summarize the main features of the New World trade order of the 17th century as follows:

> The Chinese traded their silks to the British and the Dutch who bought them with Spanish pesos that had been minted by African slaves in what is today Mexico and Bolivia and mined by indigenous peoples recruited through adapted forms of Incan and Aztec labor tribute.
>
> (Pomeranz and Topik 2006, 18)

It was in this context that the regional trade with soy sauce took off. Beginning in the 1620s, the union of merchants in Japan started to export *shōyu* (soy sauce) through VOC's trading post in Deshima/Nagasaki. The earliest surviving documentary source concerning soy sauce is dated June 11, 1637. It is a handwritten letter from the VOC official Jeremias van Vliet, writing from Siam (Thailand) to Nicolaes Coeckenbacker, head of Dutch office in Nagasaki, requesting provisions which included 10 kegs (*balien*) of "Murasaki", an ancient poetic synonym for soy sauce which literally means "purple" in Japanese. Another handwritten letter (dated 1647) from a VOC merchant at Deshima to his VOC-counterpart in Formosa, registers provisions for 10 kegs of soy sauce (Shurtleff and Aoyagi 2021b, 137). The VOC archives in the Netherlands house many more handwritten original letters, from the 1640s onwards, requesting provisions of soy and other goods (Roessingh 2022). Thus allowing us to map out the different alternative commercial lines of intra-Asian trade.[4]

Over the first half of the 17th century, Dutch trade expanded and the Netherlands colonized the Dutch East Indies (Indonesia) and took over Colombo (Sri Lanka) from the Portuguese in 1656; consolidating an important trade link along which soy sauce moved from Japan to Ceylon (Shurtleff and Aoyagi 2014a, 176–82). And, once more, all of it is reflected in the sources. For example, the governor of Ceylon and Governor-General of the VOC, Rijckloff van Goens, wrote a letter to the Deshima factory, in June 1680, requesting that sake and soy sauce be poured into well-made casks before shipping, to avoid wasting them because of broken pots (van Goens et al. 1680). In an appendix to the same letter there is a list of goods to be sent from Japan to Ceylon and to Coromandel (Southeastern India), including miso and soy sauce (Shurtleff and Aoyagi 2014a, 174–78). By the second half of the 17th century, VOC shipped more and more soy sauce to a wide network of Asian trade hubs: Ceylon, Batavia (Jakarta), Coromandel, Bengal and Malacca (Malaysia), Surat (Gujarat, in western India) and Cambodia (Shurtleff and Aoyagi 2014a, 178–80, 193–94; Roessingh 2022). While soy sauce was never the most important trade item in the intra-Asiatic trade system, it became important in the exchange between Japan, Korea, Indonesia and India (Mintz 2008, 60). The fact that all known 17th-century sources on soy trade are lists and letters related to VOC is no coincidence; the company had become a hegemonic force in the intra-Asiatic trade by the mid-17th century (Shurtleff and Aoyagi 2014a, 162–63). While Asian traders had probably dealt with soy sauce for much longer, their records, as far as we know, have not survived. But an interesting aspect in early Western sources on soy trade is linguistic. The names of soy sauce in the VOC letters vary constantly between *soij*, *zoije*, *soija*, *soije* and *shoyu* (Roessingh 2022) – all are cognates of the Japanese *shōyu*. *Soija* became the dominant form in provision letters from the 1660s onwards (Shurtleff and Aoyagi 2014b, 48). Later, during the 18th century, the English began referring to soy sauce as catch-up, catsup, ketjap, ketchup, etc. It seems that the latter must have adapted the Malay word *katsiap*, or confused soy sauce, with *ke-tsiap*, a Chinese pickled-fish sauce (Shurtleff and Aoyagi 2021a, 169–75).

Soy became a globally traded commodity during the 17th century, mainly through European penetration into Asian markets and, the critical role played by VOC's monopoly position in Japan. It is nevertheless clear from qualitative sources that soy sauce from China was present in many of the international markets and that soy sauce from Japan and Tonkin (Vietnam) could often be found in Chinese markets – particularly after the Kangxi emperor reopened the coast to trade in 1684–85, after a complete closure of coastal trade between 1662 and 1683 (Marks 1998, 166–67; Perez-Garcia 2021, 77). It is worth mentioning that the

Chinese also had an official trade permit in Nagasaki, Japan. It is estimated that as many as 190 Chinese ships a year visited Nagasaki and one-sixth of the town's population hailed from the East Asian mainland (Shurtleff and Aoyagi 2021a, 1475). Probably soy sauce and other soy products were included in this trade. The role of Chinese soy trade is not quantitatively recorded, but, all the same, several other documents register it. Noteworthy among them is the 1711 book *Account of the Trade in India*, written by the Englishman Charles Lockyer who had traveled in Sumatra, Canton and India observing different markets. In his words:

> Soy comes in Tubs from Jappan, and the best Ketchup from Tonqueen; yet good of both sorts, are made and sold very cheap in China. Buy none but what is right, which you are likelier to meet with among the Merchants than Shop-keepers. The best way is to agree by the Catty; for the Tubs are seldom or never full: But if they will not hearken to it, try which are the heaviest, and refuse all that are not likely to contain the Quantity they ought; draw it off immediately, and secure it in Bottles: Therefore in your Passage thither save as many as you can; for I know not a more profitable commodity.
>
> (Shurtleff and Aoyagi 2021a, 201)

While Europeans evidently came to dominate the intra-Asian trade routes and managed to monopolize certain strategic circuits, Asian traders continued to be competitive in many places and managed to play an important commercial role until the 19th century. Asian merchants did not complement commercial ventures with the exercise of violence; so they had lower transaction and/or overhead costs and could therefore undersell the Europeans wherever monopolistic positions had not yet been secured through military force (Pomeranz and Topik 2006, 164). Complicating a full understanding of actual trade flows are, it must be said, smuggling, piracy and plunder, which were rampant in both maritime and Inner Asia (Colla 2008, 124; Nakajima 2018).

Trans-oceanic trade: soy sauce arrives in Europe

During this regime phase, the world's transoceanic agricultural economy expanded significantly and European markets took more and more East and Southeast-Asian products. A wide range of new institutions to facilitate international trade, such as commercial law, emerged during the 17th century. Improvements in transport technology and better sailing routes reduced transport costs. Some bulky goods like sugar began to appear in transoceanic trade, with the Dutch ships delivering slaves from Africa to Brazil and sugar from Brazil to Europe (Kiple 2007, 164). Soon the commercial success of sugar plantations was followed by slave plantations of tobacco, coffee, cotton and indigo in the land-abundant and fertile "New World".

While less spectacular than in the case of sugar, vegetable ingredients and food traditions from Asia also moved West. The first known soy sauce cargo reached England as early as 1647 (Shurtleff and Aoyagi 2014a, 9, 176–82). While this commodity was never among the top five most important trade items, by 1680 it was important enough to be mentioned in the British Parliament by the barrister and political propagandist William Petyt (1680). Petyt, a member of the protectionist and mercantilist Whigs, expressed his fears that England's trade balance would worsen, driven by the domestic craving for foreign exotic goods:

> [...] in this alteration of the course of our Trade, our Importers will find out new trifles and gewgaws for our silly people: How suddenly do we find all the Women and Children of any account in England, in Amber Necklaces? Which at the rate they are sold at, must cost England of a Nation: of Excise, fyc. England at least 1000007. And now we have a new Sawce called Catch-up, from East-India, sold at a Guiney a Bottle.
>
> (Petyt 1680, in; Jewell 2011; Mun n.d., 490–91)

The earliest documents found concerning soy sauce imported to the Netherlands, in turn, come from the tri-weekly newspaper *Amsterdamse Courant.* On January 10, 1715, the *Courant* reported that a shipment of soy (among other imported goods, such as porcelain and tea) had made it to 's-Gravenhague (Hague) the day before (KB 2022). The same newspaper mentions soy again, in a cargo containing porcelain, mirrors and coffee, on May 27, 1719 (Ibid). The next time soy is mentioned in any Dutch publication – judging by the Dutch digital archive, Delpher – is May 31, 1724, when the tri weekly *'s-Gravenhaegse Courant* included an advertisement about a cargo brought by East Indian Merchants which included soy among other goods (KB 2022). From then on, several similar mentions can be found in both *Amsterdamse* and *'s-Gravenhaegse Courant* when searching for "soja" (Ibid). Numerous VOC letters also mention the arrival of soy shipments into Holland. Among them, the first known written record of a soybean cargo entering Europe is a shipment of soy sauce delivered from Deshima to the Dutch ports of Delft and Rotterdam in 1739 (Shurtleff and Aoyagi 2016a, 35). According to a translation of the 1992 Japanese book *Edo jidai shôyu no kaigai yushutsu [Exports of Shoyu from Japan During the Edo Period]*, by Yamawaki Teijirô and made available by *SoyInfo Center*, approximately 46,000 liters of soy sauce were exported from Dejima to the Batavia headquarters and 15,600 liters were then shipped from Batavia to the Netherlands during the 24 years between 1737 to 1760 (Shurtleff and Aoyagi 2021a, 1474).

In England and Holland, the traded soy sauce seems to have come mostly from Japan. As Denis Diderot noted in his 1765 *Encyclopédie*, Japanese soy sauce was much sought after by the peoples of Asia and the Dutch, adding that while the Chinese also make soy sauce, the Japanese type was regarded as superior (Ibid, 253). The Swedish doctor and professor of botany at the University of Uppsala, Carl Peter Thunberg – a physician at Deshima who traveled extensively in the region during the 1770s – echoed Diderot's impressions. As he wrote, Japanese tea was seen as inferior to Chinese, but Japanese soy was actually much better than its Chinese counterpart. That was the reason, according to Thunberg, why soy was not only exported from Japan to Batavia, but also sold thence to Europe and to every part of the East Indies (Thunberg 1776).

As the 18th century unfolded, Britain displaced the Netherlands in terms of naval supremacy and overseas reach to become the world's commercial and political hegemon. Additionally, European powers had begun shifting their presence in Asia beyond fortifications, small settlements and trading posts to full-fledged territorial empires, marked by the conquest of Bengal by EIC and VOC's acquisition of Java (Gelderblom 2009, 13; Barbier 2011, 257). The English also managed to grow commercial links with China, through which private British businessmen availed themselves of the EIC's protection to set up an extensive network on the mainland to bring home tea, quicksilver, vermillion, china-root, rhubarb, raw and wrought silk, copper, sugar candy, fans, pictures, lacquered ware, porcelain, soy and lapis lazuli, among other things (Shurtleff and Aoyagi 2014a, 227). Europe had shifted to the large-scale exploitation of the world's global frontiers. By the end of this period, 1650–1780 of, other actors had entered the lucrative Asiatic trade as well. Sweden, for instance, started to import Chinese soy sauce through the Swedish East India Company (*Svenska Ostindiska Companiet*, SOIC), founded in 1731. According to SOIC's statistical records of imports from China between 1777 and 1808, soy sauce was among the top 20 most important items (Nyström 1883). In December 1750, soy sauce (known there as "India Soy") also arrived in British North America (Shurtleff and Aoyagi 2014a, 235–36). While some retail stores in port cities along the North Atlantic coast advertised soy sauce in the following decades (Ibid, 276), it remained little more than a curiosity until well into the 20th century (Kiple and Ornelas 2000, 423) – as we will explore in the next chapter.

Unlike earlier times, international trade now affected the consumption of relatively ordinary people over a vast area (Gordon and Morales 2017, 34). Notwithstanding the limits imposed by climate and transport technologies, European powers wanted more trade and were increasingly frustrated with Chinese restrictions, using negotiations, piracy and smuggling to try to circumvent them and further "open up" Chinese commerce. The Qianlong emperor was not willing, however. As he had told the British: "there is nothing we don't have that we need from you" (Perez-Garcia 2021, 127). However, many officials, the gentry and the business elites in Fujian and Guangdong (Canton) provinces in fact desired many things from foreign powers. The introduction of silver and Western goods had progressively changed the patterns of local consumption, with European clocks, crystal glasses, mirrors, wines and liquors becoming part of the lifestyle of the Chinese elites. The three Manchu emperors (Kangxi, Yongzheng and Qianlong) issued several bans and decrees on foreign trade aimed at restricting trade networks led by local elites and gentry, but the increased demand for foreign goods and the influx of American silver meant that merchant communities were acting outside the officialdom in Southern China provinces (Ibid, 87–88, 101–02). Corruption had been a problem already during the Ming Dynasty, but it accelerated during the Qing state, due to the combined effects of the massive influx of American silver, the inefficient administrative systems and the low salaries paid to officials. The growing contradictions between the stiff principles of the tributary trade system, the actual extension of foreign trade outside of the framework of the tribute system (Perez-Garcia 2021, 21–22), as well as the increasing frustration among the European powers, particularly the British, would play an important role in the rupture of this regime. But before it all falls apart, we will further explore the remaining main functions commodified soy products during this regime: cuisine and soy foods, fertilizer, animal feed and exotic flavors.

Soybeans as a central pillar of the East Asian cuisine

In China, by the 17th century soy had clearly moved away from representing the ultimate hardship food into becoming a central element of everyday food traditions at all levels of society. Consuming and sharing food played a significant role in all parts of Chinese culture and aesthetics (Chang 1977) and because of China's dominant position in the region, Chinese culinary culture, including many types of fermented and non-fermented soyfoods, gradually spread and was locally adapted. In this way, soybean foods are mentioned in manuscripts from and/or about east Java, Vietnam, Japan, Moluccas and Korea (Shurtleff and Aoyagi 2012, 78–193; 2021a, 122–80; Shurtleff, Huang, and Aoyagi 2014, 178–210). Notwithstanding this wide diffusion, Japan and China stand out with an exceptional number of records on soyfoods from this period (Chang 1977, 77, 80–81; Huang 2000, 299–317; 2008; Shurtleff and Aoyagi 2021b). Second to China, Japan was probably the place where soyfood consumption rose the most in relative terms during this period, not least because the ruling Tokugawa shogunate promoted soybean production and made it more widely available (Isett and Miller 2016, 85–87). The availability of soy rose particularly after a law was passed in 1639 – the *Kanno Jorei* (or *Keian no Ofuregaki*) – which compelled all farmers to plant soybeans and adzuki beans between their rice fields and farms, in an effort to reduce food shortages (Shurtleff and Aoyagi 2014a, 184). While Japanese cookbooks from this period include extravagant feasts and many side dishes, Japanese food culture, since medieval times, was strictly regulated in accordance with hierarchy and most city dwellers lived on a relatively austere diet. The typical breakfast consisted of rice, miso soup (from soybeans) and pickles. Lunch and dinner looked similar, with the addition of one side dish of simmered vegetables, tofu or simmered or grilled fish (Rath 2010, 101). In times of food scarcity, however, diets became poorer. For example, in 1642 a Bakufu edict restricted polished rice and beans to tribute payments, while peasants were prohibited from purchasing wheat noodles and tofu (Rath 2010, 97–107). The resulting lack of vitamin B1 gave rise to beriberi – referred to as Edo-sickness (Ibid, 27). When harvests did not fail, ordinary people of course ate better, and then soyfoods featured prominently. Peasant food habits appear less in the written sources, but their diets were regulated with the most severe restrictions – most ingredients and dishes appearing in cookbooks were completely forbidden (Ibid, 75, 117–18). As they planted rice and soybeans, these ingredients were probably at the centerpiece of their diets.

Many different forms of soyfoods were important in both China and Japan, such as fermented black soybeans (*kuki* or *shi*), *natto*, miso soup (only in Japan – the Chinese had a soup like miso soup, but it never became popular), soy milk, the "skin" of soy milk, tofu, soy oil, soy flour, soy wine as well as *jiang* and soy sauce (*hishio, shoyu*) (Shurtleff and Aoyagi 2014a, 86–93). Both pressing soybeans for oil (wedge press) and milling (stone mill) for flour became more common (Ibid, 59). Alongside more sophisticated technologies and processed forms of soyfoods, soybeans also continued to be prepared and consumed in traditional ways, such as whole dry or boiled soybeans. Sprouted soybeans and whole young soybean pods were also greatly appreciated by common people (Bray 1984, 6: 423, 514). Moreover, during famine, it was recommended to boil and eat the leaves which were described as having a sweet flavor (Shurtleff and Aoyagi 2021a, 131). In Japan, the practice of consuming boiled green soybeans in the pod – *edamame* – also became widespread (Shurtleff and Aoyagi 2014b, 66). Of all the different soyfoods, however, the soybean curd (tofu) and soy sauce – along with miso soup in Japan – were by far the most frequently mentioned and important components of diets and cuisine (Chang 1977, 77, 80–81; Huang 2000, 292–317; 2008; Shurtleff and Aoyagi 2021b).

Tofu: popular and protein-rich

> The people of the Xi district are skilled in making tofu. They use mill stones made of very fine purple stone. Each pair of stones is worth 2–3 pieces of gold. They are of the quality of inkstone. When the soybeans are ground, the cakes of tofu are completely smooth, without dregs. When you cook them, you do not have to season them with salt or fermented black soybeans; they have a natural, sweet flavor. On this mountain lived old Mr. Wang. He used a clay pot to cook his tofu and the flavor was superb. Legend has it that a scholar, Mr. Xu, was unsuccessful in his state examinations. So he threw down his pen and said: "How much time does one have in a lifetime? Why not return to my village, heat up my pot, and make tofu?" His product became famous as the Tofu of the Xu Pavilion.
>
> (Shurtleff and Aoyagi 2021a, 150–51)

The quote comes from *Penglong yehua* [*Night Discourses by the Penglong Mountain, or Night Dialogue Under the Shade*] written by Li Rihua [or Li Jih-Hua] in 1610. The poem is the first recorded mention of fermented tofu, and it reflects the centrality and great variety of different forms of tofu during this period (Shurtleff, Huang, and Aoyagi 2014, 136). It also reflects that at this time there was movement both towards standardization of recipes and of diversification through the development of local specialties and variations that started to be picked up in the textual sources. Tofu preparation techniques have basically looked the same ever since – soaking the soybeans, milling them, filtering and cooking the soy milk produced, stirring in the coagulant and pressing the curd into solid blocks and draining off excess water (Huang 2000, 329; Tan 2008, 102). At the same time, a great variety of local tofu specialties had emerged.

Aside from the advances in food processing technologies and urbanization (leading to shops and restaurants selling soyfoods), the spread of Buddhism, with its emphasis on vegetarianism, probably played a role in promoting tofu and other high-protein soyfoods as vegetable alternatives to meat. As we saw in Chapter 2, many Confucian texts extolled the virtues of vegetarianism. Although there is a scholarly debate about the exact degree of actual meat consumption in both China and Japan, we know that diets were predominantly vegetarian, especially among the poor. Several Chinese and Japanese texts praising vegetarianism, often mention high-protein tofu and natto. Meat eaters also appreciated tofu, both for its subtle flavor (suitable in savory and sweet dishes) and for its ability to absorb other flavors, allowing it to be used in almost endless combinations (Huang 2000, 299; Tan 2008, 110).

Chinese cookbooks from this period bear witness the impressive variety of tofu. One of the many variants to emerge during this period was *fu chu*, or fermented tofu salted and cured with microorganisms. The *Shih Hsien Hung Mi* [*Guide to the Mysteries of Cookery*] by Chu-I-Tsun from 1680, states that after cooking tofu could be placed on a bed of rice straw and covered with rice husk for five to six days. As the tofu acquired a hairy growth (after being exposed to natural air-borne fungi), it could be salted, heated in the sun, marinated in a soy sauce mix until all the sauce evaporated and finally soaked in a jar with wine for six months (Huang 2000, 326). Another recipe in the *Shih Hsien Hung Mi* is frozen tofu, claiming that freezing it overnight would completely do away with any beany flavor (Huang 2000, 324–25). It also points out that tofu can be given a longer shelf life by pressing the curd as hard as possible and then air-drying it (Huang 2000, 325). Smoked tofu also has a longer shelf life (Huang 2000, 129, 325–26). In short, variations in preparation methods gave rise to a range of tofu products: frozen (*Tung you fu*), deep fried (*Tou fu phao*), dried (*Tou fu kan*) and smoked (*Hsûn tou fu*).

Tofu had its breakthrough in Japan – flavored already in the late 16th century, taking on a variety of names and forms. The first Japanese-Portuguese dictionary, *Nippo Jisho – Vocabulario da Lingoa de Iapam,* – compiled and published in Nagasaki by Portuguese Jesuit missionaries in 1603, contains many tofu entries: deep-fried, skewered, simmered, thinly sliced served with sauce, spread with miso and broiled, made into a soup with finely sliced daikon radish (*Misoyaqijiru*) or in noodles (*vdondôfu*) (Rath 2010, 181; Shurtleff and Aoyagi 2014a, 147). These variations were typically mentioned in domestic cookbooks. Recipes of grilled, boiled, lemon and sweetened tofu on a stick were also published (Rath 2010, 97–107). The Japanese also made a fermented tofu (*no misozuke)* pickled in miso (Huang 2000, 328).

Besides the dictionary published by the Jesuit missionaries, other Europeans wrote about soyfoods in the region. One early characterization of the vast quantities of tofu consumption in China, for example, comes from the travel journal of Domingo Fernández Navarrete, a Spanish Dominican missionary who lived in China between 1658 and 1669:

> Alone it is insipid, but very good so dressed and excellent fry'd in butter. They have it also dry'd and smok'd, and mix'd with caraway-seeds, which is best of all. It is incredible what vast quantities of it are consumed in China, and very hard to conceive there should be such abundance.
>
> (Shurtleff and Aoyagi 2021a, 173–75)

Fernández Navarette described tofu as one of the most remarkable things in China, noting that it was readily available everywhere for everyone. He pointed out that it was the cheapest and most common food and that everybody, rich or poor, ate it. Fernández Navarette also wrote about the process of making tofu from soy milk. These observations are of course illustrative of the fact that tofu had become a central pillar of East Asian cuisine, and they also show how surprised Europeans were by the many ways in which the soybean – hitherto unknown to them – was consumed (Ibid, 173–75). In a similar vein, Georgius Rumphius, a German scientist working for VOC in the Maluku Archipelago (Eastern Indonesia) in the 1650s noted that soybeans were used ubiquitously – fresh and dried – as well as for tofu or as roasted flour for noodles (Shurtleff and Aoyagi 2014a, 229–32).

Although a novelty for Westerners, tofu consumption appears as something natural in East Asia. A Japanese woodblock print from 1657 by Tosa Mitsunobu entitled *Tofu to somen uri* [*Sellers of Tofu and Somen Noodles*] shows a woman wearing a dark kimono and a white head-tie, seated on her heels on the floor. In front of her, there is a short table with five cakes of tofu on it (Shurtleff and Aoyagi 2014b, 49–50). Another woodblock print from 1685 by Hishikawa Moronobu, *Wakoku shoshoku e-zukushi* [*Various Japanese Occupations Described in Pictures*], shows a man carrying two water-filled tubs of tofu suspended by four ropes at each end of a shoulder pole, and at his side a woman with a similar wooden tub atop her head – most likely tofu sellers. The caption reads: "Please eat tofu. I came all the way from Nara" (Ibid, 57). Another clear illustration is the *Jinrin kinmô zui Jinrin kinmô zui* [*Illustrated Encyclopedia of Life in the Edo Period*] from 1690, which includes a picture entitled *Yakidofu-shi* [*Grilled Tofu Man*], showing a man grilling tofu over a rectangular brazier. In his right hand is a fan to feed the coals beneath several tofu cakes. In his left hand is a skewer with two prongs piercing another tofu cake. Another illustration of a street seller of *natto* – whole fermented soybeans (Ibid, 59–60) – is included in the same source. Yet another is entitled *Tofu-shi* [*Tofu maker*], featuring the text: "[A]mong the craftsmen, tofu makers rise the earliest each morning". The illustration shows a tofu maker in his shop pressing the soy milk out of the *okara* (the byproduct) in the pressing sack. Generally, these publications indicate that soyfoods were sold as inexpensive, fast food and that walking street vendors were fairly common. Soyfoods appear to have played a crucial role in improving the diet of the poor, especially since the most common food staple - rice - lacks important key nutrients. Soyfoods thus provided an important complement to fill the nutritional gaps in the diet (Shurtleff and Aoyagi 2014a, 5–6).

The importance of tofu in Japanese cuisine is also clearly illustrated by the 1695 *Pen chao shih chien* [*A Mirror of Food in this Dynasty*] by Hitomi Hitsudai, which argued that the tofu in Edo was "nowadays" quite good, but that it could not compete with the tofu made in Kyoto (Shurtleff and Aoyagi 2021a, 191). The many local variations of tofu, such as the boiled kind served in restaurants near Nanzenji, one of Kyoto's most famous Zen temples, and the sweetened tofu on a stick served in two Buddhist tea houses in the area of Gion, were also highlighted in the book from 1782 *One Hundred Favorite Tofu Recipes* and its sequel in 1783 by Ka Hitsujun, which was followed by the 1786 book, *A Hundred More Tricks with Tofu*, this time published by an unknown author (Rath 2010, 76–85).

In short, tofu had by this time become an important source of protein in both China and Japan, often used in the same way as meat or fish, and is commonly understood to have filled the same roles that meat and dairy (particularly cheese) played in Europe (Chang 1977, 77, 80–81; Huang 2000, 329–30; 2008; Shurtleff and Aoyagi 2021b). Although it was many times cheaper than meat, it was highly prized by emperors and peasants alike (Huang 2000, 329).

Soy sauce as flavoring – and Japanese miso soup

While tofu became an increasingly important source of protein, soy sauce became a principal condiment. As we explored in the section on trade, soy sauce was also the first soy item that entered the evolving transoceanic trade. In China, during this period, soy sauce (*jiangyou*) started to rival its precursor, *jiang*, the disintegrated paste of fermented black soybeans mixed with flour, salt and water, which as we have seen, had been essential even for the humblest peasant family (Bray 1984, 6:5). By the early years of the Qing Dynasty, soy sauce was already more widely used than *jiang* (Huang 2000, 373). The variety and ubiquity of soy sauce in China during this period is clearly illustrated in the *Tiaodingji* (or *T'iao Ting Chi* [*The Harmonius Caldron*], a collection of recipes of unknown authorship and date, assumed to originate in this period. It is the most comprehensive of the premodern food canons. By way of illustration, 58 recipes include *jiang* and around 350 recipes include different types of soy sauce. The book also contains many recipes involving fermented black soybeans, roasted whole soy flour, soy oil and soy milk (Shurtleff and Aoyagi 2021a, 252–53). Many recipes from this time include dishes in which fish, vegetables, pickles or sweet potatoes are simmered in soy sauce. Other recipes show ingredients boiled in soy sauce, covered in soy sauce or flavored by some splashes of soy sauce.

In Japan, by this time, soy sauce seems to have replaced both vinegar dressing and fish sauce as the most common food condiment (Rath 2010). Several European travel stories also register the remarkably broad use and culinary variety represented by soy sauce in Japanese cuisine. One example, from almost 100 years earlier, comes from the Florentine "merchant-traveler-narrator" Francesco

Carletti, who gave a detailed account of Japanese customs from his visit to Japan between June 1597 and March 1598, including the universal usage of soybean sauce there:

> They prepare various sorts of dishes from fish, which they flavour with a certain sauce of theirs which they call misol. It is made of a sort of bean that abounds in various localities, and which cooked and mashed and mixed with a little of that rice from which they make the wine already mentioned, and then left to stand as packed into a tub-turns sour and all but decays, taking on a very sharp, piquant flavor. Using this a little at a time, they give flavor to their foods, and they call shiro what we would call a potage or gravy.
>
> (Shurtleff and Aoyagi 2014b, 144)

While Carletti wrote that the sauce was called miso, it seems clear that he was actually referring to soy sauce. At the same time, it is perhaps not so strange that Europeans would mistake soy sauce and miso, as both were very popular. In addition, according to some Japanese myths, the first soy sauce in Japan actually came from miso, as a 13th-century Buddhist monk (Kakushin) returned from China where he had learnt to make fermented miso. In the story, Kakushin discovered that the liquid which gathered on the bottom of the vats while fermenting the miso could be used as a tasty seasoning. This liquid byproduct of miso fermentation – *tamari* – is considered Japan's first soy sauce (Ibid, 31–32). *Tamari* nevertheless continued to be consumed – and still is – in its own right. Independently of the exact origins of soy sauce in Japan, however, both these fermented soy products became extremely popular during the Edo period. The aforementioned Japanese-Portuguese dictionary also indicates that it was common to buy miso in Japan at this time. The dictionary even contains a proper entry for the word *Misoya* – a term referring to a specific miso-shop (Shurtleff and Aoyagi 2014a, 147) – suggesting that commercialization of miso was important, something which can also be confirmed by a passage in Alessandro Valignano's 1586 book about the history of a Jesuit mission in Japan, in which he mentions buying rice, miso and dried fish (Ibid, 137).

Besides shops, restaurants became more common at this time (Rath 2010) and several specialized soy sauce and miso companies emerged in Japan. Some of these still brew soy sauce today, three centuries later (Shurtleff and Aoyagi 2021a, 172). Nevertheless, most peasant families continued to make their own miso at home for many more centuries to come (Rath 2010, 27). According to the aforementioned 1695 *Pen chao shih chien*, soy sauce was also made by hand in the large majority of houses in all regions and could be purchased anywhere (Shurtleff and Aoyagi 2021a, 191). Similarly, the 1686 *Yôshû fushi* by Kurokawa Dôyû, published in Kyoto says that soy sauce was made and could be purchased in all of Japan. It also states that soy sauce was made by hand in the large majority of homes (Shurtleff and Aoyagi 2014b, 57–58). Moreover, the 1690 *Jinrin kinmô zui Jinrin kinmô zui* [*Illustrated Encyclopedia of Life in the Edo Period*] includes an illustration titled *Miso-ya*. It shows two men mixing or pounding something in a wooden mortar using long spatulas. Behind the men to the left is a vat of fermenting miso with stones on top for pressure. The text says: "Miso makes a good seasoning and helps keep people healthy. A day cannot go by without it" (Ibid, 59). In short, regardless of whether the soy sauce or miso soup was commercial or homemade, they appeared everywhere, and all soups were presumed to be miso unless otherwise specified (Mintz 2008, 60).

As we saw in the trade section, the principal food condiment in many East Asian countries at this time was soy sauce. Large quantities of soy sauce were traded within Japan, brought from the Kyoto-Osaka area to Edo by ship. But this trade was not in VOC's hands, and we only have indirect references to domestic trade from sources like the 1690 *Jinrin kinmô zui Jinrin kinmô zui* which states that *shōyu* was a famous product of Sakai and Osaka and was shipped from thence to various places (Shurtleff and Aoyagi 2014b, 59). Typically, soy products were exported together with other land-based products (rice, timber, horses) to core regions selling manufactures, particularly textiles.

The German botanist and physician Engelbert Kaempfer, who worked in the VOC factory at Deshima in the 1690s, published a book in 1712, where he wrote:

> [T]his legume supplies to the Japanese kitchen vital elements, for they make from it the following: 1–A kind of pap that they call miso, which is added to dishes instead of butter. Butter is unknown under this strip of heaven. 2–And then the famous so-called shoyu, a sauce which is poured over if not all dishes, at least over all cooked and fried meals.
>
> (Shurtleff and Aoyagi 2014a, 204–207)

Kaempfer also included detailed information about the preparation methods and steps involved in making miso and soy sauce, adding that those who make the former are held in high esteem and sell it ready-made. While other Europeans before Kaempfer had talked about soy sauce, miso and tofu, he was the first to show a detailed understanding for the plant – the soybean – from which all these dishes derived (Du Bois 2018, 39).

Kaempfer's successor in VOC's Deshima plant was the Swedish doctor and professor of botany Carl Peter Thunberg. In his abovementioned 1776 book, he described the process of making both the sauce and miso soup and observed that it was eaten throughout the day in Japan (Thunberg 1776). Another Swede, Karl Gustaf Ekeberg, a captain of the Swedish East India Company, described tofu as a dish similar to sweet cheese, but prepared with Chinese beans. Ekeberg would return to China and write about the Chinese way of making soy sauce, hoping that perhaps it might be produced in Europe (Ekeberg 1764).

Other soy foods

Besides tofu and soy sauce (and miso in Japan), there are a myriad of ways of using soy for different food purposes, providing proteins, fats, vitamins, minerals and flavor. Oftentimes, specific local varieties emerged (Huang 2008, 50; Mintz 2008,

60). One prominent example is *tentsuyu* – the traditional Japanese dipping sauce for all kinds of tempura. Another example is Indonesian *tempeh* – a cake made out of fermented whole soybeans, a good source of vitamin B12 (Huang 2000, 345; Du Bois 2018, 32). Yet another fermented Indonesian specialty from this time is *oncom*, sometimes made with dregs from the first stages of tofu production (Du Bois, Tan, and Mintz 2008, 33).

It is impossible to go into all soyfoods and their functions, but let us cover the medicinal functions of soy briefly. The principle that "medicine and diet derive from the same origins" had become a dominant view in both Chinese and Japanese culture and remained strong throughout the early modern period. The Chinese masterpiece *Bencao Gangmu* [*Compendium of Materia Medica* or *The Great Pharmacopeia*] by Li Shizhen from 1596, describes most soybean forms and varieties as sweet, neutral and nontoxic. In Li's words:

> When soybeans are raw they are neutral, but when they are roasted they become hot, and when they are boiled they become cool. When made into fermented black soybeans they become cold. When used to make jiang or soy sprouts they are neutral. Soybean oil is described as pungent, sweet, and hot and slightly toxic. When cattle eat them, they are warm. When horses eat them, they are cold. So even though it is one substance, when it is eaten in different ways, it has different effects.
>
> (Shurtleff and Aoyagi 2021a, 131)

The text also points out, however, that prolonged ingestion would make one's internal organs feel heavy (Ibid, 143). The *Bencao Gangmu* also refers to a type of soy wine called *tou-lin chiu* [translated as bean soak wine] which is described as a sake-like fermented alcoholic beverage made from black soybeans. Other texts about the medical properties of food, for example, the 1609 *Sancai tuhui* [*Illustrated Encyclopedia of the Three Realms*] by Wang Qi and the 1665 *Xiaxiaozheng jie* [*Explanation of the Lesser Annuary of the Xia Dynasty*] by Xu Shipu say similar things about soybeans, and several others point to soybean congee as a means of dispelling "the humidity of the body" (Shurtleff and Aoyagi 2014a, 148–76). Moreover, during the Qing era several new books emerged discussing the pleasures, health benefits and dietary habits associated with food consumption (Kuo 2013). In Japan, the *Pen chao shih chien* also covered its medicinal properties: inducing calm, relaxing the stomach and having a good overall effect on the intestines. Miso is described as something "one should not be without", while *natto*, again, makes one feel calm, conditions the stomach, enhances good appetite and detoxifies poisons (Shurtleff and Aoyag 2021a, 191). Soy sauce, finally, is said to "[I]nactivate any poisoning from eating food, drinking alcohol, or taking medicine" (Ibid, 191–92). In Korea, fermented soybean products also appear to have been used as medicine, and they are talked about as such in the *Dongui bogam* [*Principles and Practices of Eastern Medicine*], compiled by Heo Jun in 1613 (Shurtleff and Aoyagi 2014a, 192).

Less celebrated, but still in high demand during the regime phase, was soy oil for cooking. Knowledge of oil presses spread and the most common way to grind the beans and press the oil was either of two different types of stone mills driven by ox or buffalo. One important work in this respect is the 1637 Chinese book entitled *Tiangong kaiwu* [*Exploitation of the Works of Nature*] by Song Yingxing, which claims that the oil of yellow soybeans is among the best for eating and includes graphic illustrations of soybean pressing. The reader learns the quantity of oil obtainable from each measure of yellow soybeans is precisely 9 catties of oil from each *tan* of yellow beans (Shurtleff and Aoyagi 2021a, 160). In another work, *Wuli xiaoshi* or *Wu Li Hsiao Shih* [*Mini-encyclopedia of the Principles of Things*], by Fang Yizhi, 1664, we read that when pressing yellow soybeans, it is possible to get 18 catties of oil from one picul, or even 22 catties if a wedge press is used (Shurtleff and Aoyagi 2016b, 40). Further, even if soy oil was mainly used for cooking, it was also used for lubrication and lighting. In Japan soy oil is also mentioned as used for food (Shurtleff and Aoyagi 2014a, 147). Whatever its uses, at any rate, it is clear that more and more soybeans were being pressed, and that the residue from pressing oil – the presscake, or bean cake, was in very high demand (Perkins 1969).

Another snowballing soy byproduct is the residue of dietary fiber that comes when making soy milk or tofu, often referred to as *okara* (in Japan) or *zhi* (in China). This could be used as starvation food, but was also wanted for other purposes. In other words, as considerable amounts and numbers of byproducts were created by the production of tofu, soy milk and oil extraction, new functions of the soybean emerged and were developed.

Soybean cakes as fertilizer

It takes millennia to generate fertile soils, but they can be degraded or eroded after only a few years of poor management. Legumes (such as soybeans) have been used in rotations since ancient times to fixate nitrogen and enrich soils. Another traditional source of nutrients was animal and human excrement. Both of these sources were used in China and Japan at this time (Tan 1986, 115), but there is also some evidence that soybeans were being used as fertilizer elsewhere in Asia as well. For example, the aforementioned Georgius Rumphius noticed that the soybean was abundantly cultivated both for food and to be plowed under as green manure to enrich the soil in Java, Bali and other Malayan islands (Shurtleff and Aoyagi 2014a, 232).

However, specialization in cash crops and the intensification of farming depleted the soils faster and required larger inputs of fertilizer than before. With urbanization and trade, most of what plants extracted from the soil by farming resulted in displaced nutrient supplies. As McNeill notes "with cities, human societies systematically exported nutrients from farming" (McNeill 2001, 23). In China, in the densely populated Yangtze Delta, and south of the Yangtze, many farmers had increasingly adopted the practice of double-cropping rice with wheat in order to raise total output to feed growing populations. Thus, the peasants who farmed paddy rice added an additional crop of winter wheat (which further depleted soils) and skipped fallow farming (which enriched soils). This farming regime exhausted soils, which in turn threatened to reduce yield (Isett 2007, 236). Even worse than the paddy rice and wheat double cropping was the continuous cropping of cotton. Moreover, Columbian Exchange crops such as tobacco, corn, peanuts and mulberry all depleted soils.

To counter these effects, peasants needed to fertilize the land more, but traditional sources were scarce. Both Chinese and Japanese farming systems included few manure-producing animals. Since meat consumption was almost nonexistent, there was a constant scarcity of manure from livestock and a need for other sources of protein in diets (Myrdal 2022, 15–16). Soybeans played important roles in responding to both needs – nutrients for soils and humans directly. Besides excrement, peasants and state officials (in the gazetteers) experimented with fertilizers from all kinds of origins, from canal sludge to the ashes of burned compost (Bray 1981, 45; Isett and Miller 2016, 31, 110) to pressed soybean cake (Lander and DuBois 2022, 35). In this way, the byproduct from extraction of soy oil (as well as *okara* from the making of tofu and soy sauce) became highly valued on account of its high quantity of nitrogen, phosphoric acid and potassium (Wells 2018). Further, soybean cake was lighter than manure fertilizers which made it easier to store and spread and it was thus, labor saving. Soybeans now started to provide nitrogen in two ways: the way it had been done since ancient times, grown in cultivations or plowed under as green manure, and in a new way, bean cake spread on the soil.

Besides playing an important role in fertilizing the paddy rice fields, soybean cakes were highly demanded for wheat cultivation. Note that together, rice and wheat constituted the lion's share of imperial tax revenues (Barbier 2011, 181). Bean cakes were also used to boost intensive cotton and sugarcane cultivation throughout the Yangtze Delta, particularly in the densely populated Jiangnan region (Isett 2007, 235–36; Choi 2014, 46). Bean cake was one of the most advanced commercial enrichers available before modern chemical fertilizers were invented. However, as demand for soybean cakes grew, the amount of available soy byproduct diminished (Tan 1986, 116).

It is unclear when soybean cake as fertilizer made its first appearance, but in one early document from the Chang-chou prefecture, a 16th-century magistrate instructs farmers to mix finely chopped hog bristles with ashes, soybeans and other legumes and to soak the mixture for several days before applying it as fertilizer in the rice field (Rawski 1972). Another early mention comes from an encyclopedia of the Wanli emperor of the Ming period (1573–1620). In it, cake-shaped fertilizers (noted as "the basis of fields" or just "cake fertilizer") are categorized into different kinds depending on the source: hemp, vegetable, tung tree or cotton seeds on the one hand, or beans on the other (Sakai, Yoshinobu, and Yōichi 2000, 318–20). According to Bray, by 1500 many rice farmers "were purchasing commercially produced lime and soybean waste for fertilizer or imported nightsoil (human manure) from the cities, raising annual output to two or three tons per acre in some double-cropping regions" (Bray 2004, 14–20). A single cake could fertilize an entire *mu* of young rice (Bray 1984, 6:294–95).[5] According to the 1621 *Qunfang Pu* [*Assembly of Perfumes* or *Monographs on Cultivated Plants*] by

Wang Xiangjin, soybean press cake was already widely used as fertilizer in the province of Shandong during the late Ming period (Katō 1953).

But how important was the bean cake, really? – a scholarly debate

While there is no doubt that the introduction of bean cake was a major technological change, its relative importance is still debated – often within the realm of the *Great Divergence* debate. The Chinese economic historian Li Bozhong (echoed later in Pomeranz's work) has argued that the soybean cake created a "fertilizer revolution", allowing for a substantial rise in agricultural productivity (with yield increases in rice of 47%) in the lower Yangtze Delta between 1620 and 1850 (Li 1998, 114; Huang 2002). According to Bozhong, around 20 million *shi* of imported soybeans arrived in the South for this specific purpose, and from this around 26 million piculs of bean cake could be made. While it was also used as feed for the pigs, it still had a very important impact on the demand and supply balance of fertilizer in Jiangnan (Li 1998, 114). Further, according to Li, vast imports of bean cake fertilizer increased the regional specialization of the South in high-yielding double-cropped wheat and rice cultivation as well as the output of mulberry and cotton.

In addition, it is argued that the substantial release of female labor from agriculture that accompanied the spread of double cropping allowed a vast expansion of textile production and exports (as mentioned in the previous section on cultivation) which in turn paid for the bean cake and other imports. Altogether, these interlinked and reciprocal changes underlie the booming economic prosperity widely remarked upon by both Chinese and European observers during the early 18th century (Marks 1998, 285). The widespread use of bean cake as fertilizer in Jiangnan and elsewhere in the lower Yangze Delta – and its benevolent effects on yields – is one of the central arguments of Kenneth Pomeranz's claim in his seminal book *The Great Divergence: China, Europe, and the Making of the Modern World Economy* that the levels of development in this area and England were comparable at this time. In this view, bean cakes contributed to a significant rise of agricultural output, feeding the rapidly growing population in the Yangze Delta – which, circa 1750, was estimated to be between 31 and 37 million people (Pomeranz 2021, 7). Indeed, Pomeranz says that the widespread use of bean cake suggests that China was not suffering an agricultural involution (in disagreement with Patrick O'Brian and others):

> [T]he sharp increase in purchases of soybean cake fertilizer—which cost much more than manure (especially, of course, self-supplied manure) but could be applied with much less labor—is quite suggestive. One could, in fact, infer from wage and price data that households that bought beancake were implicitly valuing male labor at a rate that was roughly in line with the market wage. Finally, since estimates of the number of labor days used to cultivate one mu (or one-sixth of an acre) of rice in the Yangzi Delta are virtually identical for the 1600s, the 1800s, and the 1930s, while output per mu rose and rents as a percentage of output probably fell, we actually have less

> indication of agricultural involution for at least this part of China than for early modern Europe, with its increasing working hours and arguably falling real returns for unskilled labor.
>
> (Pomeranz 2021, 98–99)

By contrast, other scholars argue that the position of Li, Pomeranz et al. is exaggerated (Xue 2007; Deng and O'Brien 2015, 255). In their view, while demand for bean cake soared, supply was limited and import of bean cake was not enough to sustain high yields in Jiangnan since the prices of this (and other fertilizers) were too high (Deng and O'Brien 2015, 255; Isett and Miller 2016, 110–12). Drawing on secondary literature, Deng and O'Brian observe that imported soybean cakes cost the same as rice in the Yangtze Delta (Deng and O'Brien 2015, 254–56). Consequently, only rich households could afford commercially available and nitrogen-rich soybean cakes as a basic fertilizer, while poor households used compost, night soil and manure from pigsties and goat pens (Isett and Miller 2016, 31, 110–11). Others could perhaps use it in combination with cheaper sources, or only to stimulate crop growth in the final sprint of the growing season, but not throughout its entirety. Their point, in short, is that the high prices of bean cake taken together with the low price of labor, hampered the spread of this innovation. Instead, peasants could only afford to put more labor into the land to maintain the same yields and growth stagnated. The patterns of adoption nevertheless differed between crops. Peasants growing sugarcane in the far South, for example, could better afford to buy bean cake imported from North and Central China (Ibid, 112).

The discussion of the magnitude and multiple contexts of utilizing bean cake among scholars writing in English (Huang 2002; Fan 2007; Xue 2007) does not appear among scholars writing in Chinese. The narrative that bean cake was a revolutionary fertilizer in Jiangnan from the 17th century onwards, is generally accepted without divergent opinions among scholars writing in Chinese (Katō 1953; Tan 1986; Choi 2014). For them, a growing population and the need for cash crops required more agricultural skills and innovation in farming to maintain the fertility of farmlands (Long 1997). In such a social context, adds Choi digging sediments from rivers and transporting excrement became uneconomic due to the mass labor investment required, so the use of bean cake soared (2014). Choi argues further that as bean cake emerged in the market, traditional fertilization techniques became inefficient both in terms of manpower and prices (Ibid, 46). The significant role of bean cake and the remarkable scale of its trade are frequently mentioned in the Chinese historical literature, but detailed figures from the Qing period are not available, probably due to the lack of reliable quantitative sources from this period. Choi has nevertheless made some estimates based on available data and concludes that, on average, wheat farmers in the Yangtze Delta applied around 10 *jin* of ground bean cake to one *mu* of wheat field as basal dressing (Ibid, 51–53). There are no clear records of how commonly these bean cake treatments were adapted by ordinary Chinese peasants; their utility in different provincial contexts is not documented. What we can be certain of, in short, is that such common approaches to fertilization as using river sediment, ash or excrement required mass labor. At the same time, bean cake became available as a more efficient – but also more expensive – option. Soy also played an additional, though indirect, role as fertilizer; as feed to the increasing number of pigs which in turn increased the amount of available manure (as we will explore in the coming section on feed) (Pomeranz 2021, 226).

Bean cake as fertilizer in Japan

As in China, soyfoods in Japan generated considerable amounts and numbers of byproducts, used for different purposes, especially in the form of bean cake as fertilizer. According to Isett and Miller, soybean cake actually became relatively more widely used in Japan than in China (2016). In their view, high land taxes pushed out many peasants from their land and turned them into tenants or landless wage laborers, which spurred a rapid process of agricultural innovation and intensification (Isett and Miller 2016, 85–87). Landlords tied the wages of hired laborers to land productivity, so tenants needed to demonstrate an ability and willingness to experiment and innovate in pursuit of higher productivity to successfully compete for leaseholds. This worked as a strong incentive – or pressure – for tenants to experiment with superior fertilizers, such as bean cakes, in pursuit of higher productivity. Moreover, in great contrast to China, the prices of the commercially available fertilizers remained stable despite rapidly increasing demand. This indicates that the manufacturing sector managed to scale up production in ways that kept costs down. Isett and Miller argue that one reason was the reduction in demographic growth rates, which created increased competition for labor, driving up wages substantially. As labor became scarcer (and thus more expensive), technology and innovation became more important to compensate for labor. Isett and Miller thus understand the use of bean cake as illustrative of the Japanese transition to capitalism, which in turn, they argue, is best seen as the unintended outcome of the struggle among elites and between elites and peasants to take a greater share of the rural surplus (Ibid, 85–89).

Soybeans as animal feed

Soybeans and soybean residue had been used as fodder in China since ancient times, and as many texts attest, the practice continued. For example, the *Nongzheng quanshu,* from 1639, states that vegetables and soybeans were fed to cattle, horses and water buffaloes before they prepared the ground for planting (for a detailed treatment of the source see Kuo 2013, 82). Another section of the book, entitled "opening up new fields", records the role of soybeans as a recommended feed for cattle and horses and describes a mixture for pig feed consisting of one *dou* (about 20 liters) of yellow soybeans (*huang dou*), one *sheng* (about two liters) of hemp seeds, as well as various other ingredients. Feeding this to pigs for 12 days is claimed to make them fat. This work integrated the main points of previous agricultural treatises, including many points found in the great agricultural treatise *Qimin Yaoshu* from the 8th century (Ibid). According to Louis D. Lecomte, a French Jesuit missionary in China writing his memories from China in 1697, all provinces of the North and West use black and yellow peas in place of wheat to feed horses (Shurtleff and Aoyagi 2021a, 192). In a similar vein, the previously mentioned 1637 *Song Yingxing* reads: "[I]t is customary when taking a long journey north of the Huai river, to feed the horse with black beans in order to make it strong" (Shurtleff and Aoyagi 2021b, 160). In general, black soybeans are described as even more hardy than yellow soybeans and therefore suitable to be fed to livestock – or as food backup when all other crops failed (Lander and DuBois 2022, 35).

Soybeans were of course not the only source of feed for animals, but a variety of natural provisions available in the given landscape were used – with important changing specificities from one region to the next (Kuo 2013, 84–88). Already in *Qimin Yaoshu*, Jia Sixie had remarked that pigs were not to be given fodder (for example, whole beans) but should instead be put out to pasture as long as possible, storing the feed for the winter (Kuo 2013, 62; Bray 2019, 368). This way of raising pigs was reproduced in several agricultural books and seems to have represented the most common practice for centuries. However, besides the traditional techniques, animal feeding had started to undergo a transformative shift, particularly regarding pig raising. With the increased amount of soybean residuals (byproducts from the production of tofu – *okara* – and the extraction of oil – bean cakes) in the 17th century, the old rule of not feeding pigs started to become outdated. In the 1621 *Qunfang Pu* by Wang Xiangjin, the use of soy milk or tofu byproducts (*okara* in Japan, here called *zhi*) is mentioned as pig feed and, indeed, as human food in times of famine (Shurtleff and Aoyagi 2021a, 154–55). Besides *okara* (or *zhi*), the bean cake was emerging as an important source of feed for the more commercialized and concentrated pig farmers. The earliest known text alluding to defatted soybean cakes as feed for pigs is the 1637 *Tiangong kaiwu* by Song Yingxing (Shurtleff and Aoyagi 2021a, 157–60).

These changes also marked a qualitative shift in pig farming, since the pressed soybean cake is extremely protein rich (Shurtleff, Huang, and Aoyagi 2014, 133). As a consequence, pig husbandry in China's early modern era gradually shifted from semi-nomadic to a profit-oriented non-nomadic practice and pork production and consumption increased significantly (Kuo 2013, 39). A dynamic pig industry began to flourish during this period in the South and pork became the most widely consumed meat (Ibid, 47–48, 53). This practice seemed to have expanded, helping to shape a radical transformation in Chinese pig husbandry by the 18th century. Roughly half of the soybean cakes exported to the South from Manchuria were used as feed. As with fertilizers, there is controversy about just how important the role of soybean cakes was since prices of bean cakes were relatively high (Choi 2014, 44).

Alongside the high-protein bean cake, other forms of soyfeed persisted. One example comes from the Qing-era treatise *Binfeng Guangi* (1740) by Yang Shan, which argued that agricultural businesses in the northern Shaanxi area in North

China, gave piglets high-quality feed such as cooked wheat and cooked beans to ensure that they would receive sufficient nutrients (Kuo 2013, 89–92, 99).

Soy enthusiasm in Europe and the United States

Soy sauce entered Europe within a pre-existing social and psychological registry of acquired meanings and associations. Soy sauce (catch-up) emerged as a new ingredient in European cuisine, enjoyed by a European upper class with a taste for "exotic" flavors from the Far East. One of the first texts mentioning soy sauce in Europe comes from a 1679 journal by the famous philosopher John Locke (not published until 1829): "Mango and saio [*shyu*] are two sorts of sauces brought from the East Indies" which one ought to enjoy in London (Shurtleff and Aoyagi 2015). The context suggests that soy sauce was widely available in the British capital already in 1679 (Shurtleff and Aoyagi 2014a, 7). "Ketchup" was also mentioned in the 1701 *The Mysteries of Opium Reveal'd* as one example of a diet that could counterbalance the effects of opium intake. Soy sauce was also referred to as medicine in a pharmacology book written in Latin, *Pharmacologiæ, seu manuductionis ad materiam medicam supplementum* [*Pharmacology, or a Food as a Supplement to Medicines*]. It was published in 1705 in London, noting that a pharmacist's laboratory should have soy sauce made out of the Japanese small white bean (Shurtleff and Aoyagi 2014a, 196, 199). In 1712, there was an advertisement for soy sauce published in the London-based newspaper *The Daily Courant*, and in 1739 there was another ad for soy sauce in the *General Evening Post*. Moreover, several cookbooks in England emerged containing recipes with soy sauce, for example, Elizabeth Mixon's 1748 *English Housewifery* (Ibid, 233, 235).

Thus, a taste for soy sauce spread among the upper classes in Europe in the early 18th century. The Swedish Royal Court, for instance, consumed it in 1732 (Hovförtäringen 1732, 939). Some decades later, in 1755, the Swedish cook Cajsa Warg published what would become a seminal cookbook *Hjelpreda I Hushållningen För Unga Fruentimber* [*Guide to Housekeeping for Young Women*]. The fifth edition of this book includes a recipe for a domestic version of soybean sauce based on boiled mushrooms, spices and salt, noting that it tasted just like foreign soy sauce (Warg 1770, 71–72). Warg was not alone among European cookbook authors to provide recipes for soy sauce that did not include soybeans as an ingredient. British cookbooks promoted a less expensive mushroom-based *kejap*, as an alternative to the pricy "ketjap" imported from Asia (Shurtleff and Aoyagi 2014a, 212–13). In addition, since soy sauce arrived in Europe well before the soybean plant, it was generally not understood that the soybean was one of its main ingredients. Thus, European soy sauce consumers had, for the most part, no idea of how it was made; they only knew it was expensive, but tasty, and so the race was on to see if it could be made using domestic ingredients, such as mushrooms, vinegar or walnuts.

Knowledge about the soybean and its relation to soy sauce was nevertheless beginning to spread in Europe, at least among scientists and Europeans who had direct experience of soyfoods in Asia. Engelbert Kaempfer, mentioned before, was the first European to publish a detailed illustration of the soybean plant and a description of its uses in his seminal *Amoenitatum Exoticarum Politico-Physico-Medicarum* in 1712 (Hymowitz 1970; Shurtleff and Aoyagi 2014a, 204–07). This book marked the beginning of the European academic fascination with the versatile soybean.

Another important scientific contribution, with a good description and illustration of the soybean plant, was the 1747 publication of *Herbarium Amboinensis* [*The Flora of Amboina*] by Rumphius. The six-volume treatise was originally penned while working for VOC in Amboina (now Ambon Island, Indonesia) during the 1650s. The company found its detailed contents commercially too sensitive, however, and the book was not published until several decades after his death, between 1741 and 1750 (Shurtleff and Aoyagi 2014a, 229–32; 2014b, 167). The earliest document concerning the cultivation of soybeans in Europe, however, was written by the world's most famous botanist, the Swede Carl von Linné (or Linnaeus). In 1737, he described a sample he obtained from the first soybean plant growing at Clifford's Garden [*Hortus Cliffortianus*] in Hartecamp, Netherlands (Shurtleff and Aoyagi 2014a, 7; 2021a, 218–19). Ten years later, Linné mentioned the soybean again in *Flora Zeylanica*. The publication covers plants collected by Paul Hermann between 1670 and 1677 in the island of Ceylon and builds on Kaempfer's previous work among others (Shurtleff and Aoyagi 2021a, 227–28). Linné's description in this publication differs from his great *Species plantarum*. In vol. II, from 1753, he described two supposedly different plants, mistakenly assuming that the soybean from Clifford's Garden, which he named *Phaseolus max*, was a different vegetal from the real soybean, which he named *Dolichos soja* in accordance with his new binomial system (Ibid). In addition, Linné wrote that the native country of his *Dolichos soja* was India. This was likely incorrect as Linné probably based his description of the plant on a soybean specimen collected one year earlier in China by his student Pehr Osbeck who, according to his own writings, had never been to India (Osbeck 1757; Shurtleff and Aoyagi 2014b, 172).

Others followed in the footsteps of Linné. In 1757, Karl Gustaf Ekeberg of the Swedish East India Company, released *Kort Berättelse Om den Chinesiska Landt-Hushåldningen*, published in English in 1771 as *A voyage to China and the East Indies*. In it he describes the soybean and its many uses in China. Ekeberg also wrote about soybeans in *Berättelse om Chinesiska Olje-fröet och dess Trefnad i Sverige*, in which he details how the Chinese made soy sauce, with the explicit intent that Swedes might start cultivating soybeans and make soy sauce themselves (Shurtleff and Aoyagi 2014a, 252). Moreover, we have already mentioned the work of Carl Peter Thunberg (1776, 121).

In a more strictly scientific genre, the Swedish botanist Peter Bergius wrote *Beskrifvning på Soja-bönan* [*Description of the Soybean*], which was published by the Royal Swedish Academy of Agriculture and Forestry (KSLA) in 1764. Bergius drew on the works of Linné and Kaempfer and improved their description and illustrations of the soybean (Bergius 1764, 271–72). He also drew on work by Ekeberg, taking his suggestion that soybean cultivation should be tried in Europe – Ekeberg's own attempts had ended in failure, resulting in stunted plants with tiny flowers. In a passage that echoes the English Whig William Petyt eight decades earlier, Bergius also argued that soy sauce had become such a popular item that it was draining money out of the country. This, he claimed, was a good economic reason to bring the soybean to Sweden and adapt it to the local climate and soil (Ibid, 271–72). As Bergius knew, however, it was doubtful that soybean adaptation to Swedish conditions could ever be possible. When he carried out his own trial, most of summer passed before the plant bloomed, leaving no time for maturation (Ekeberg 1764, 272–73). This left Bergius wondering if all the knowledge accumulated about the preparation of soy sauce was of any real value, reasoning that perhaps other beans could be used to make a similar sauce. In any case, the soy sauce hype in Europe, combined with botanical enthusiasts taking an interest in the bean, resulted in multiple attempts to make soy thrive in European soils. Such attempts were, again, impelled by the mercantilist ideals of the time, with its fear of negative trade balances and wish to avoid imports and relatively high transport costs (Hymowitz 1970). It turned out that Europe was poorly suited for soybean production (Liu 1997).

Rupture (1780s–1860)

Throughout the "roots" and "regime" phases (1000–1780s) of the second soy cycle, China was the unquestionable hegemonic power in the region, but, as we have seen, the global balance of economic and political power had been gradually changing as Europe took off and China lagged behind (Perez-Garcia 2021, 74). According to some scholars, the "Fall of the East" preceded the "Rise of the West" – facilitating a relatively easy takeover by the Europeans, who then built on a preexisting system (Barbier 2011, 40, 202). Scholars such as Pomeranz, however, remark that China still had a trade surplus with Europe and remained the end market of silver and gold flows until the late 18th century and that the remarkable economic development in Europe of the 19th century was, in part, shaped by its privileged access to overseas resources (Pomeranz 2021, 4–7). Irrespective of exactly when, it is clear that by the late 18th century China was far behind the world's leading economies (in Northern Europe) in technology, commerce, military power and labor productivity. Soybeans clearly remained central in China even after this global shift. In fact, there was a widespread promotion and development of soybean agriculture in new areas well into the 19th century, which can be observed in several gazetteers (Huaian Fuzhi 2008). However, all major soy-related discoveries and innovations (from cultivation practices and seed breeding to food preparation techniques and bean cakes for fertilization) from now on originated in Europe and the United States.

One reason for the rupture of the Sinocentric soy regime was agrarian. The frontier expansion model had begun to reach its limits in China. Whereas population tripled over the course of Qing, cultivated acreage only doubled and gains in land yields, though significant, were ultimately insufficient to secure food supplies (Isett

2007, 252; Deng and O'Brien 2015, 257–61). If Isett's figures are correct, by 1750, the Chinese population had reached 200–250 million and the average cultivated area of land per capita had fallen to less than half a hectare, or 6 *mu*. By the late 19th century, the population had reached 410 million, while cultivated land per capita had fallen below a quarter hectare, or 3 *mu* (Isett 2007, 293). Thus, Chinese peasants each had much less land from which to secure subsistence. The three Manchu emperors (Kangxi, Yongzheng and Qianlong) were too incapacitated by revenue scarcities to deal with the challenge of population growth. This became obvious in the early 19th century, as the enlargement of areas for settlement was reaching its endpoint. Apart from the mountainous terrain of Yun-nan in the Southwest, new frontiers now lay only in Manchuria. The infusion of new lands and new resources declined steadily and Chinese society began to exhibit signs of resource and land scarcity. At the same time, energy, in the form of fuelwood, became expensive and hard to come by. While China was rich in natural resources, the continuously increasing population came up against hard resource limits, leading to a worsening environmental crisis (Richards 2003, 112–47). With less available "new" land to incorporate and a growing population, the Chinese would have needed to increase yields per hectare (productivity) for food production to keep up. As we have seen, soy played a role during the regime phase to improve yields in the lower Yangtze (providing better fertilizer) and productivity of husbandry (providing better feed), but supplies of better fertilizer and feed were limited. Moreover, apart from the use of bean cake, there were no other major innovative solutions for the agriculture of this period. According to Isett, this ultimately resulted in a Malthusian trap, hampering sustained economic growth (Isett 2007, 301). Consequently, food deficits grew beyond the organizational capacities of the state.

The environmental and resource economist Edward Barbier follows a similar line of thought, arguing that, despite the abundance of resources and trade, Chinese institutions and policies hindered a true scientific revolution. This constrained economic growth and industrial development, which in turn hindered the establishment of a modern economy in Qing China. Moreover, frontier expansion had led to massive deforestation, which resulted in waterway siltation, increasing the risk of bursting banks and resultant flooding and loss of water control also for transport. Farmers also brought hillsides and upland forests under the plough and axe, which had disastrous effects, especially soil erosion and flooding (Barbier 2011, 271–72). Even if some of the new land was fertile, much of it was less productive and the quality of land declined over time (Perkins 1969, 18). Quantitative data on rural China is, as previously mentioned, scarce, but many scholars point out that regional resource depletion was often serious, particularly regarding forest cover and fuel supply (Pomeranz 2021, 225). The combination of dramatic population growth with only modest improvements in yields due to stagnant labor productivity and environmental degradation, resulted in persistent food shortages. According to Barbier, the core of land emperors in China failed to translate their dominance of the world economic system into a successful strategy of sustained economic development based on natural resources. Reversely, "the rise of the West" was, at least in part, due to its ability to expand extraterritorial exploitation of natural resources and frontiers (Barbier 2011, 27–30).

Paradoxically, one of the reasons for the long-term decline of China was the dramatic inflow of American silver. This influx created strong economic networks of Shanxi bankers, Huizhou traders, the so-called Sangleyes in the Philippines, the Hong traders of Canton and their Chinese and Western partners in Macau. Qing officials, however, were not able to supervise and properly tax any of these groups (Sng 2014, 108; Perez-Garcia 2021, 96–102). American silver ended up accumulated predominantly in private family fortunes which escaped official control or sanction. These strong commercial networks – mostly Han Chinese – were perceived as a major threat to Manchu leaders (Perez-Garcia 2021, 89). The economic expansion in the South, combined with the demographic expansion, had exacerbated the problems of administrative control (Sng 2014, 108). At the same time, the state needed more revenue to sustain its burgeoning bureaucracy, including the big military force needed for constant territorial expansion. It was against this backdrop that the emperor Qianlong imposed the Canton System in 1756, which meant, in essence, closing all ports of foreign trade, with the single exception of Canton where taxes on maritime and foreign trade could be more easily levied and monitored and widespread corruption and smuggling curbed (Perez-Garcia 2021, 117).

The Canton system was severely criticized by the Europeans who at this time were violently imposing their model of "free trade" across the world. While both the EIC and VOC were by this time struggling (they had disappeared by the 1830s), their home states remained interested in free trade (Pomeranz and Topik 2006, 165). In addition, the British had finally found an export commodity – opium – that the Chinese wanted to import and which their Indian colonies could supply. Consequently, EIC expanded cultivation of opium in its Indian Bengal territories. Between 1729 and 1800, opium exports to China had grown more than twentyfold, offering a solution to Britain's negative trade balance (Ibid 2006, 91, 160–61). Opium became India's most important export item and China was the by far biggest buyer (Linter 2021, 36–40). With the opium trade snowballing, China's trade surplus with Great Britain gradually disappeared and soon the flow of silver reversed its course. Ending the negative trade balance with China was extremely important to Britain, since it had a even deeper trade deficit with the Atlantic world, despite being the world's leading industrialized nation. The surplus trade via India was thus necessary to compensate for the deficits with the Americas and to get adequate foreign exchange.

However, as opium addiction soared, China prohibited its import in 1839 and ceremonially destroyed some of the incoming cargo. This, together with the general discontent of the British with the Canton monopoly and the "tribute trade", was motivation enough to send troops and go to war. British "gunboat diplomacy" was a success, and the First Opium War (1839–42) ended with the Treaty of Nanjing, which put an end to the Chinese tribute system. It granted Europeans greater access to China with four more ports opened to foreign trade (putting an end to the Canton system), an exchange of ambassadors, the end of ritual obedience and extraterritoriality for their subjects (Pomeranz and Topik 2006, 91–92). In addition, a fixed tariff on imports and exports replaced the indeterminate charges levied previously and China lost its external tariff autonomy. China was also forced to cede Hong Kong and pay reparations both for the financial losses incurred by British traders as well as for damages to Britain during the fighting (war reparations amounted to 21 million silver dollars or roughly \$800 million in 2022). Britain managed to impose all of its initial demands and more and only came short of obtaining the legalization of the opium trade, which remained illegal, at least on paper. The actual trade, however, even flourished, generating fortunes for European and American merchants as smuggling soared (Linter 2021, 52–79).

The Chinese refer to the Treaty of Nanjing as the first in a series of "Unequal Treaties" between China and Western powers. British diplomacy demanded from the Qing government strict adherence to the agreement and to the expansion of free trade, but it also supported Qing efforts to suppress domestic unrest. The year 1850 was a "threshold" year in Chinese history, demarking the close of the early Qing period and the beginning of the late Qing, and simultaneously those of China's ancient and modern periods, respectively (Shi 2017, 21). By this time, Qing China was also engaged militarily in fighting off the Taiping Rebellion – one of the bloodiest civil wars in world history, with over 20 million dead. The Qing Dynasty prevailed, although at an enormous cost to its fiscal and political structure. In 1854, the *Imperial Maritime Customs* or *Chinese Maritime Customs Service* (CMCS) was established to collect customs duties in new ports freshly opened to foreign trade (Keller, Andres Santiago and Shiue 2017, 31; Lyons 2003). The staff was dominated by "great British administrators", such as the Inspector General Sir Robert Hart from 1863 to 1911, but there were also large numbers of German, US and French staff (Foster Hall 1977). Hart created an organization that served the interests of British free trade policies and which, at the same time, conspicuously supported the Qing authorities (Horowitz 2006).

As China lost its external tariff autonomy with the treaty of Nanjing, it installed internal transit duties. These duties, taken together with the old prohibition and newer restrictions on opium trade, motivated the British (this time in alliance with the French) to wage a new war against Qing China: the Second Opium War (1856–60). The result was an even greater set of concessions from China. The Treaty of Tientsin (1858) and the ensuing Convention of Peking (1860) forced China to open up even more cities and ports to foreign commerce, legalized the opium trade and granted foreign merchants and missionaries the right to travel within China. Qing China was also again forced to pay enormous reparations to Britain and France (Linter 2021, 93–117).

A weakened China entered negotiations and signed treaties with an increasing number of rival states other than the United Kingdom and France, especially the United States, Japan and the Russian Empire. In fact, Russia gained 600,000 km^2 in territorial concessions in Manchuria as part of the Treaty of Aigun (1858) – later confirmed in the Treaty of Peking (Shao 2017). Soon the Japanese (who in

1854 abolished its policy of self-imposed national isolation) also started to make territorial claims in Manchuria (Barbier 2011, 258). Trade in soybeans already flourished in the local markets when the CMCS treaty port was established at Newchwang, Manchuria, in 1864. For centuries Manchuria and North China had shipped soybeans, bean cakes and soy oil by boat to Southern Chinese ports. And already by the 1820s, Manchuria alone had sent 40 million *shi* of soybeans to Jiangnan, Shanghai, Fujian and Guangdong – more than double of what it had been exporting to this region only three decades earlier (Li 1998, 114; Fan 2007). Notwithstanding the importance of Manchuria for the entire country (not least as the homeland of the ruling Manchus), China was gradually losing control over this territory. Soybeans in Manchuria, however, would play an increasingly important role over the following century, but in an entirely new geopolitical and technological context.

It was in this context of the mid–19th century that the soybean became globally disseminated through trade and imperialism, as clipper ships and then steamships knit the world more tightly together. The rupture period of the second soy cycle, between the 1780s and 1860s, was not only framed by the long decline of Chinese hegemony, but also by an increase in Europe's agricultural output (both domestically and due to privileged access to overseas resources), marking an important step away from Malthusian pressures (Pomeranz 2021, 17). The dismantling of medieval institutions, abolition of the feudal system and the expansion of capitalist relations in European agriculture and increasingly commodified land and agriculture created larger productive units, allowing for economies of scale. Moreover, beginning in England by the mid-18th century and spreading all over Europe, human and animal labor were replaced by machines powered by fossil fuels (Roberts 2004, 70–75). This, combined with new technologies and cheaper production and diffusion of knowledge, made yield grow explosively: estimates show that from around 1800 Europe's agricultural productivity grew at a rate of about 1% a year (Federico 2010, 23–27).

While there is a significant amount of debate over when, how and why Europe underwent its "agricultural revolution" (Pomeranz 2021, 75–76), clearly the process of agricultural modernization drew on new technologies, innovations in seed breeding (new higher-yielding varieties), machinery (steam engines and cheap coal), infrastructure (railways) and systematized knowledge diffusion (new research centers, cheap printing, institutionalization of information sharing and systematized knowledge). Moreover, Europe also could tap huge larders and frontiers in other parts of the world. In this way, the increasing scientific and experimental interest by Europeans in the soybean (and other "exotic" plants) soon spread to new geographical areas in other continents. As world agricultural production took off, so did population (Federico 2010, 21–22).

Like many other foreign plants, soy was grown in botanical gardens for botanical and taxonomical purposes (Peruchi Moretto, Nodari, and Nodari 2022, 23). Soybeans were introduced in many European countries and European colonies, but the place where they would become truly important was in what would later become the United States. In 1765, soybean seeds reached the Southern colonies of British North America, brought over by Samuel Bowen, an Englishman who had sailed to China with EIC some years earlier and who later acquired land in Savannah, Georgia, where he planted soybeans – most likely using slave labor (Cobb and Inscoe 2009). Bowen used some of his soybean harvest to make soy sauce, a technique he had picked up while in China (Hymowitz and Harlan 1983, 371–79). Only one year later, on May 28, 1766, the *Georgia Gazette* reported that ten bottles of "Bowen's Patent Soy" had been registered at the Custom House in Savannah as exports to London (Shurtleff and Aoyagi 2021a, 254). This was the first time that soybeans grown outside Asia were traded. On June 5, 1766, the *General Evening Post* of London reported that Samuel Bowen had received a gold medal by the *Society of Arts, Manufactures and Commerce* "for introducing several Chinese manufacturies in his Majesty's province of Georgia" (Shurtleff and Aoyagi 2021a, 255). The possibility for England to import soy sauce from Georgia (still under British colonial rule at the time) was much appreciated (Shurtleff and Aoyagi 2014a, 8, 257). While Samuel Bowen had high hopes in the commercial and productive potential of soybeans, the crop would be largely ignored by farmers and scientists until the early years of the 20th century (Du Bois, Tan, and Mintz 2008, 4). This is a thread we will take up again in Chapter 4.

Reflections on the soybean and its functions

The present chapter has covered the ways in which multicausal processes propelled an increased agricultural intensity in which the soybean emerged as a vital part of everyday food and cuisine, while also playing important roles in soil fertility and in feeding animals. Between 1000 and 1800, China was the most powerful and populous state in the world (Bray 2004, 10–15). China's hegemonic power in the region was reflected in the spread of Chinese agricultural practices and culinary culture to neighboring regions. When the regime stood in full bloom 1650–1780, the soybean played a central role in a broad range of arenas throughout East Asia. At the same time, the newly installed Tokugawa shogunate government (1603–1868) in Japan promoted soybeans, and it is possible that soy became even more important for this small island-nation of constant land scarcity.

While China at the beginning of this period was by far the most developed region of the world (Ponting 2001, 377, 382), it saw a long-term reversal of fortunes with the rise of the West and the relative decline of the East. Some of these significant global shifts were reflected in soybean history: through the increasing amount of texts written by European travelers, missionaries and merchants mentioning the many varieties of Asian soybeans and soyfoods; in the trade lists of the European merchants penetrating intra-Asiatic trade (including commerce in soy sauce) and, by the end of this period, texts talking about imported soy sauce in Europe, as well as in the enthusiastic "discoveries" of the properties of the soybean by European scientists.

Notwithstanding the substantial geopolitical, commercial and technological shifts taking place between 1650 and 1780, from the vantage point of the soybean, this century marks a relatively stable and coherent set of roles and functions. For example, soybeans remained important in both subsistence peasant agriculture and as a cash crop (Li 1998). Peasants in the North typically continued to cultivate soybeans in line with its traditional roles (to enrich soils and for food and fodder), but cultivation for commercialization was rising. As demand increased due to urbanization and geographical specialization, soybean flows moved from more sparsely populated zones to core areas in China and Japan, as did the majority of land-intensive commodities (especially forms of energy). As soyfoods became a vital part of the diet of the increasing number of people living in the cities, soy became an increasingly important cash crop.

The growing cities and the geographical differences and complementarities between North and South China spurred trade. Trade expanded during this period and specialized soy guilds and merchants appeared, becoming increasingly powerful. In addition, in spite of the critical attitude of Chinese rulers toward foreigners and international commerce, a wide stream of silver began pouring into the country from the ports in the South. This allowed a deep marketization of the economy. An increasing number of people purchased soybeans or other soyfoods on the market: soy sauce had become a necessity at most meals and tofu now occupied a central position in East Asian cuisine. The trade with soy, from the North to the South not only sustained markets, shops and restaurants with ingredients and food (selling whole soybeans, soy oil, milled soy meal, as well as tofu, soy sauce and other soyfoods), but also other soy products, e.g. *miso*.

Yet, soy trade was not only a Chinese phenomenon, soy sauce was demanded throughout Asia, and as the Europeans took over important routes of the intra-Asian trade, soy sauce began to emerge in European records. In this way, the aggressive European expansion came hand in hand with a boom in the written history of the soybean. The evolution of trans-oceanic trade also made possible a global diffusion of agronomic and gastronomic knowledge about the soybean.

Another important characteristic of this period was the fact that technologies for extracting oil from soybeans had become popularized in China and East Asia. At the same time, the byproduct of oil extraction – bean cake or soybean meal – as well as the byproduct from making tofu – *okara* – were praised as a high-caloric and effective feed in large-scale commercial pig farming in China and Japan. Soybeans and other legumes were thus increasingly grown to fill direct human caloric needs and provide a savory taste, as well as to feed the burgeoning pig industry.

At the same time, rice and wheat, tobacco, corn, sugar cane, peanuts and mulberry began to be intensively cultivated as cash crops, which caused severe soil degradation and deforestation. Rice was increasingly double-cropped with wheat and triple cropping even occurred. The increased land use intensification alongside the low use of livestock in Chinese agriculture, resulted in rising demands for fertilizers. Soybean cake played a significant role here as a potent fertilizer to the intensive agricultural systems in the South, as well as being used widely in Japan. The growing use of soybeans as fertilizer thus enabled the simplification of Chinese and Japanese landscapes and farms. In time, however, the reliance on bean cake for fertilizer became a top priority for the Japanese, partly accounting for Japan's growing interest in Manchuria. This would have enormous implications for soy history as we shall see in the next chapter.

The soybean enabled simplification at the production end of the commodity chain, while the entire agrofood system became ever more complex. Ultimately, the technological, economic and regulatory shifts which propelled the global spread of capitalism would create new, social-ecological effects where soybeans would play a new, central role.

Notes

1 North here refers to the following provinces and regions: Zhili, Fengtian, Shandong Shanxi, Henan, Shaanxi, Gansu and Xinjiang.
2 South here refers to the following provinces: Jiangsu, Anhui, Jiangxi, Zhejiang, Fujian, Hubei, Hunan, Guangdong, Guangxi, Sichuan, Yunnan and Guizhou.
3 It is nevertheless probable that soy products were included in Southeast-Asian maritime trade much earlier. While the spice trade rendered the biggest profits, it seems clear that metalware, forest products and bulk foods – such as rice, grains, beans and pickled and dried fish – filled an important space in the merchant ships already in the 16th century (Reid 1993).
4 In VOC's archives, the originals of these hand-written letters, mentioning trade with *shoyu* (soy sauce), can be found. The inventories for the whole period are now digitized: *1.04.21 Inventaris van de archieven van de Nederlandse Factorij in Japan te Hirado [1609–41] en te Deshima, [1641–1860], 1609–1860 | Nationaal Archief.* The scanned handwritten letters are nevertheless difficult to read, but fortunately the *Soyinfo Center* hired a Dutch researcher (Herman Ketting), to rewrite and translate many of these documents.
5 one *mu* was approximately 666.5 square metres, or 0.067 hectares

References

Banister, Judith. 1992. "A Brief History of China's Population." In *The Population of Modern China*, edited by Poston and Yaukey, 51–57. Boston, MA: Springer US.

Barbier, Edward B. 2011. *Scarcity and Frontiers: How Economies Have Developed through Natural Resource Exploitation.* Cambridge: Cambridge University Press.

Bergius, Peter Jonas. 1764. "Beskrifning På Soja-Bönan." *Svenska Vetenskapsakademiens Handlingar*, KSLA, 25 (25): 271–75.

Bray, Francesca. 1981. "Select Bibliography on Chinese Agricultural History." *Journal d'agriculture Traditionnelle et de Botanique Appliquée* 28 (3): 369–74.

———. 1984. "Science and Civilisation in China. Volume 6, Biology and Biological Technology. Part 2, Agriculture", In *Science and Civilisation in China*, edited by Needham. Cambridge: Cambridge University Press.

———. 2004. "Rice, Technology and History: The Case of China." *Education about Asia* 9 (3): 14–20.

———. 2018. "Where Did the Animals Go?" In *Animals through Chinese History: Earliest Times to 1911*, edited by Sterckx, Siebert, and Schäfer, 118–38. Cambridge: Cambridge University Press.

———. 2019. "Agriculture" In *The Cambridge History of China: Volume 2: The Six Dynasties, 220–589*, edited by Dien and Knapp, 2:355–73. The Cambridge History of China. Cambridge: Cambridge University Press.

Bray, Francesca, and Georges Métailié. 2001. "Who Was the Author of the Nongzheng Quanshu?" In *Statecraft and Intellectual Renewal in Late Ming China*, edited by Jami and & Blue, 322–59. Leiden: Netherlands: Brill.

Chang, Kwang-chih. 1977. *Food in Chinese Culture: Anthropological and Historical Perspectives*. New Haven, CT: Yale University Press.

Chaudhuri, K. N. 1978. *The Trading World of Asia and the English East India Company, 1660–1760*. Cambridge: Cambridge University Press.

Choi, Duk-Kyung. 2014. "明末清初豆饼的出现和江南农业的发达。农业考古，年第 4 期，页 The Origin of Bean Cake in the Late Ming and Early Qing Period, and the Development of Agriculture in the Jiangnan Region." *Agricultural Archaeology*, no. 4: 41–57.

Cobb, James C., and John C. Inscoe. 2009. "Georgia History." In *New Georgia Encyclopedia*, edited by Inscoe, Hatfield, and Forrester. Georgia: Georgia Humanities, University of Georgia Press.

Colla, Elisabetta. 2008. "16th Century Japan and Macau Described by Francesco Carletti." *Bulletin of Portuguese – Japanese Studies* 17: 113–44.

Crosby, Alfred W. 1972. *The Columbian Exchange: Biological and Cultural Consequences of 1492*. Westport, CT: Greenwood Press.

Deng, Kent, and Patrick O'Brien. 2015. "Nutritional Standards of Living in England and the Yangtze Delta (Jiangnan), circa 1644–circa 1840: Clarifying Data for Reciprocal Comparisons." *Journal of World History* 26 (2): 233–67.

Du Bois, Christine M. 2018. *The Story of Soy*. London: Reaktion Books.

Du Bois, Christine M., Chee-Beng Tan, and Sidney W. Mintz, eds. 2008. *The World of Soy*. Singapore: NUS Press.

Ekeberg, Carl Gustaf. 1764. *Om Chinesiske Soyan*. Stockholm: Kongl. Svenska Vetenskaps Academiens Handlingar. https://hosting.devo.se/kvah/searchResult.htm.

Fairbank, John King, and Merle Goldman. 2006. *China: A New History, Second Enlarged Edition*. Boston: Harvard University Press.

Fan, Jin-Min. 2007. "清代前期淮安关税收的盈绌原由。安徽历史，第1期 Reasoning the Causes of the Fluctuation of Taxation in the Huai-An Guan (Huai-An Customs) during Early Qing Dynasty." *Anhui History*. 范金民 (1): 35–41.

Federico, Giovanni. 2010. *Feeding the World: An Economic History of Agriculture, 1800-2000*. Princeton: Princeton University Press.

Foster Hall, Basil E. 1977. The Chinese Maritime Customs: An International Service, 1854-1950. Greenwich, England: National Maritime Museum.

Frank, Andre Gunder, and Barry K. Gills. 2000. "The Five Thousand Year World Systems in Theory and Praxis." In *World System History: The Social Science of Long-Term Change*, edited by Denemark, Friedman, Gills, and Modelski, 3–23. London and New York: Routledge.

Gelderblom, Oscar, 2009. *The Political Economy of the Dutch Republic*. Farnham, England; Burlington, VT: Ashgate.

Gordon, Peter, and Juan José Morales. 2017. *The Silver Way: China, Spanish America and the Birth of Globalisation, 1565–1815*. London: Penguin Books.

Gordon, Stewart. 2008. *When Asia Was the World: Traveling Merchants, Scholars, Warriors, and Monks Who Created the "Riches of the East."* Pennsylvania, PA: Da Capo Press.

Han, Bei-Zhong, Frans M. Rombouts, and M. J. Robert Nout. 2001. "A Chinese Fermented Soybean Food." *International Journal of Food Microbiology* 65 (1–2): 1–10.

Horowitz, Richard S. 2006. "Politics, Power and the Chinese Maritime Customs: The Qing Restoration and the Ascent of Robert Hart." *Modern Asian Studies* 40 (3): 549–81.

Hovförtäringen. 1732. "Soja Af Schampinjoner." Stockholm. Handskrift i Slottsarkivet. https://www.saob.se/artikel/?seek=soja&pz=1.

Hsu, Tan. 1998. 明清时期山东商品经济的发展。北京：中国社会科学出版社/ *The Development of Commodity Economy in Shandong Province during Ming and Qing Dynasties*. Beijing: Social Sciences Press.

Huaian Fuzhi. 2008. "The Collection of China's Gazetteers." In 吳昆田、高延第纂，孫雲錦修，. Vol. 7. Kun-Tian, Gao and Sun 54. 淮安府志（光緒）。中國地: 方志集成，第.

Huang, Hsing-Tsung 2000. "Science and Civilisation in China. Volume 6. Biology and Biological Technology. Part 5. Fermentations and Food Science". In *Science and Civilisation in China*, edited by Needham. Cambridge: Cambridge University Press.

———. 2008. "Early Uses of Soybean in Chinese History." In *The World of Soy*, edited by Du Bois, Tan, and Mintz, 45–55. Singapore: NUS Press.

Huang McBeath, Jenifer, and Jerry McBeath. 2010. *Environmental Change and Food Security in China*. Advances in Global Change Research. Dordrecht: Springer.

Huang, Philip C. C. 2002. "Development or Involution in Eighteenth-Century Britain and China? A Review of Kenneth Pomeranz's 'The Great Divergence: China, Europe, and the Making of the Modern World Economy.'" Edited by Pomeranz. *The Journal of Asian Studies* 61 (2): 501–38.

Hymowitz, T. 1970. "On the Domestication of the Soybean." *Economic Botany* 24 (4): 408–21.

Hymowitz, T., and C. A. Newell. 1981. "Taxonomy of the GenusGlycine, Domestication and Uses of Soybeans." *Economic Botany* 35: 272–88.

Hymowitz, T., and J. R. Harlan. 1983. "Introduction of Soybean to North America by Samuel Bowen in 1765." *Economic Botany* 37: 371–79.

Hymowitz, T., and W. R. Shurtleff. 2005. "Debunking Soybean Myths and Legends in the Historical and Popular Literature." *Crop Science* 45 (2): 473–76.

Isett, Christopher. 2007. *State. Peasant, and Merchant in Qing Manchuria, 1644–1862*. Stanford, CA: Stanford University Press.

Isett, Christopher, and Stephen Miller. 2016. *The Social History of Agriculture: From the Origins to the Current Crisis*. Lanham: Rowman & Littlefield.

Jewell, William. 2011. *The Golden Cabinet of True Treasure. London: John Crosley, 1612*. Ann Arbor, MI: Text Creation Partnership.

Katō, Shigeshi. 1953. "Trade between Manchuria and China during the K'ang-Hsi 康熙 and Ch'ienlung 乾隆 Periods / 支那經濟史考證（下卷），康熙乾隆时代に于ける满洲と支那本土と通商について，東京都: 東洋文庫。." *Studies in Chinese Economic History* 2: 595–616.

KB *Koninklijke Bibliotheek* (National Library of the Netherlands) 2022. "Delpher Database." Online Database Books and Periodicals from the Seventeenth Century until 1995 in the Netherlands: The Hague. https://www.delpher.nl/ (Accessed 2022-08-26).

Keller, Wolfgang, Javier Andres Santiago, and Carol H. Shiue. 2017. "China's Domestic Trade during the Treaty-Port Era." *Explorations in Economic History*, Special Issue: A new economic history of China, 63 (January): 26–43.

Kiple, Kenneth F. 2007. *A Movable Feast: Ten Millennia of Food Globalization*. Cambridge: Cambridge University Press.

Kiple, Kenneth F., and Kriemhild Coneè Ornelas, eds. 2000. *The Cambridge World History of Food*. Cambridge: Cambridge University Press.

Krech, Shepard, J. R. McNeill, and Carolyn Merchant, eds. 2004. *Encyclopedia of World Environmental History*. 1st ed. Berkshire.

Kuo, Chunghao Pio. 2013. *Pigs, Pork, and Ham: The Practice of Pig-Farming and the Consumption of Pork in Ming-Qing China.* Doctoral thesis, New York University. ProQuest Dissertations Publishing.

Lander, Brian, and Thomas David DuBois. 2022. "A History of Soy in China: From Weedy Bean to Global Commodity" in *The Age of the Soybean: An Environmental History of Soy during the Great Acceleration*, edited by da Silva and de Majo, 29–47 Winwick, Cambridgeshire: The White Horse Press.

Li, Bozhong. 1998. *Agricultural Development in Jiangnan, 1620–1850.* London: Palgrave Macmillan UK.

———. 2019. "1524: The End of the Silk Road." Plenary Keynote Speech presented at the The IV Congress of the Asian Association of World Historians, Osaka, Japan.

Linter, Bertil. 2021. *Opiumkrigen.* Stockholm: Historiska Media.

Liu, KeShun. 1997. *Soybeans: Chemistry, Technology, and Utilization.* Gaithersburg, MD: Aspen Publisher.

———. 2008. "14- Food Use of Whole Soybeans." In *Soybeans*, edited by Lawrence A. Johnson, Pamela J. White, and Richard Galloway, 441–81. Urbana, IL: AOCS Press.

Long, Deng-Gao. 1997. *The History of the Development of Chinese Traditional Markets* 中国传统市场发展史。北京：人民出版社. Beijing: People's Press s. 龙登高.

Lyons, Thomas P. 2003. *China Maritime Customs and China's Trade Statistics, 1859-1948.* Trumansburg, NY: Willow Creek of Trumansburg.

Mann, Charles C. 2011. *1493: Uncovering the New World Columbus Created.* New York, NY: Random House.

Marks, Robert. 1998. *Tigers, Rice, Silk, and Silt: Environment and Economy in Late Imperial South China.* Studies in Environment and History. Cambridge: Cambridge University Press.

McNeill, John Robert. 1998. "China's Environmental History in World Perspective." In *Sediments of Time: Environment and Society in Chinese History*, edited by Elvin and Liu, 31–49. Cambridge: Cambridge University Press.

McNeill, John Robert. 2001. *Something New under the Sun: An Environmental History of the Twentieth-Century World.* First published as a Norton paperback. New York: W. W. Norton.

Mintz, Sidney W. 2008. "Fermented Beans and Western Taste." In *The World of Soy*, edited by Du Bois, Tan, and Mintz, 56–73. Singapore: NUS Press.

Mun, Roberts. n.d. *A Select Collection of Early English Tracts on Commerce.* Accessed August 13, 2022.

Myrdal, Janken. 2020. "Agricultural Treatises in Eurasia before ca. 1500: The Iceberg Method, a Survey, and a Grand Hypothesis." *The Medieval Globe* 6 (2): 1–47.

——— 2022. "A Theory Kit for World History." Science report 7. *Agrarhistoria.* Uppsala: Department of Urban and Rural Development, Swedish University of Agricultural Sciences

——— Forthcoming. *The Agricultural History of East and Southeast Asia. Global Agricultural Systems.* Uppsala: Department of Urban and Rural Development, Swedish University of Agricultural Sciences.

Nakajima, Gakusho. 2018. "The Structure and Transformation of the Ming Tribute Trade System." In *Global History and New Polycentric Approaches*, edited by Perez-Garcia and De Sousa, 137–62. Palgrave Studies in Comparative Global History. Singapore: Palgrave Macmillan.

Needham, Joseph. 1971. "Science and Civilisation in China. Volume 4. Physics and Physical Technology. Part 3. Civil Engineering and Nautics". In *Science and Civilisation in China*, edited by Needham, Cambridge: Cambridge University Press.

———. 2004. *Science and Civilisation in China. Volume 7. The Social Background. Part 2. General Conclusions and Reflections*, In *Science and Civilisation in China*, edited by Robinson. Cambridge: Cambridge University Press.

Nyström, Johan Fredrik. 1883. *De Svenska Ostindiska Kompanierna: Historisk-Statistisk Framställning.* Göteborg: Göteborgs Kongl. Vetenskaps- och Vitterhets Samhälles Handlingar.

Osbeck, Pehr. 1757. *Dagbok Öfwer En Ostindisk Resa Åren 1750. 1751. 1752. Med Anmärkningar Uti Naturkunnigheten, Främmande Folkslags Språk, Seder, Hushållning, m.m.* Stockholm: KSLA.

Perez-Garcia, Manuel. 2021. *Global History with Chinese Characteristics: Autocratic States along the Silk Road in the Decline of the Spanish and Qing Empires 1680–1796*. Singapore: Springer Singapore. 3.

Perkins, Dwight H. 1969. *Agricultural Development in China, 1368–1968*. Chicago, IL: Aldine Publishing Company.

Peruchi Moretto, Samira, Eunice Nodari, and Rubens Nodari. 2022. "Soybean as a Critical Resource: Domestication, Dissemination and Introductions (Glycine Max)." In *The Age of the Soybean: An Environmental History of the Soyacene during the Great Acceleration*, edited by de Majo and da Silva, Chapter 1:19–28. Winwick, Cambridgeshire: White Horse Press.

Petyt, William. 1680. *Britannia Languens: Or, A Discourse of Trade Shewing, That the Present Management of Trade in England, Is the True Reason of the Decay of Our Manufactures, and the Late Great Fall of Land-Rents; and That the Increase of Trade, in the Method It Now Stands in, Must Proportionably Decay England. Wherein Is Particularly Demonstrated, That the East-India Company, as Now Managed, Has Already Near Destroyed Our Trade in Those Parts, as Well as That with Turky, and in Short Time Must Necessarily Beggar the Nation. Humbly Offered to the Consideration of This Present Parliament.* Printed for Tho. Dring, at the Harrow at Chancery-Lane end in Fleetstreet and Sam. Crouch in Popes-Head-Alley near the Royal Exchange in Corn-hill, London.

Pomeranz, Kenneth. 2021. *The Great Divergence: China, Europe, and the Making of the Modern World Economy*. The Princeton Classics edition: Princeton University Press.

Pomeranz, Kenneth, and Steven Topik. 2006. *The World That Trade Created: Society, Culture, and the World Economy, 1400 to the Present*. 2nd ed. Armonk, New York: M.E. Sharpe.

Ponting, Clive. 2001. *World History: A New Perspective*. London: Pimlico.

Rath, Eric. 2010. *Food and Fantasy in Early Modern Japan*. Berkeley: University of California Press.

Rawski, Evelyn Sakakida. 1972. *Agricultural Change and the Peasant Economy of South China*. Cambridge: Harvard University Press.

Reid, Anthony. 1993. *Southeast Asia in the Age of Commerce, 1450–1680*. Vol. 2. New Haven, CT: Yale University Press.

Richards, John F. 2003. "Internal Frontiers and Intensified Land Use in China." In *The Unending Frontier: An Environmental History of the Early Modern World*, 1st ed., edited by Richards, 112–47. Los Angeles: University of California Press.

Roberts, John M. 2004. *The New Penguin History of the World*, 4th ed. London: Penguin Books.

Roessingh, M. P. H. 2022. "Inventaris van de Archieven van de Nederlandse Factorij in Japan Te Hirado [1609–1641] En Te Deshima, [1641–1860], 1609–1860." Inventarissen. 1.04.21. Haag: Nationaal Archief, Den Haag 1964. Nederlandse Factorij Japan.

Sakai, Tadao, Sakade Yoshinobu, and Ogawa Yōichi, eds. 2000. *Santai Wanyong Zhengzong.* Tokyo: Kyūko Shoin.

Shao, Dan. 2017. "Manchuria in Modern East Asia, 1600s–1949." In *Oxford Research Encyclopedia of Asian History*. Oxford Research Encyclopedia.

Shi, Zhihong. 2017. *Agricultural Development in Qing China: A Quantitative Study, 1661–1911*. Leiden, The Netherlands: Brill.

Shurtleff, William, and Akiko Aoyagi. 2012. *History of Soy Sauce (160 CE to 2012): Extensively Annotated Bibliography and Sourcebook*. Lafayette, CA: Soyinfo Center.

———. 2014a. *Early History of Soybeans and Soyfoods Worldwide (1024 BCE to 1899): Extensively Annotated Bibliography and Sourcebook*. Lafayette, CA: Soyinfo Center.

———. 2014b. *History of Soybeans and Soyfoods in Japan, and in Japanese Cookbooks and Restaurants Outside Japan (701 CE to 2014): Extensively Annotated Bibliography and Sourcebook*. Lafayette, CA: Soyinfo Center.

———. 2015. *History of Soybeans and Soyfoods in the United Kingdom and Ireland (1613–2015): Extensively Annotated Bibliography and Sourcebook*. Lafayette, CA: Soyinfo Center.

———. 2016a. *History of Soybeans and Soyfoods in the Netherlands, Belgium and Lxembourg, 1647–2015*. Lafayette, CA: Soyinfo Center. https://www.soyinfocenter.com/pdf/186/NBL.pdf.

———. 2016b. *History of Soybean Crushing, Soy Oil, and Soybean Meal (980–2016): Extensively Annotated Bibliography and Sourcebook*. Lafayette, CA: Soyinfo Center.

———. 2021a. *Early History of Soybeans and Soyfoods Worldwide (1024 BCE to 1899): Extensively Annotated Bibliography and Sourcebook*. Revised from 2014. Lafayette, CA: Soyinfo Center.

———. 2021b. *History of International Trade in Soybeans, Soy Oil and Soybean Meal, and Trade Policy (1859–2021): Extensively Annotated Bibliography and Sourcebook*. Lafayette, CA: Soyinfo Center.

Shurtleff, William, Hsing-Tsung Huang, and Akiko Aoyagi. 2014. *History of Soybeans and Soyfoods in China and Taiwan, and in Chinese Cookbooks, Restaurants, and Chinese Work with Soyfoods Outside China (1024 BCE to 2014): Extensively Annotated Bibliography and Sourcebook, Including Manchuria, Hong Kong and Tibet*. Lafayette: Soyinfo Center.

Sng, Tuan-Hwee. 2014. "Size and Dynastic Decline: The Principal-Agent Problem in Late Imperial China, 1700–1850." *Explorations in Economic History* 54: 107–27.

Spence, Jonathan D. 1993. "Book Reviews: Nourish the People: The State Civilian Granary System in China, 1650–1850. By Pierre-Etienne Will and R. Bin Wong, with James Lee" *The China Quarterly* 133: 162–63.

Tan, Chee-Beng. 2008. "Tofu and Related Products in Chinese Foodways." In *The World of Soy*, edited by Du Bois, Tan, and Mintz, 99–120. Singapore: NUS Press.

Tan, Qi-Yu. 1986. *The Draft of The History of Chinese Crops and Agriculture* 北京：中国作物栽培史稿. 北京：中国作物栽培史稿. Beijing: Agriculture Press.

Thunberg, Carl Peter. 1776. *Travels in Europe, Africa, and Asia*. London: Printed for F. and C. Rivington.

Tremml-Werner, Birgit. 2015. *Spain, China, and Japan in Manila, 1571–1644*. Amsterdam: Amsterdam University Press.

van Goens, R., Cornelius Speelman, Balthasar Both, Anth Hurrt, Dirk Blom, W. van Outhoorn, and Joan Camphuijs. 1680. "Re: Sending Sake and Soy Sauce in Casks and Pots." Letter Unpaginated. Handwritten, with signature. Den Haag. Nationaal Archief, De Archieven van de Nederlandse Factorij Japan (NFJ).

Verschuer, Charlotte, and Wendy Cobcroft. 2016. *Rice, Agriculture, and the Food Supply in Premodern Japan*. 1st ed. London: Routledge.

Warg, Cajsa. 1770. *Hjelpreda I Hushållningen För Unga Fruentimber*. Femte upplagan. Med Kongl. Maj:ts Allernådigste Privileg. Stockholm: Grefingska Tryckeriet.

Wells, Richard Evan. 2018. *The Manchurian Bean: How the Soybean Shaped the Modern History of China's Northeast, 1862–1945*. Doctoral dissertation, University of Wisconsin–Madison.

Werner, Dietrich, and William E. Newton. 2005. *Nitrogen Fixation in Agriculture, Forestry, Ecology, and the Environment*. Springer Science & Business Media.

Xue, Yong. 2007. "A 'Fertilizer Revolution'? A Critical Response to Pomeranz's Theory of 'Geographic Luck.'" *Modern China* 33 (2): 195–229.

4 The roots of the third soybean cycle (1860–1949)

> The Commissioner of Chinese Customs at Niu-chwang refers, in his report for the past year, to the important part played by Japan in the recent rapid growth in the foreign trade of Manchuria. This growth in trade with Japan may be said to have commenced with the war between Japan and China, which probably opened the eyes of the Japanese to the wealth of Manchuria and the peculiar value of its products to themselves. The Soya bean is consumed in enormous quantities in Japan, the better qualities for food and the inferior ones as manure. In former years the supply was obtained from Korea, but the Manchurian bean is cheaper. Beancake is used as a food for cattle and for manure, and its value in both respects is well known. Japanese shipping at Niu-chwang naturally shows a great increase likewise.
>
> (Shurtleff and Aoyagi 2022, 61)

This quote comes from the article "Japanese Commercial Interests in Manchuria" which appeared in *The Times* of London on October 5, 1900. Most of the key processes that constitute the roots of the third (and present) soy cycle – the object of this chapter – are encapsulated herein. For instance, the post of Commissioner of the Chinese Maritime Customs Service (CMCS) mentioned in it had been established in 1864 at the Manchurian port of *Niu-chwang* [Newchwang – today Yingkou] as a consequence of the "unequal treaties" that followed China's defeat in the Opium Wars, (Horowitz 2006) as discussed at the end of the previous chapter. The CMCS (and thus, ultimately, the British) oversaw all operations at Newchwang (the northernmost commercial treaty port), which by that time had become the most important hub in the soy trade. In this piece, *The Times* also reflected another key development essential to this phase: the mounting inter-imperial territorial rivalries and, specifically, the Sino-Japanese War of 1894–95. After Tokyo's successful invasion of Manchuria, Japan's dependence on this region's soy imports soared. Although they are not mentioned in the piece, two further essential trends also began to take shape at this time. First, Heilongjiang, Fengtian/Liaoning and Jilin – the three Northeastern provinces of China – as well as parts of Inner Mongolia would soon take a central position in the nascent, but fast growing, European agro-industrial sector. Indeed, about 90% of

DOI: 10.4324/9780367822866-4

globally traded soy at this time was Manchurian. Second, soybean production had also begun taking root in the US Corn Belt and within a mere few decades it would become large enough to fill the void left by the eventual implosion of the Manchurian soy economy after the defeat of Japan in World War II.

This chapter is exclusively devoted to the roles and functions of soy during the period between 1860 and 1949; in other words, it only considers the roots phase of the third soy cycle. By the mid-19th century, soy was not yet the key global commodity it is today. However, a set of transformative forces during this phase (new technologies, commodity chain activities, uses and power configurations) would ultimately usher in the post-World War II regime phase. After this introduction, we first critically reflect on empirical sources before we delve into three main Chapter sections. The first section analyzes the roles of soy within the processes of frontier expansion and agricultural intensification as they occurred in Manchuria during 1860– 1939 and will then situate the Manchurian soy boom in the wider context of global agrofood rivalries and imperialism. In the second section, we turn to the roles of (predominantly Manchurian) soybeans in European agroindustries, during roughly the same period. Our main focus is on the technological advances that enabled soy to supplant other fats and to be used in products such as lamp oil, soaps, synthetic wool and leather, as well as new foods like margarine. We then cover the gradual consolidation of soybeans in the United States, where it would pave the way for consolidation of a new regime phase. The final section summarizes and reflects on the great changes between 1860 and 1949. As in other chapters, we begin with a discussion of our sources.

A critical note on sources

During the period under consideration in this chapter, the modern scientific canon takes off; scores of scholarly books and journals about agriculture, geography, politics, history and a whole host of other disciplines will be penned, published and critically discussed in Europe and many other places (Myrdal 2020). Moreover, there is a noticeable "scientific" proliferation of specialized and systematized approaches to examine the various dimensions of agriculture. Seeds and animal breeds, for instance, which had already been bred to develop specific characteristics for millennia, were now being subjected to a more technically rigorous and selective breeding process, and all of it was being extensively documented. Soy is no exception – shelves of articles began to appear exploring, in particular, different aspects of various soybean seed types. Simply put, what in Chapters 3 and especially 2 had been a problem of scarcity of sources now becomes, in this and the following chapter, a problem of selection. However, a notable exception to this, in our opinion, is that some of the accounts of the Manchurian soybean complex around the turn of the century probably underestimate Russian, and later Soviet, involvement in the region, including approximated figures for soy demand and trade.

Even with the abovementioned abundance of materials, however, quantitative data on agricultural trade and (especially) production in this period remains problematic. While most countries in the world had developed national statistical offices in charge of gathering information on production and trade, these were often incomplete. It became possible to construct fairly representative indices of agricultural output from statistics drawn in Western Europe, the United States and Australia, but data on agricultural production from the rest of the world remained patchy and uncertain until at least 1950 (Federico 2010, 16–17). Throughout this period, in fact, world agricultural trade was actually much better documented than agricultural output (Ibid, 28).

In the present chapter, we have proceeded by compiling available data on soybean, oil and cake/meal production as well as imports and exports in weight since the 1860s. For Manchurian production and trade in the 1890s, we have mostly relied on reports from the Chinese Maritime Customs Service, as well as the trade section of the 1901 book *Manchuria: Its People, Resources and Recent History* by Alexander Hosie, responsible for the British Consulate at Newchwang between 1894 and 1900 (1904).[1] For the period between 1905 and 1948, we have also used published international agricultural statistics and trade data from the International Institute of Agriculture (IIA) and US Department of Agriculture (USDA). Some other sources are the *Farmers' Bulletins*, the *American Soybean Association Digest* and, just as for all the other chapters, the integrated bibliographies and sourcebooks produced by the *Soyinfo Center* published by William Shurtleff and Akiko Aoyagi. Whenever we were not able to check specific primary sources or materials mentioned in *Soyinfo* we have used their annotated summaries and aimed to triangulate these as our work progressed.

Frontier expansion in Manchuria

Following the terms of the Treaty of Tianjin (or Tientsin) of 1858, China gradually opened ports to foreign trade and adopted a 5% *ad valorem* tariff limit (O'Rourke and Williamson 1999, 54). Only one such port – Newchwang (Yingkou) – was open to international trade in Manchuria (Hu 2002; Zhang 2008:, 526–527) – from where soybeans, oils and cakes were shipped out. The British Consulate and the Chinese Maritime Customs Service (CMCS) were also established (in 1861 and 1854, respectively). Detailed information about Newchwang appears in the comprehensive *The Treaty Ports of China and Japan: A Complete Guide to the Open Ports of Those Countries*, edited by the British civil servant, journalist and Sinologist, Nicholas Belfield Dennys, from 1867:

> The first impression on the eye when approaching Newchwang is dreary in the extreme, and the place possesses no advantages to correct this feeling subsequently. The muddy river winds through a plain of mud, without a single natural elevation to break the dismal monotony of the scene, and houses built of mud are all that constitute the human habitations included in the survey. Filth and squalor in an unusual degree are the characteristics of the native town of *Ying-tz'*; [Yingkou] whilst the foreign residences which have been erected above the town are primitive, bare, and isolated. The British

> Consulate is established in a building, formerly a temple, somewhat better than the rest. Except for a few weeks during the summer, the region in which the port is situated is little more cheerful than an arctic swamp. The only noteworthy features of the town are the large enclosures in which the native dealers or warehousemen carry on the business of storing and manufacturing Pulse and Pulse-cake, upon which staples the sea-borne trade of Ying-tz' depends… The number of British and German vessels visiting the port is, notwithstanding, very considerable, but, as will be shewn below, their freights are carried exclusively on native account.
>
> (Dennys 2012, 540–41)

"Pulse" and "Pulse-cake" refer to soybeans and soybean cake and the cargo sailed southwards from what, in 1867, was still a modest town enlivened by a vibrant soy trade. Again, in Dennys' words, the commerce of pulses, which afforded "employment to a large number of foreign vessels, is almost exclusively conducted by native Chinese, principally from the southern provinces, who have gained complete control of the local markets" (Ibid, 543). At the start of the period we are surveying, trade volumes were relatively modest and the attendant documented sources were relatively scarce. We do know, however, that in 1865, 119,168 tons of bean cakes were exported from Newchwang (Shurtleff and Aoyagi 2022, 22). Fairly soon, other foreign commercial interests were making themselves known alongside those of the British. The city, always under the oversight of the British Consulate, would quickly grow into one of the most important of the treaty ports, centering on the soy trade and consolidating its position as the main exit point for Manchuria's voluminous soybean production (Ibid, 28). The beans arrived at Newchwang where they were pressed into oil and cake and then shipped off in iron hoops to the Southern ports of China for different purposes.

While soy cultivation and exports were important and expanding, much of the rich resources in the vast Manchurian hinterland remained "untapped". As we saw in the previous chapter, the ruling Manchus had tried to limit Han-peasant immigration and, while some settlers managed to remain, the region was still sparsely populated, comprising large tracts of land used mainly for hunting (Dennys 2012, 543). In 1880, however, the central Qing government removed mobility restrictions and offered Han peasants land for sale or lease, often even adding capital loans to the offer (Chang 1963; Hymowitz 1970, 417; Bix 1972). Predictably, a wave of migrants rolled into Manchuria after this drastic policy turn. Though some see it difficult to explain, the sudden decision to change tack and bring more land into cultivation was motivated by the need, politically, to stabilize and even pacify the country. Over the previous years, famines, floods and a severe drought in Northern China had ignited revolts against the government. Between 1850 and 1864 alone, up to 20 million people had perished during the Taiping Rebellion (Isett and Miller 2016, 116–17). In addition, the Qing Dynasty was concerned about declining levels of imperial revenues: clearly, selling or leasing land and then taxing it was one way of addressing this problem (Langthaler

2020, 248–49). Finally, there was also an obvious geopolitical calculation at play: by populating Manchuria with Chinese peasants, the central government hoped somewhat to counter foreign (and perhaps particularly, Russian) imperialist ambitions (Isett 2007, 9). After this reversal in migration policy, millions of Han Chinese abandoned their homelands (mostly in Shandong and Zhili), in an attempt to escape poverty and hunger and soon found themselves living in a far richer land (Shan 2014, 4). The settlement transformed the landscape, where natural grasslands and forests were now cleared and plowed (He et al. 2008). Manchuria's black soils, extensive lands and temperate climate were considered perfect for the cultivation of soybeans and production (as well as profits) increased rapidly (Lander and DuBois 2022, 38). A booming Manchurian soy industry took off, attracting even more incomers, including seasonal peasant workers for the soybean harvest (Shurtleff and Aoyagi 2022, 105).

The importance of soy was underscored in the book *Manchuria: Its People, Resources and Recent History*, by Sir Alexander Hosie (Hosie 1904, 40–41). In one illustrative passage, Hosie noted that "[T]he virgin soil gave forth abundantly; there was a great wealth of agricultural products to be disposed". As he also added: "I shall have much to say hereafter in regard to this insignificant looking bean, which is the subject of a considerable industry throughout Manchuria and Northern China" and which was "worth much more than all other exports combined" (Ibid, 181). Hosie's testimony is a long, vivid and personal account of Newchwang's metamorphosis from a "mud village", consisting of "a few semi-Chinese cottages", to "a rich and populous town, with many fine shops, houses, and temples, and with something of a modern look, due to the tall chimneys of the steam beancake factories". Such transformation took place over a period of just a few years (Ibid, 70–77). Hosie's individual field report chimes perfectly with a 1880 US consulate report confirming that exports of soybeans, oil and bean cakes from this previously "neglected area" had taken off and were changing it radically (US Bureau of Foreign Commerce 1880, 1:814).

While observers from the United Kingdom and the United States described the transformation of Newchwang in mostly positive terms, Chinese guilds were not so keen on foreign merchants and sought continuously to keep them out of the trade in soy. In 1887, the US Bureau of Manufactures (Department of Commerce and Labor) published another report, *The Chinese guilds. Monthly Consular and Trade Reports*. With regard to the Shantung guild, the document was clear:

> [the guild's] last struggle to maintain its supremacy was shown in its procuring insertion in the regulations of trade appended to the Tientsin treaty a special clause prohibiting the export of pulse and beancake from Tungchow and Newchwang under the British flag.
>
> (Shurtleff and Aoyagi 2022, 45)

While efforts were made to revive the prohibitory enactments, the Chinese were eventually forced to forfeit this stand as well. In 1869, the prohibition was withdrawn and the clause was rescinded (Ibid, 487). Thus, while Britain made no territorial claims on Manchuria (and was, at this time, not particularly interested in soybeans), it played an important role in overseeing that China kept all ports open, extending full access to British (and other) merchants. According to a dispatch written on behalf of the Associated Chambers of Commerce in London by Lord Beresford in 1899, two of the largest British trading companies – Butterfield & Swire and Jardine & Matheson Steamship Co. – made about 250 trips in and out of Newchwang in 1897. In Lord Beresford's boastful words: "the backbone of the Chinese coasting trade under the British flag is the Newchwang coasting trade– soya beans, bean-cake, pease, and kindred stuffs" (Ibid, 55). Predictably, the economic potential of the region and its trade brought increasing foreign interest; with France, Germany, Britain and the United States all engaged in the area. But it would be the inter-imperialist rivalry between Japan, tsarist Russian and the disintegrating Chinese Qing Empire, however, that would flare up into an open territorial conflict.

Imperial rivalries and railway imperialism

Before the mid-19th century, Japan had avoided direct contact with Qing China, but tensions would quickly build up as the Tokugawa shogunate disintegrated and the Western oriented *Meiji* era was proclaimed in 1868. Japan, aware of its food insecurity and dependence on foreign lands, began developing its commercial and territorial ambitions in Korea and Manchuria (Seth 2010, 445). On August 1, 1894, war broke out officially between China and Japan: it was the first Sino-Japanese War, in which Japan invaded Manchuria by land and sea, destroying every city under occupation (Hosie 1904, 39). The conflict neared its end as the Japanese took the city of *Kai-p'ing Hsien*, as well as Newchwang, on March 6, 1895, dealing a fatal blow to Chinese forces only a few days later at *Tien-chuang-t'ai*, 13 miles north of Newchwang (Ibid, 40). The Treaty of Shimonoseki forced China to cede Taiwan and an important part of Manchuria – the *Fêng-t'ien* Peninsula, including the port of Newchwang – to Japan. The treaty also forced China to pay 200 million taels of silver and certified the loss of Korea as a tributary state. Meantime, Russia, France and Germany made representations to Japan arguing that "the permanent possession of the ceded districts by the Empire of Japan would be detrimental to the lasting peace of the Orient" (Ibid, 40–41). A convention for the retrocession of all the territory occupied by Japan in Manchuria was signed on November 8, and the evacuation of Japanese troops was soon complete. But what followed it, however, was a new commercial treaty, signed in 1896, which forced China to open yet more ports to trade and concede Japan the same privileges enjoyed by Western powers; including the control of manufacturing outfits in treaty ports (Shurtleff and Aoyagi 2022, 52). While Manchuria remained Chinese (at least for the time being), China was again humiliated. As far as soy exports were concerned, the implications of Japanese victory were felt immediately (Akio Saito has extensively documented this in his 1985 *Daizu bunka-shi nenpyô* [*Chronology of Soybeans*, 122 BCE to 1984] – see Shurtleff and Aoyagi 2022, 58–59). Again, in Lord Beresford's words:

> Until recent years the bulk of the beans and bean-cake trade was carried on with the south of China; but since the Chino-Japanese war an extensive trade has been carried on with Japan; in fact Japan has outstripped China altogether. [...] Exports to Japan increased from 460,000 taels in 1891 to 5,079,000 taels in 1897, from 2,727,000 taels in 1891 to 2,438,000 taels in 1897. Exports to Canton increased from 1,751,000 taels in 1891 to 2,338,000 taels in 1897. [...] Beans are sent to Hong Kong and Canton for food, and bean-cake is sent to Swatow for manure.
>
> (In Shurtleff and Aoyagi 2022, 55–56)

These numbers provide a similar picture to that offered by Alexander Hosie, who carried out the first comprehensive attempt to provide reliable statistics on the emerging production and trade in Manchuria and who had firsthand information from all vessels entering and exiting the port, from his position in the CMCS and the British Consulate in Newchwang. According to Hosie's estimates, Manchuria's total production of soybeans was around 600,000 MT in 1899. Of these, 567,019 MT passed through the CMCS treaty port at Newchwang – 280,428 MT as beans, 260,798 MT as bean cake and 9,512 MT as soy oil (Hosie 1904, 242–43). Moreover, the total value of these soy products from Newchwang was 2.5 million British pounds sterling, of which Japan imported almost half – 1.2 million British pounds sterling, followed by the ports of Shanghai (0.95), Swatow (0.51), Canton (0.4) and Amoy (0.27). These figures did not include the native junk trade, however, which was instead conducted under the supervision of the Native Custom House (i.e. not under supervision of the CMCS). According to Hosie, this trade could be considerable, but there existed no reliable data on it (Ibid, 243). As well as forcing China open to international trade via gunboat diplomacy, the British took great pride in "modernizing" and advancing Chinese administration and trade statistics. Again, following Hosie, Newchwang (under the supervision of the Chinese Maritime Customs Service – i.e. Hosie himself) was the only place where one could obtain reliable trade statistics (Ibid, 243). Hosie's accounts show clearly that Japan was the first buyer of beancake already in 1899 – the fact is confirmed by other sources and in literature (Reardon-Anderson 2005, 200; Shurtleff and Aoyagi 2022, 46–47).

While the soybean business was "booming" as never before, China's wider political and economic crisis, combined with its loss of exclusivity of the waterways and ports, led to a series of protests and upheavals, especially the anti-imperialist and anti-foreign Boxer Rebellion during 1899–1901 (Hymowitz 1970, 417; Shurtleff and Aoyagi 2021, 11). Tsarist Russia did not lose the opportunity to take advantage of China's weakness after its loss against Japan. Russia was an important geopolitical actor, a land-based empire surrounded by abundant frontier resources and keen on conquering and settling neighboring territories. Russians began gaining Manchurian territory in 1858: a treaty signed by China and Russia in 1860 recognized the Tsar's sovereignty over the space between the Sea of Japan and the Ussuri River (a tributary of the Amur) to the west. Although located in Manchuria, it marked the boundary between China and the Russian Maritime Province (Zatsepine 2014). Russia offered a Sino-Russian alliance against Japan in exchange for a railway concession in Northeast China. Once more, in Alexander Hosie's words:

> In the autumn of 1896 some surprise was caused in China by the publication in the Shanghai English press of a translation of what purported to be a Convention (known as the Cassini Convention) between Russia and China for the construction of a railway through Manchuria to connect the Trans-Baikal and Southern Ussuri sections of the Siberian Railway, and for the leasing to Russia of certain ports in Manchuria and China. (…) The general terms of the agreement are that the shareholders of the said company shall be Russians and Chinese only; that the gauge shall be the Russian gauge of five feet; that work shall be begun within twelve months.
>
> (Hosie 1904, 43)

Construction of the Chinese Eastern Railway (CER) began in 1897 and the rail line intensified the colonization ambitions of imperial powers over Manchuria. Another important hallmark in Russian expansion was the 1898 pact with China over the Kwantung Peninsula and its southernmost portion Liaodong (Liao-tung), which included an extension of the railroad to Port Arthur and Dairen (Dalian). According to this agreement, the Qing emperor agreed to lease Port Arthur (Dalny) and Ta-lien-wan, together with the adjacent seas, to the Russian Empire for at least 25 years. The Russians opened a new commercial port on the Peninsula and linked Port Arthur by railway to Harbin, which in turn was connected with the CER and the Trans-Siberian Railway. Writing in 1901, Alexander Hosie, expressed high hopes for these Russian investments in railways and ports, which he expected to alleviate transportation problems and make Manchurian agricultural products even more competitive (Ibid, 45–49). On their part, the Russians further extended the Trans-Siberian Railway to CER, which opened in 1903 and began construction on the South Manchuria Railway (SMR) (Lyons 2003, 27; Zatsepine 2014). While soy was an important driver behind the investments in infrastructure, its production was further spurred by decreasing transport costs, enabling the transfer of goods from Port Arthur to Vladivostok (Figure 4.1). The city of Harbin sprang up almost instantly where the railway lines joined: Harbin became the central transport node in the Heilongjiang Province and would play a central role for the coming decades (Wolff 2000, 244).

Other international players did not stand by idly. In 1902–03 (only one year after Hosie's book was published) Japan allied with Britain to stamp out the Russians and continue their own colonization of Manchuria. As Thomas Lyons has argued, railroad imperialism accelerated the Russian and Japanese colonization of Manchuria and greatly contributed to the dismemberment of Chinese territorial fragmentation (Lyons 2003, 23–24). The Chinese collapse, however, would have to wait until the full-scale Russo-Japanese War of 1904, which was triggered by essentially two strategic objectives: the control of Port Arthur and

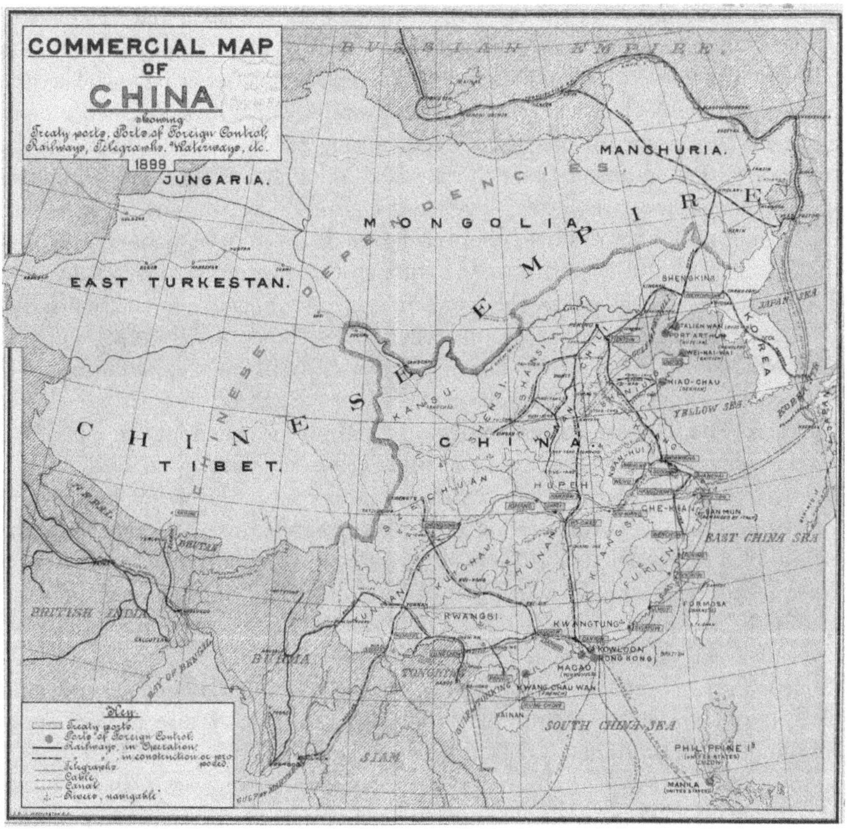

Figure 4.1 Map of commercial infrastructure in China in 1899 by the US Bureau of Statistics, Treasury Department. The map shows treaty ports, ports of foreign control, railways, telegraphs and waterways. Note that Dzungaria, East Turkestan, Manchuria, Mongolia and Tibet are separate from China.

the control of soy (Du Bois 2018, 7). Moreover, soy was not only an objective in the war; it fueled it (Wolff 2000, 145). Fermented soy and dried frozen tofu were the two most important sources of protein for the Japanese army; and, though soy played only a minor part in the diet of Russian soldiers, it was important as fodder for Russian horses. Meanwhile soy oil was used for soap and axle and artillery lubricant on both sides (Wolff 2000, 145; Du Bois 2018, 45). The Russo-Japanese War 1904–05 ended with another surprising victory for Japan, sealed by the peace Treaty of Portsmouth in September 1905. This treaty was brokered by the United States, who "aimed at preserving the bean-related balance between Russia and Japan in Northeast Asia" (Wolff 2000, 248). In addition, and as further explored below, the USDA's Division of Plant Exploration and Introduction intensified exploration in Manchuria, collecting and sampling soy seeds to be grown in the United States (Ibid).

In the aftermath of the Treaty of Portsmouth, the rail section linking Port Arthur and Dairen to the Yellow Sea came under Japanese control. The South

Manchurian Railway (SMR) network was further extended, becoming the most important Japanese company ever organized. Not entirely dissimilar to a modern vertically integrated agrobusiness giant, the SMR took on the construction of port facilities (particularly in Dairen – which replaced Newchwang as the central Manchurian hub), channeled investment in mining operations, supplied gas and electricity, established hotels and, of course, emerged as a central player in the transport of soy. Meantime, Manchuria was thus left practically divided into two spheres of influence: the North remained mainly under Russian control, with CER, and Russian businessmen continued to trade soy from Manchuria also after the defeat in the Russo-Japanese War through the Siberian port of Vladivostok (Du Bois 2018, 47). However, the Japanese sphere of influence in the South had the largest network and was the most important (Prodöhl 2013, 470). Further, the SMR continued to invest heavily in storage, mills and export infrastructure, greatly expanding its capacity to keep, handle and crush great volumes of soybeans. The Chinese Eastern Railway (CER) also invested and acquired two regular steamship lines on the Sungari River (Shurtleff and Aoyagi 2022, 258).

The amount of Manchurian bean cakes imported by Japan for fertilization and for feeding cattle rose from 216,198 tons (representing 58% of all fertilizers) in 1903 to 367,210 tons (45% of all fertilizers) in 1907 (Shurtleff and Aoyagi 2014), reflecting "a substantial transfer of nutrients from a frontier to the imperial metropole" (Lander and DuBois 2022, 40). Meanwhile, Japan's imports of whole soybeans from Manchuria rose from 146,971 tons in 1903 to 177,365 tons in 1907 (Shurtleff and Aoyagi 2014, 492). As soy became cheaper and more easily available in Japan, it played an important role for food security; both directly, for human consumption, and, indirectly, as fertilizer for their thin soils and as feed to hogs and horses (Du Bois 2018, 8). It should be remembered, however, that while a growing share of soybean output went to Japan, production was also expanding; South China continued to buy large amounts of Manchurian whole beans, bean cake and oil for food, illumination, cooking, lubrication, waterproofing, fertilization and animal feed (Hosie 1904, 183; Wolff 2000, 244; Prodöhl 2013, 466).

Manchurian soy entered a new phase in late 1908 to early 1909, as trial shipments of whole soybeans were transported successfully to Britain by the Japanese firm Mitsui Bussan Kaisha (the largest soy trader/exporter by far). In total, 19 steamers left Vladivostok and arrived at the port of Hull in England. In a report from 1909, the US Bureau of Manufactures (Department of Commerce and Labor) noted that Mitsui Bussan Kaisha had plans to take another 200,000 tons of soybeans from Vladivostok and around 400,000 tons via Dairen (previously Dalny) to Europe before the end of that year (Shurtleff and Aoyagi 2022, 133–34). Indeed, directly after the first soy shipments reached England, European imports began increasing dramatically, reaching 449,000 tons in 1910 (Shurtleff and Aoyagi 2007). The first Manchurian soy cargo to the United States reached its destination also in 1910, and, soon after, from 1915, the Dairen branch of Mitsui Bussan Kaisha began shipping bean oil via steamer directly to North America.

International demand for Manchurian soy soared hand in hand with the Chinese population's discontent with both their own rulers and foreigners. The immediate effect was the final collapse of the Qing Dynasty in 1911–12, and the inauguration of the Republic of China (1912–49), which, through our lens, resulted in more soy exports being released onto the world market. As the *Wall Street Journal* reported on November 20, 1911: "[R]evolutionary conditions in South China have closed that portion of the market for soya beans and made available much larger quantities than usual for western market" (Ibid, 219). In a similar vein, the *Daily Consular and Trade Reports,* from the US Bureau of Manufactures (Department of Commerce and Labor), reported on September 11, 1912 that "the Chinese revolution did not greatly disturb the peace of Manchuria" and that the export of bean cakes was rising (Ibid, 220). The revolution did not seem to have a great impact on Manchurian soy business, although according to the *San Francisco Chronicle* (August 7, 1913), it had lessened the already "vague and shadowy authority exercised by Chinese officialdom over the population" (Ibid, 261). However, the Chinese Revolution made Japan's dominance in Manchuria even stronger.

As the Bureau of Foreign and Domestic Commerce (US Department of Commerce), reported on August 4, 1917, total soybean exports from Manchuria had grown from 1,724,292 to 2,076,688 tons between 1911 and 1915. During the same time, soy oil exports rose from 65,919 to 81,863 tons, while soy cake exports grew from 911,821 to 1,112,661 tons (Shurtleff and Aoyagi 2022, 330). The main firms engaged in the export business were all Japanese (Shurtleff and Aoyagi 2014, 492). By 1915, almost 80% of all Manchurian exports were soy products and 90% of world soy exports came from Manchuria (Wolff 2000, 246). By way of illustration, the *Far Eastern Review* – a China-based English-language commerce journal – reported that the SMR had carried 2,000,000 tons of soybeans in 1923. Of that, 751,000 tons were shipped as whole beans by steamers from Dairen and the remaining 1,250,000 tons "were consumed in oil factories called by the Manchurians *Yu-Fang*, factories which flourish in Dairen and Yingkou [previously called Newchwang] for the manufacture of oil and cakes". By 1923, Manchuria exported 140,000 tons of oil and 1,300,000 tons of bean cake (Shurtleff and Aoyagi 2022, 480). Europe had become the main buyer of whole soybeans, while soy oil imports were quite equally distributed between Europe, the United States and Japan. By far, the main destination of bean cakes was still Japan, followed by South China (Ibid, 290). According to *Manchuria: a Survey*, published in 1925 in New York, "Japan found in soya beans the savior of her fast-failing rice fields. That must mean something to a people of 57,000,000 hungry mouths which can not get along without rice three times a day" (Ibid, 485). The soybean had now moved from a basically regional to a global commodity. As the *Far Eastern Review* noted in March 1928, a permanent conference for soybean investigations was to be established in Harbin: "[I]t may be said without exaggeration that in our days soya beans are among the most widely distributed agricultural products in the world" (Ibid, 518). Manchuria would soon become the first world soy frontier.

Manchuria under extraction and control

Manchuria had fallen under the control of Japan and, to a lesser extent, Russia: firms and public officials from Britain, the United States, Germany and France were also present in varying degrees trying to gain access to the region's resources. The forces of colonization, capitalism and occupation eventually restructured the social and economic landscape of the region. As Alexander Eckstein, Kang Chao and John Chang (1974) have pointed out, Manchuria became one of the world's specialized zones: a frontier area specializing in the production of soybeans. The increasing infrastructure and trade with soybeans and other commodities attracted millions of people and investment into the Manchurian frontier. This mass migration and frontier settlement boom totally transformed the social-ecological system (Barbier 2011, 257). Manchuria went from being a sparsely populated area of some 3 million people practicing hunting, extensive animal husbandry and agriculture in the 1850s, to being inhabited by over 19 million in 1914, occupied mainly in intensive farming (Wolff 2000, 144–45; Barbier 2011, 228). Its population would continue to sharply increase over the following two decades. In fact, between 1846 and 1940, Manchuria would receive a wave of mainly Chinese and Russian immigrants almost as significant as that of the great transatlantic exodus (Barbier 2011, 228). Such huge migration flows taking place in other parts of the world that characterized what is often described as the "Golden Age" of Resource-Based Development between 1870 and 1914.

The combination of cheap labor, decreasing transport costs and abundant (cheap) land, implied that this export-oriented system, a specific form of frontier-based development, continuously yielded high profits (Bix 1972, 431–33; Barbier 2011, 251). Between 1906 and 1921, soybean production grew from 0.6 to 4.5 MMT (Piper and Morse 1923, 7), while cultivated acreage went from 1.7 Mha in 1872/73 to 15.3 Mha by 1940 (Prodöhl 2013, 471). Most of the increase in soybean output was due to extensification, or the enlargement of cultivated acreage: but as a result, gradually only poorer soils would be incorporated, while the land already in use became degraded through nutrient mining (Langthaler 2020, 247). Yields per unit of land actually decreased by an astonishing 21% in the period between 1924 and 1931. Moreover, the cultivation practices, regardless of farm size, were labor intensive and centered on the use of rather simple and traditional technologies – using a single-handed plough pulled by draft animals, meticulous weeding, harvesting by hand and separating the seeds from the pods using a stone roller drawn by a mule (Ibid, 250–51). The small scale and simple technology of most production units was noticed in the 1924 article, "Manchuria: Home of the soy bean", published by *The Bean-Bag*, a bulletin released by the National Geographic Society:

> Aware of its enormous soy bean crop, and its heavy yield of sorghum and millet, the visitor expects to see great farms of these products. The Manchurians, like most Chinese, are gardeners rather than farmers. The vast aggregate of these grains and vegetables come from patches of an acre or so. Tens of

> thousands of farmers with tens of thousands of the primitive plows, shaped like crude shovels, turn up millions of ridges, and drop seeds into the loose earth. Then they run a roller over their patches and wait until it is time to spread their beans or their grain on clay floors. They thresh the grain with flails or with oxen to trample it.
>
> (Shurtleff and Aoyagi 2022, 474)

While most soybean cultivation was produced by landowning family farmers or tenants working in rather small plots, there were also big owners engaging in exploitive relations with their workers – typically landless young men (Langthaler 2020, 251–52). Different institutional arrangements, including debt obligation, were used to keep wages down (Barbier 2011, 259).

According to a 1909 report by the US Bureau of Manufactures (Department of Commerce and Labor), farmers could make good money from planting soy since this crop could: "always be relied upon to afford the farmer ready cash" (Shurtleff and Aoyagi 2022, 104). That soybeans earned better pay than many other land uses can also be traced in the fact that many farmers abandoned their previously diverse and long rotation schemes in favor of simplified and shorter ones, with a specialization in soy (Shan 2014, 70). Soy was, for the first time in its long history, grown as a monoculture; in a model which resembled the plantation economies of sugar, cotton, rubber and tobacco from a century earlier (Lander and DuBois 2022, 29, 38). Manchurian soybean growers used the scant, but nitrogen-rich, plant residues, including stalks and husks from processing, as fertilizer in combination with planted seeds (Ibid, 37–38). Meanwhile, the majority of farmers continued to produce soybeans in traditional ways – in three- or four-year rotation cycles with other crops, not least, sorghum, or *kaoliang*, an increasingly important part of the Manchurian diet of millet, wheat and corn (Langthaler 2020, 250). Sorghum actually represented the greater portion of the Manchurian's daily food and alcoholic drink and was often combined with soy, benefitting from its service of providing nitrogen to the soil. In this way, soy continued playing its traditional role as fertilizer, whether directly and locally, through the byproducts of processing and fixation in soils, or as exported bean cake to farms far away.

Manchuria was of course not the first frontier area in the periphery that became inserted as a commodity provider to core areas, and soybeans were definitely not the first highly specialized or monocultural cash crop grown for mass consumption somewhere else. The preconditions for the new frontier economies lay in the impressive and interconnected advances in chemical, transport and agricultural technologies stemming from earlier centuries (O'Rourke and Williamson 1999, 43–47). By the 1800s, fossil fuels enabled hundreds of millions of years' worth of photosynthesis to augment human power to increase production using steam engines (McNeill 2001, 13). Huge investments in infrastructure, ports, railways and steamships were part and parcel of this global food order, allowing for an increasing division of labor and specialization. In this way, by the end of the 19th century, transport costs plummeted, allowing, for the first time in history, an industrialized food supply and long-ranging commercial expansion and significant trans-oceanic price convergence (O'Rourke and Williamson 1999, 47). Oceanic transport's real cost in 1900 was a seventh of what it had been just a 100 years earlier (Roberts 2004, 710). In addition, international monetary relations in this period were framed by a stable system – the gold standard – which gradually included more and more countries and further reduced the (transaction) costs for trade. Moreover, the command over resources from faraway lands all over the world depended on control of vastly superior means of violence, e.g. machine guns and battleships, but also the control of improved means of communication, steamships, railways and telegraph (Buzan and Lawson 2013). Europe and Japan became the most important food importing regions, while the "New World" and India were the most important food exporting regions (O'Rourke and Williamson 1999, 74). The new global food order emerging was a key component of an international system organized along a firm pattern of core and periphery. In the periphery, the frontier economy manifested itself as successive waves of commodity export booms, in which the core country actors played a dominant role. Accordingly, land use patterns in large parts of the world changed considerably as a result of the process of commodifying productive agroecosystems in the periphery.

The drive to intensify and transform landscapes was now global and the experiments with soy agriculture in European colonies formed part of these efforts. The Dutch and French were the first to try and cultivate soybeans in Indonesia and Indochina (Du Bois 2018, 46). French agronomists started with soybean trials in Algeria in 1894 and sent seeds to the other French colonies in 1898. The Belgians introduced soy in Congo in 1908. Meanwhile, between 1909 and 1913, Britain systematically tested the yielding capacity of soy in what is now Gambia, Ghana, Kenya, Tunisia, Malawi, Zambia, Nigeria and Sierra Leone (*Bulletin of the Imperial Institute; L'Economiste Francais* 2009 April 9; *Wall Street Journal* 1910 April 16). Between the years 1885 and 1904, attempts were also made to cultivate soybeans in several states in India, but the results were not encouraging (Werner and Newton 2005, 45). In face of relatively poor results from growing soy in the colonies, and in a context of steeply falling transport costs, efforts to produce soy in European colonies were largely abandoned. Thus, during the period spanning from 1860 and over the first three decades of the 20th century, soybean production was still largely confined to the Orient. China proper, Japan, Korea and Indonesia (Dutch East Indies) were all major soybean producers, but only Manchuria produced a large exportable surplus (Hymowitz 1970, 408).

Instead, Europe relied overwhelmingly on Manchurian soy exports and on more successful cultivation efforts of other crops in the colonies. And, while soy received an astonishingly favorable reception in the West, the lion's share of the cargo was traded by a handful of Japanese crushing and trading firms (e.g. Mitsui, Mitsubishi, Honen Oil Milling and Nisshin Oil). These firms were also increasingly vertically integrated, investing heavily in the milling, crushing, storing and transport stages of the soybean chain and entering in strategic partnership with SMR (Bix 1972; Noguchi and Boyns 2013; Qi 2020). SMR in turn, as already mentioned, enlarged its field of activities to mining, agriculture, manufacturing and trade (Ginsburg 1949, 402). As soy history scholar Ines Prodöhl has said: "Japan's interest in China became more and more complex. It affected many economic and political questions for which bean cake and the SMR formed a strong anchor" (2013, 468). One of the reasons for the Japanese engagement in the soy trade was of course that it was heavily reliant on foreign lands for food security for its rising population (Wolff 2000, 248). Soybeans were absolutely central for the Japanese kitchen, with soy sauce, miso and tofu considered "the three daily articles of diet for all classes" (Shurtleff and Aoyagi 2022, 301). Besides this food, soy was also considered vital for fertilization of rice fields.

While the power of the Manchurian soy commodity chain (e.g. revenues, ownership, governance) was increasingly in the hands of a handful of Japanese investors, inexpensive Han-Chinese labor was not only doing the actual soy farming, but also a lot of work with transport (as carriers) and crushing (as workers in the mills, which to a large extent still relied on human muscle power). In the midst of all the "modern" investments railways, steamships and mills, technologically backward and intensive man-powered technologies also persisted. For example, Chinese carriers remained vital for shorter distances. A news article, "New Food from China" published in New Zealand in 1909, recounted:

> marvelous stories of Chinese carriers who will cover thirty miles a day regularly with a load of 120 lb or 130 lb upon their backs, and these laborers keep up their strength for this work by eating the soja bean.
>
> (Shurtleff and Aoyagi 2022, 103)

Human labor power also remained important in the processing industries, with thousands of small-scale mills worked by hand screws or by driving wedges with heavy stone mallets suspended from the roof of the mill. Even when soybeans could also be crushed by steam-powered cylinders, the crushed bean wafers had to be placed in iron-screw presses turned by capstan bars by hand. More modern mills eventually emerged, driven by hydraulic power or by steam, but, as the US Bureau of Manufactures reported in 1909, the most common kind of "manufacture of bean cake" was conducted in a "very primitive way", by hand, which opened "an opportunity to introduce some American machinery to replace the antiquated manner of forming and pressing" (Shurtleff and Aoyagi 2022, 91). This dual development seems to have remained a central feature of the Manchurian soy complex even many years into the "soy boom". According to *Manchuria: a Survey*, Manchurian oil-extraction mills could vary from hand presses or traditional mule- and donkey-powered mills (*yufang*) to the gigantic steam and electric presses at Dairen. Moreover, a new method of chemically extracting bean oil had been developed by the SMR and then turned over to the *Suzuki Bean Mill* (Shurtleff and Aoyagi 2022, 484–85). For a time, the most modern capital investments could thus operate side by side with traditional, cheap, labor-intensive forms of milling and pressing. But the trend would soon change: by the 1920s, the center of the bean oil industry had begun to shift to Europe. Traditional mills would become uncompetitive and Manchuria would come to specialize merely in the supply of raw materials (Shurtleff and Aoyagi 2022, 530).

While soy exports continued to rise, so did tensions over the territory. China was divided between different warlords. Zhang Zuolin (backed by Japan) had controlled Manchuria between 1916 and 1928, but was ultimately defeated by the National Revolutionary Army. The new leaders tried to restrict foreign control over the soy business. As *The New York Times* reported on the November 4, 1928, a new mandate had just been issued which harmed foreign firms dealing in Manchurian soy:

> With millions of Chinese in the fields new harvesting the largest crops Manchuria has ever produced, the Mukden government has astounded the grain trade by forbidding foreign buyers from purchasing direct from the farmers. The farmers are ordered to sell their grain only to 'authorized government agents,' and foreigners who wish to buy soya beans and grain for export must then deal with these agents.
>
> (Shurtleff and Aoyagi 2022, 526)

In 1929, the new Chinese leadership moved to end Soviet management of the Chinese Eastern Railway (CER) with a view to reclaim control over some of its foreign concessions. The Soviet Union, however, responded with a military invasion and successfully forced a return to the *status quo ante* of joint administration. Things did not end here. The Japanese saw the strengths of the Red Army and decided to accelerate plans to fully conquer the Northeast in 1931 – probably this decision was also influenced by the fact that geologists had discovered new minerals in Manchuria (Ginsburg 1949, 405). With this conquest, Japan set up the puppet-state *Manchukuo* in 1932, into which the three Northeastern provinces of Manchuria were incorporated (Wolff 2000, 243).

After the Japanese takeover, Manchuria continued to be the world's greatest soybean exporter (Fletcher 1950, 116). In fact, all but a negligible amount of all traded soybeans in the world came from Manchuria. In 1934, for instance, Manchuria exported 2.5 MMT while the next biggest exporter, Korea sold only 0.2 MMT. The leading importers the same year (in MMT) were Germany (0.9), Japan (0.7), Denmark (0.3), United Kingdom (0.2), Netherlands (0.1) and Sweden (0.09). These figures reflect only trade in beans, although meal and oil were also traded. The export structure of processed soy goods such as soy oil differs from that of beans in very telling ways, however: while Manchuria was still the main exporter of soy oil in 1934, with just over 97,000 tons, the second and third soy oil exporters were now European – Denmark and Netherlands exported around 16,000 and 11,000 tons, respectively (IIA 1939, 69–76).

In March 1935, the Japanese bought the rights to the Chinese Eastern Railway from the Soviets (Ginsburg 1949, 405). The railway expansion was regarded as key "to long-term regional domination and vital to the position of Japan on the continent, which otherwise was weak strategically" (Ginsburg 1949, 409). By 1936, the annual production of soybeans in Manchuria was 5 MMT. Soy sales benefitted from a trade agreement with Germany, to alleviate that economy's shortage of vegetable oils (Prodöhl 2013, 473). But the tide was changing: several European countries had put import tariffs on soy oil and the United States was close to becoming self-sufficient.

In control of essential Manchurian raw materials, the Japanese carried out their ambitious plans for conquest of Southeast Asia, which would eventually trigger the second Sino-Japanese War of 1937–45 (Bix 1972). The Japanese military took the Chinese capital and other major cities, and the Chinese government had to go into exile in the mountainous Southwest, with soybeans feeding both troops and refugees (Lander and DuBois 2022, 41). Millions of civilians perished in the conflict, which included the Nanjing Massacre, among other episodes of infamy and destruction. Even then, the Manchurian soy economy did not stand still; exporting 2.8 MMT in 1939. However, the outbreak and evolution of World War II finally disrupted the soy trade between Manchuria and Europe. After the nuclear strikes on Hiroshima and Nagasaki, Chinese communist forces seized Manchuria as Japan's empire disintegrated. Severely damaged by the fighting, the Manchuria's railway system lay in ruins; and as it retreated and evacuated Manchuria, the Japanese destroyed much of the infrastructure, including bridges, telegraph lines and, critically, oil factories. China sold what was left of the soy complex to the Soviet Union in return for military aid (Lander and DuBois 2022, 42). As we will see, US soy production would prove capable of covering European demand. But, before considering this, let us turn to the ways in which the Manchurian soy economy began to transform the European agrofood industry.

Soybeans in Europe and its colonies

The soybean was first introduced to Europe in the late 18th century (see Chapter 3) and, over time, many European scientists became enthusiastic about the plant and its potential roles for European agriculture and its colonies. A wide range of burgeoning new Western sciences (especially nutrition, microbiology, chemistry and agronomy) were applied to the study of soybeans. For example, in Vienna, the Austrian botanist Friedrich Haberlandt gathered many soybean varieties, shared samples and initiated hundreds of soybean field trials throughout Europe. Haberlandt was enthusiastic about the soybean and its many potential uses as food. His seminal book, *Die Sojabohne* (1878), provided state-of-the-art knowledge of this crop and inspired several other European researchers to write about the various health benefits of consuming soy (Du Bois 2018, 42). Haberlandt envisioned that the transfer of the soybean plant to Central Europe would have a huge impact and become an important ingredient for food and a central crop in agriculture. However, outside of agro-scientific circles, soy remained largely unknown. One of the reasons was that yields were highly disappointing. However, the completion of European scientists understanding of the mechanics of symbiotic nitrogen fixation in legumes in the late 19th century, offered a promising step towards increased soy yields outside of Asia. These scientists discovered that in order to fix nitrogen from the air, legumes needed to form a symbiosis with specific bacteria (*Rhizobium leguminosarum*) present in Asian (but not European) soils and in soybean root nodules (Eaglesham 1989, 29; Laranjo, Alexandre, and Oliveira 2014; Shurtleff and Aoyagi 2018, 18). In this way, samples of Asian soils were gathered to inoculate soils of the West (Shurtleff and Aoyagi 2018, 54–55). In 1896, a prepared inoculum was developed, patented and commercialized in Germany (*Nitragin* or *Germ Fertilizer)* and quickly spread around the world (Shurtleff and Aoyagi 2021, 12). This artificial inoculant was based on the use of pure cultures of rhizobia for legume seed treatments and came into common practice during the early part of the 20th century (Werner and Newton 2005, 4:16–17).

This was during the time when an important segment of North European agriculture became more knowledge-intensive, professionalized and capitalized. This led to the emergence of several new technological innovations for land use intensification, of which one was the integration of legumes in rotations for nitrogen soil fixation and livestock feed improvement. European family farmers at the time generally farmed in mixed systems and a basic principle was to provide a substantial part of the inputs from the farm. Fundamentally a closed circuit, practices were essentially guided by the concept of self-sufficiency with the rotation of pastures with legumes to improve the quality of the soil (Thompson 1968). The most common rotation crops were clover and rapeseed, mustard and linseed. Soybeans were not at all common since yields were still low. As the London-based *Botanical Journal* laconically put it in March 1918: "The story of soya bean cultivation in this country, with a single brilliant exception, may be summed up in one word–failure" (Shurtleff and Aoyagi 2022, 342). But Britain had not given up its attempts to grow soy. As the *National Food Journal* reasoned in piece published the same month:

> [N]umerous experiments had been made before the war [World War I] in the cultivation of the soya bean, but without success. A North Manchurian variety has, however, now been discovered which appears to be suitable for cultivation in Great Britain, and some specimens of the plant grown in Regent's Park were shown at the recent Food Economy Exhibition of the Ministry of Food held at the Institute of Hygiene. The soya bean is a hardy plant which does not demand a specially good soil, and if it should prove practicable to grow it on a considerable scale in this country there would be a considerable increase in the home resources of oil.
>
> (Shurtleff and Aoyagi 2022, 342)

Notwithstanding the still hopeful tone of many European soy enthusiasts, interest in engaging in the arduous work of adaptation to make soy produce better soon faded. A main reason was, of course, that cheap Manchurian soy had begun flowing into the region (Prodöhl 2013, 471). Eastern Europe constituted an exception, however, there research in adaptation and experiments with alternative soybean cultivation and processing techniques continued (Du Bois 2018, 48; Ryzhova 2022).

Manchurian soy enters the burgeoning European agro-industry (1909–30)

> Low ocean freights and the low price of silver are working some noteworthy changes in the world's market. Among them is to be noted the springing up of some new lines of trade between the Far East and the West. Such trade connections are greatly encouraged by the high level of prices for food products, in particular in the United States. For instance, the industrial and commercial depression in the Far East has driven capital and labor to agricultural industries. In Manchuria, where labor is being more generally restored to its normal agricultural uses, the cultivation of the soya bean, a rich and nutritive product with a high oil content, has begun to bring prosperity to the growers and to open new sources of supplies for the seed crushing industries in such centers as Hull and Liverpool.
>
> (Shurtleff and Aoyagi 2022, 103)

Above quote comes from *The Wall Street Journal* published March 15, 1909. It reflects one of many news texts throughout the world that commented on the arrival of around 46,000 tons of Manchurian soybeans to England, to be crushed in oil mills at Liverpool and Hull, primarily for industrial uses (Du Bois 2018, 46). That the first large soybean cargo arrived to Great Britain is not so surprising since Britain at this time was by far the world's largest agricultural importer and had been so for many decades (Magnan 2012). In fact, since the abolition of the *Corn Laws* in 1846, Britain and to some extent other European countries were increasingly engaged in long-distance supply chains allowing for cheap food imports for its urban and industrial workforces – a central element of what food regime scholars have called the first international food regime (Friedmann 1982; McMichael 2013, 26–30). As we saw in Chapter 3, Dutch and British imperial trade in tropical commodities had been going on for centuries centered around luxury consumption, such as soy sauce. Now, however, even bulky low value products could be moved over vast distances and become everyday products. While European industrialization was still quite limited outside of Britain until at least 1860 (Pomeranz 2021, 16), by the turn of the century it was in full swing. Thus, while the elements of relatively free trade and low transport costs were already in place by the late 19th century, soybeans were not yet in great demand. But by the early 20th century, fats and oils started to take on a new importance in Western economies. At the same time, failed linseed harvests in Argentina and cottonseed harvests in the United States, drove up demand for substitute seed oils (Wolff 2000, 246). Thus, in the context of persistent demand for vegetable oils, mainly for industrial purposes, coupled with the high interchangeability among the oils, the importance of soybeans soon became considerable. As *The Scotsman* reported in March 5, 1909:

> The arrival at Hull of the first cargo brought into this country of Soya beans, marks the beginning of what promises to be an important new industry in this country, and it is intimated that several other large cargoes of the same material will follow forthwith. [...] Hull is not to get the whole of these cargoes. Already one vessel of over 9,000 bags of the beans has arrived at Leith, and further supplies are expected at that port during the spring. Considerable interest is manifested in this new development as it will have an important bearing upon the future prosperity of the oilcake industry. For a long period of time the Soya bean has been one of the principal feeding grains grown in certain parts of the East, notably in China, Japan, and Manchuria, in which countries it is a staple food. [...] The outstanding features in its composition are its exceptionally high percentages of oil and albuminoids, and its lack of starch. Whereas the beans, peas and lentils grown in this country contain only about 2 per cent of oil, the Soya bean contains from 15 to 20 per cent. As to albuminoids, while our beans contain from 20 to 25 per cent, the proportion in Soya beans rises as high as 30 to 35 per cent. On the other hand, the soya bean contains practically no starch, which is the largest constituent in British beans. The Soya bean is thus a particularly rich feeding material, really more suitable for mixing with other foods than for consumption by itself. Stockowners in this country will watch the development of this new branch of the feeding stuff industry with deep interest, and for their sakes it is to be hoped its progress will be all that is expected of it. It is understood that the oil pressed from the Soya bean is valuable for soap manufacture. It is therefore probable that from Soya beans used in the soap industry useful feeding material might be obtained as a by-product (Shurtleff and Aoyagi 2022, 102–3).

Scientific and technological advances in food preparation techniques were rapidly reshaping the agrofood system, providing oils with more solid consistency and longer shelf life (hydrogenation). In Europe, solvent extraction techniques, first patented in 1795 by Joseph Bramah, began to replace pressing systems in the early 20th century (Sheikh and Zakiuddin 2019). Oil and cake extraction in Manchuria was still mostly performed by mechanical pressing; after the soybeans were ground and steamed, the oil was pressed in primitive wooden wedge presses, which were later replaced by screw oil presses. In contrast, the technologically more sophisticated oil mills in the West – already in use for processing a wide range of other vegetable oil crops – could also handle soybeans without any major difficulties (Prodöhl 2013, 469). With rising demand for fats and oils, and with these innovations in place, the soybean was quickly accepted throughout Europe (Prodöhl 2010, 111–12; Du Bois 2018, 47). According to the *Oil, Paint and Drug Reporter* on June 21, 1909:

> Within the last 6 months, there have been shipped to the United Kingdom, to Scandinavia, and to Germany upwards of 400,000 tons of the soya bean, which have been converted into oil, and from oil into soap, and the cake has found a useful place in the fodder markets of these countries. It is seldom indeed that in the course of one short season an unknown or rather an untried substance has ever forced its way into a market so cautious and conservative as that of Great Britain, where the manufacturer and consumer alike are so wedded to established formulas and customs.
>
> (Shurtleff and Aoyagi 2022, 107–08)

At the same time, the availability of relatively cheap Manchurian soybeans also sparked new investments in scientific research, looking for ways to make plastics, soap, fuel and food products from soy (Lander and DuBois 2022, 39). While the amount of oil in soy (c. 15–20%) was small compared to most other oil-bearing crops (e.g. linseed c. 35–40%, rapeseed 42%, mustard seed c. 40–50%, cottonseed c. 15–20%, peanuts c. 45–52%), the versatile qualities of soybean oil, and the increasing prices of other vegetable oils and fats made it a highly attractive substitute. The invention of margarine, developed by the French chemist Hippolyte Mège-Mouriès in the late 19th century, was of particular importance. As the margarine industry developed, the incorporation of new techniques pushed for a homogenization of taste, which rendered various vegetable oils interchangeable: gradually margarine's price became more important than its flavor (Prodöhl 2013, 465). The rise of the European margarine industry, moreover, pulled the demand for vegetable oil upwards with it. When dairy-exporting countries like Denmark and the Netherlands sold their butter for foreign revenues, domestic consumption could rely heavily on imported oils for margarine (Berg 2013). The Russian margarine industry also became more and more reliant on Manchurian soybean imports (Du Bois 2018, 47).

Soy oil was initially primarily an input to the manufacture of soap, margarine, candles and other industrial products, like the expanding cosmetics industry (Prodöhl 2010, 111; Prodöhl 2013). In 1909, for instance, two German scientists patented a soy-based substitute for rubber (Du Bois 2018, 46–47). The invention contributed to the end of the extremely extractive, but profitable, Amazonian "rubber boom". Soy oil rubber, however, would soon be outcompeted by a cheaper synthetic rubber substitute made from petroleum (ibid, 47).

Between 1911 and 1914, Europe imported between 30 and 40 million bushels of soybeans (or between 8.2 and 10.9 MMT) annually for processing (Fletcher 1950, 118). The main soybean crushing centers were in the large port cities in Northwestern Europe; Liverpool, Hull, Rotterdam, Amsterdam and Copenhagen (IIA 1944, 214–15, 384–85). In this way, the flows of soy entering Europe mirrored the general rise in the agricultural supply from faraway lands, which in turn contributed to keep European labor-force reproduction costs low (McMichael 2013, 26). The inpouring of agrofood products from transoceanic frontier zones heralded the triumph of both mass production and consumption (for a good statistical overview of the first decades of the 20th century see Bacon and Schloemer

(1940). While specific trade policies varied among European countries and shifted over time, the general pattern of the early 20th century was protectionist policies in favor of the domestic production of processed goods (i.e. high import tariffs on soymeal and oil to protect European oil mills), but relatively "freely" traded raw materials (whole soybeans) for cheap inputs (Prodöhl 2013, 469).

International trade in agriculture, in general, expanded continuously from the end of the Napoleonic wars until World War I. Friedmann and McMichael (1989) famously argue that the first international food regime started to fall apart after World War I, with the breakdown of the gold standard and the abandonment of free trade policies. Most agricultural trade indeed contracted during the interwar years, but this was definitely not the case for soy: on the contrary, global demand for oils and fats increased unabated (Prodöhl 2013, 464). As mentioned, oil was mostly used in the manufacture of industrial products like soap and lamp oil, however, Europe had already begun to "discover" a multitude of additional uses. By 1913, Germany and Russia were the major European soy importers, followed by tiny Denmark, who had become a relatively important margarine producer (IIA 1936, 260–62; 1940, 296–97). With the outbreak of World War I, the spread of the hydrogenation technology and, as we have seen, the relative scarcity of other oil sources, soy oil became an input in the production of several edible products, such as lard and suet substitutes. Soy oil was also mixed with cottonseed to produce a lighter oil much more amenable to hydrogenation (Shurtleff and Aoyagi 2022, 524).

At the same time, the residue from soybean crushing for oil – the protein-rich meal, flour or cake – was also becoming available in growing quantities and, spurred by shortages of many sources of food under the war economy, it begun to be used in all kinds of foods: e.g. bread, soups and as meat powder (Du Bois 2018, 47). By the end of World War I, the struggle between the Allies and the Central Powers over the control of trade routes had severely reduced Manchurian soybean imports, but demand continued to rise globally. In the words of an Associated Press (AP) article published in 1916:

> The secret of the soja bean is its universal usefulness. A British government report gives the following list of soja products: Vegetable food; soups; meat substitutes; chocolate substitute; macaroni preparation; flour; artificial milk; cheese; coffee substitute; artificial horn; biscuit and food for diabetic patients; sauce; meal for cattle; oils, oil cake for fodder; fertilizer; beancake. … [Manufactures:] dynamite and high explosives, soaps, linoleum, rubber substitute, margarine, paints, varnishes, toilet powder waterproof cloth, paper umbrellas and lanterns, salad oil, lubricants, lamp oil, preservative for sardines, substitute for lard. … [Feed:] Sweden uses large quantities of the bean cake as food for milch cows; Denmark has a large pressing factory at Copenhagen; France has a factory built in Paris by a Chinese firm; and South Africa has recently begun to grow the bean in competition with the Manchurian farmers. Germany in 1912 rescinded her former import duty and

> installed reduction plants for the far-eastern vegetable products in all her oil mills, importing the beans directly from Vladivostok by the shipload.
>
> (Shurtleff and Aoyagi 2022, 304)

Soy's versatility seemed endless. Another significant new market emerged, for instance, with the invention of lecithin in Germany in the 1920s (List 2015, 1). The variety of profitable commercial outlets for soy attracted the participation of big agrofood companies. Unilever, for example, was born out of a merger between the Dutch company "Margarine Unie" and the British soap maker "Lever Brothers". From the vantage point of soy-based ingredients, the most important Unilever brands were *Dove* soap, *Hellman's* mayonnaise and *Flora* margarine – all of them soy-based products. Trade flows expanded also due to advances in technical processing (e.g. deodorization) in the late 1920s and early 1930s, which made soy oil suitable for a wider range of food industries beyond margarine (Veraart 2022). The *Chinese Economic Monthly* discussed important technological advances in 1926:

> Originally soybean oil was used as an edible oil by the Chinese, but its strange smell has repelled Japanese and Western palates. [...] The advance of science in recent times has quickened the development of methods of refining, deodorizing, decolouring, and hydrogenating oils. As a result, the partition that used to divide food oils from industrial oils has collapsed. Whale oil and fish oil, as well as soybean oil, are now in use in Europe and America as a regular constituent of edible oils and fat.

In short, soy oil could now be used for a wide range of purposes and it was highly demanded so long as it was cheaper than other oils and fats. In parallel, soy meal moved gradually into the market for protein fodder, particularly in the years when the prices of cottonseed and linseed went up (IIA 1936, 260–62). According to soybean specialists from the USDA writing in 1916: "In Europe soy-bean cake ground into meal is used almost entirely for feeding cattle, and the low price in comparison with other concentrated feeds has made it very popular" (Piper and Morse 1916, 13). The idea was to produce meat more efficiently, and here soy was seen to have a role in different mixes with other grains. However, the bulk of fodder inputs needed for animal production in Europe and America up until the 1930s–40s were still cultivated on-farm. The animals that gave the butter and the bacon that reached British breakfast tables were reared on farms where no more than 15–20% of feed value came from purchased concentrated feed inputs, such as oil cakes (Jonsson and Pettersson 1991). Cheap and increasingly available cake and meal from crushed oil seeds were nevertheless starting to challenge mixed farming systems. Denmark (who had lost an important share of the international wheat market to the United States), pioneered the use of Manchurian soybeans as feed for livestock, which strengthened its position as producer and exporter of bacon and butter (Du Bois 2018, 48–49). Germany, by far the largest European importer of whole soybeans from the mid-1920s throughout the 1930s, was also using the soybean cake (or meal) as animal feed. In 1928, soybean cakes constituted around 32.6% of German farmers' animal feed (Prodöhl 2013, 473). Unsurprisingly, the leading soy importers in order of size in 1929 were Germany, Denmark, Britain and the Netherlands (IIA 1939, 181–364).

Statistical figures are still incomplete, but it is clear that soy imports from Manchuria were eventually interrupted during World War II, at least for the ally countries: the United States quickly took over the role as soy supplier (Fletcher 1950, 119). With food scarcity threatening Europe again, the value of soy increased again. As economic historian Ernst Langthaler notes "[N]utritionists in collaboration with state agencies and food companies promoted soy flour as a cheap *ersatz*' version of scarce foodstuffs" (2018, 4). Several European countries promoted soy as a cheap foodstuff, but Germany, in particular, made the strongest efforts to exploit soy's nutritional potential; producing soymilk for undernourished children, using soy flour in traditional dishes (e.g. soups, sausages, bread, biscuits), distributing soy-based rations among its soldiers and enforcing soybean cultivation in Southeast Europe as part of its geopolitical aims for a "Greater Space Economy" (Fletcher 1950, 118; Shurtleff and Aoyagi 2022, 799). Accordingly, Western Allies began to talk about the "Nazi bean" (Langthaler 2018), although ironically, soy cultivation outside Asia since the turn of the 20th century had only really taken off in the United States.

Soy in the United States until World War II

As we saw in the previous chapter, Samuel Bowen was the first to cultivate soybeans in North America in 1765 (Hymowitz 1990, 160–62). While his efforts had been replicated several times since then, it was not until the second half of the 19th century that the potential of soy as a North American crop really took off (Werner and Newton 2005, 4:16). The first systematic experiments were conducted by scientists from the New Jersey Agricultural Experiment Station at Rutgers College in 1878 with seeds from Europe and Manchuria. The seeds were multiplied in the field and distributed to agricultural experiment stations in other states (Ibid). Similar projects were carried out throughout the Americas in the late 19th century. In Canada, the earliest known (though unsuccessful) cultivation attempt was in Ontario in 1882 (Shurtleff and Aoyagi 2021, 10–11). The same year a similar attempt was made by the Bahia School of Agriculture in Brazil (Shurtleff and Aoyagi 2009, 7, 16, 19) and, while soy was known in Argentina already in 1880, the first field experiments date from around 1908. Of all field trials across the Americas, however, it was in the United States where the most extensive public efforts were put into introducing it as a farm crop. Many great American scientists and USDA officials were enthusiastic about the soybean and worked hard to make it thrive on US soils. For example, in 1903, the USDA distributed soybean inoculants to farmers free of charge (Williams 1897; Piper and Morse 1916; Shurtleff and Aoyagi 2018, 6), though most farmers remained uninterested and soy yields remained relatively low (Shurtleff and Aoyagi 2022, 102–03).

In 1909, only a mere 600 ha of soybeans were grown (IIA 1923; Kromer 1961, 26). However, the USDA and other actors continued to engage actively in soy breeding. One telling example is when USDA officials Howard Dorsett and William Morse collected 4,500 different soybean varieties, mainly from Manchuria, and tried them out across state experimental stations (Peruchi Moretto, Nodari, and Nodari 2022, 24). Generally, the USDA's efforts cannot be overstated (Arntzen and Ritter 1994, 466) and, as its *Weekly News Letter* of June 27, 1917 reported: "[I]n the United States much attention has been given to the breeding of pure adapted varieties, and there are now about 20 satisfactory sorts on the market" (Shurtleff and Aoyagi 2022, 327). The aforementioned William Morse, writing in the pages of yet another USDA journal – the *Yearbook of Agriculture* – a decade later noted that the:

> acreage of soy beans in the United States increased from about 500,000 acres [202,342 ha] in 1917 to over 2,500,000 acres [1,011,714 ha] in 1924. This enormous increase in the use made of soy beans in this country has been largely due to the development of better-adapted varieties.

Morse wrote further:

> Moving forward slowly through the years with new varieties, increased acreage, wider interest, greater utilization of crop and by-products, its safety and dependability under adverse conditions, more efficient methods of planting, cultivating and harvesting, its availability as a relief crop, the lowly soybean of 1907 has risen to the rank of a major crop in 1927.
>
> (Shurtleff and Aoyagi 2022, 509)

Efforts, thus, begun to pay off. Soy farmers at this time were still mainly mixed farmers who made use of its nitrogen fixation capacity to alleviate problems of declining soil fertility. By putting soybeans into their crop rotation, farmers found they could boost their yields of corn and other commodity crops at a time when commercial fertilizer was not yet widely available and affordable (U.S. Soybean Export Council 2008). Growing soybeans for green manure to enrich the soil (when plants are plowed under) was popular, but soy was initially popular as forage (hay or grazed) (Du Bois et al. 2008, 4–5). In fact, up until the 1930s, soybeans were still mainly used on-farm for seed, hay, green manure or forage and not for commercial sales (Kromer 1961, 26).

Before harvests became abundant and farmers were convinced to sell most of their produce to processors, however, United States' industrialists began importing soybean oil from Manchuria – free from tariffs (Shurtleff and Aoyagi 2022, 253). By 1916, the imports of Manchurian soy reached around 200,000 tons (Ibid, 328). Newspapers from this period often wrote about cargoes arriving at different ports and how they were used. One illustrative example comes from *Los Angeles Times*, which on July 22, 1917, reported on a record cargo of 9,373 tons of Manchurian soy that just had arrived to the Los Angeles harbor in a Danish ship (Shurtleff and Aoyagi 2022, 328). The soybeans were to be unloaded for the Globe Grain and Milling Company and crushed in the company's new oil mill in Vernon (which was normally used to crush cottonseed). The extracted oil would, in turn, be used to manufacture soap, but, according to the newspaper, there were also plans for an extensive series of experiments for the production of a palatable cooking and salad oil. It was further reported that the remaining substance from oil extraction would be converted to meal for cattle and hog feed. Later, in the same new article, it was remarked that the soybean was "creating a tremendous interest in American agricultural circles" (Ibid). It was also mentioned that the USDA had recently sent a Chinese-born woman graduate of an American college (Yamei Kin) back to China to gather facts about soy production and uses. "While the Americans have raised the bean for stock feed and eaten the meat the Chinese have taken a short cut to get the protein which is the food value of meat and milk by eating the bean itself" (Shurtleff and Aoyagi 2022, 328). As illustrated in this article, there was a kind of hype of soy and all its potential uses. Moreover, oil mills now saw an opportunity to work at full capacity all year and not only when cottonseeds were available (Shurtleff and Aoyagi 2019, 5).

Domestic farmers also started to sell soybeans to crushers. A 1916 circular from the *Paint Manufacturers' Association* (U.S. Educational Bureau), read: "[T[his year, for the first time, American grown Soya Oil has appeared on the market". The circular further noted that the entire lot of soy oil was sold to a soap manufacturer at a price considerably above offer (Shurtleff and Aoyagi 2022, 305–06). Just as in Europe then, the main commercial uses of soy in the United States, outside of the actual farm unit, were initially industrial:, paints, soaps, linoleum, glycerine, explosives, enamels, varnish, butter substitutes, lard substitutes, edible oils, salad oils, waterproof goods, rubber substitutes, printing inks and lighting and lubricating oils (Du Bois 2018, 59; Shurtleff and Aoyagi 2022, 420–26). Soon, Canada also started to import soy oil for use in the manufacture of soaps and paints.

Although grain trading and processing infrastructure was initially built for other grains, not least cottonseed, these crushers could easily shift to handle soy as well. A wave of mergers in the food processing and oil industry had significantly reduced the number of mills/crushers and concentrated this market segment further (Bedier 2018, 37–40). By 1910, ADM, for example, was already the world's largest producer of linseed oil and an important crusher of flaxseed. By the late 1920s, it began using some of its existing presses for soybeans (e.g. hydraulic presses). Initial soybean operations were small so other big crushers did not get involved for decades still. Meanwhile ADM scoured the countryside for farmers willing to grow more soybeans (Shurtleff and Aoyagi 2020a; ADM 2022). The crushers did not have an easy time sourcing domestic beans, probably because they paid farmers a low price. In the same 1927 text, William Morse addressed this tension:

> Several oil mills are now crushing domestic-grown soybeans for oil and oil meal in the Southern and Western States, and many others are being equipped for this purpose. Complaint is often made that oil mills pay too

> little for seed, making seed production for this purpose unprofitable. We must take into account, however, that the soybean is a legume. We must consider the fertilizing value, the feeding value of the straw, and not expect too much in comparison with other standard crops. Let us be fair with this oil-mill industry, and forget the high prices for seed which have prevailed with the introduction of new varieties and the large increase in acreage. To me, the production of soybean seed for oil and oil meal appears to be one of the bright spots in the future of the soybean which will firmly establish it as a major crop.
>
> (in Shurtleff and Aoyagi 2022, 510)

Morse argued that farmers should consider soy's value as fertilizer (its ecosystem service) and feed and plant more soy in spite of the low prices, but US farmers countered that what depressed prices were the rising imports of soy oil from Manchuria (Ibid, 485–86). More specifically, the claim was made that US soy oil purchases were feeding Japan's expansionism and depressing the American market by discouraging domestic production (Shurtleff and Aoyagi 2022, 392). In 1921, *the National Board of Farm Organizations* pledged a permanent high protection against "all oriental vegetable oils" and, as the *Los Angeles Times* put it on February 15 that year, "especially the soya bean of Manchuria and its products". The US government listened and imposed an emergency tariff on soy oil, at the rate of 20 cents per gallon through the Act of 1921 (Ibid, 501). While the tariff disincentivized Manchurian bean oil, imports continued up until 1930, when the still young *American Soybean Association* (ASA) managed to achieve the enactment of a very high protective tariff on soybean, meal (cake) and oil imports. It also struck a deal between farmers and processors in which the latter were guaranteed a certain supply of soy if the former, in turn, were guaranteed a minimum pay (ASA 2022). This legislative victory was but one in a long row of successful interventions by ASA, the producers' association founded in 1920, which, we will see the coming chapter, remains a powerful player within the soy regime today (Figure 4.2).

Increasing amounts of soy crushed for oil lead to an overabundance of meal byproduct that needed new markets. Significant efforts and funding went into investigating the uses of meal as a cheap input in the expanding dairy and meat industries. Indeed, the development of mixed feed from various components had already begun in the early 20th century (Brother, Murray, and Griffiths 1951, 345), but increasingly abundant soy meals now meant that much focus shifted to the specific role of soy within these feeds. For instance, new research confirmed that the nutritional value of soybean meal was enhanced when it was properly heated (Hymowitz 1990). Looking for new ways to enhance the "effectiveness" of animal farming – rearing animals to slaughter weight as cheaply and quickly as possible – the USDA also explored a number of new techniques during the 1920s and 1930s, including growth hormones, antibiotics, selective breeding, artificial insemination and, not least, soybean meal. It was soon evident that soy meal had a high-protein and low-fiber content relative to grasses and, thus, had an

— s b d —

SOYBEANS

By TAYLOR FOUTS

First pub. in Vol. II of the Association's Proceedings.

Soybeans! Soybeans! You're like a Musical Band
To the Farmer who's tuned for the "Best on his land."
Microbic Composers, on the millionth wave length,
Sing "love" to the Rootlets as they're reveling in strength.
The Chlorophyll hums in the rays of "Ole Sol,"
And frosts, rains and winds are a "medley with Soys."
The "Pop o' the Pods" is Jazz to the Pigs—
Puts pep in the Porkers—they grunt and grow big.
The "Rustle o' Leaves" to the Lamb is sublime,
He waltzes right up in full zest and on time.
The harvests of hay, pastures and seed-time
Give enchantment to tune and thrill to the chime.
The Kiddies they hop to the "Jingle o' Coin"
That Daddy rakes in while Soy Music's a goin'.
Oh! Soybeans! Soybeans! You're a symphony grand
To the Farmer who's tuned for the "Best on his Land."

Birthplace of A. S. A.

The three Fouts brothers, from left to right, Taylor, Finis and Noah, at Soyland during the first meeting September 1, 1920.

16

SOYBEAN DIGEST

Figure 4.2 The three Fouts brothers at the first meeting of the American Soybean Association in 1920 on their farm near Camden, Indiana. Taylor, on the left, penned a hymn in praise of the beans.

Photo Credit: Fouts Family and American Soybean Association. Excerpt: The Soybean Digest, September 1944, Vol. 4, Number 11, page 16.

enhancing effect on animal rations (Du Bois 2018, 36; Roth 2018). The private sector also engaged actively in the development of crushing industries and other end products for the market (Du Bois 2018, 63). ASA members, for example, promoted the sale and use of soybean meal for livestock feeding, hoping that it would have a positive impact on prices. This initiative was also keenly supported by the National Soybean Processors Association (NSPA), founded in 1930, which financed the drive to open new markets for soy flour and oil from the mid-1930s onwards (Shurtleff and Aoyagi 2019, 75–81). Soon, the United States took the lead in developing new uses for pressed soybeans for feed.

The Great Depression and the New Deal changed traditional land tenure which pushed many farmers off their lands and this period also saw the increasing specialization in both livestock production and cultivation. Soy expanded rapidly. Surpluses in wheat and cotton suppressed their prices and made soy more attractive for farmers as a cash crop (ASA 2022): indeed, American soybean production more than quintupled in the decade following the Great Depression. By adding soybeans into their crop rotation, farmers found they could boost their yields on the succeeding crop. Moreover, soy was sold to crushers when prices were high, but used as on-farm pig feed when there was an oversupply on the market. Farmers could also often buy soybean meal from the same processors to whom they sold their soybeans and use the meal as a high-protein feed supplement (Kromer 1961, 26; Roth 2018, 124). In the 1930s, the United States began to outstrip Europe as the center of soy research and innovative new uses. By this time, all the necessary links of a new soybean chain were being forged: the soybean was becoming commercially viable in the US Corn Belt and by the 1940s it would start realizing its potential as an American cash crop (U.S. Soybean Export Council 2008). Simultaneously, soy meal was gradually becoming a low-cost, high-protein feed ingredient (Roth 2018, 125–26).

The USDA, through the tireless endeavors of researchers like William J. Morse, continued to explore additional uses of soy: his 1923 book *The Soybean* – the first major American treatise on the plant – defended soy's huge potential in the food industry (Piper and Morse 1923). Morse, and other enthusiasts like him, insisted that the "miracle bean" could contribute to a nutritious, inexpensive and savory diet – inspired by its roles in Asia (IIA 1936, 260–62). Private entrepreneurs also pushed to introduce soybeans as human food and government campaigns promoted it as a healthy meat substitute. The Commodity Credit Corporation (an agency formed under the New Deal) and the Office of Price Administration established and calibrated a price ceiling for soybean meal to make manufacturers of mixed feed boost the protein content of their formulas – and to substitute for more expensive ingredients (Ibid, 124).

> The scarcest food commodities in the world today are proteins and fats. In the soybean we find a commodity which contains, per bushel, three times as much protein, ten times as much fat and three times as much valuable mineral as wheat. It contains four times as much protein, over four times as much fat, and nearly four times as much valuable minerals as corn. Soybeans are the world's cheapest source of edible oil today, and the year 1948 established a new world record in soybean production of approximately 575 million bushels (15.7 MMT).
>
> (Fletcher 1950, 116)

Still, convincing American consumers to change their eating habits and warm up to the different new soyfoods propounded by enthusiasts, advocates and lobbyists was not so easily done. The Asian community living in the United States did set up tofu shops and miso restaurants, but the adoption of soy as food in the rest of the population was slow (Du Bois 2018, 54–55). A major exception was soy oil, which began to play a much less important role in industry and began to acquire a protagonist role in human diet through its use in food processing (Ibid, 8). In 1933, for example, under a million pounds of soy oil were used for edibles, relative to a mere 22 million used in industry. Within a decade, this use had both exploded and turned entirely on its head: about 1,200 million pounds became food, while only about 20 million pounds were used in industry (Kishlar 1944). The first food uses for oil were in shortening and margarines, although this industry was actively opposed by anti-margarine laws successfully lobbied for by the US butter sector (Ibid, 9). ASA tried again and again to obtain a repeal of the margarine laws. For years, one agrarian lobby, defending the interests of dairy farmers, stood against another, one representing soybean farmers. It wasn't until

1950 that the federal tax on yellow margarine was finally repealed (Shurtleff and Aoyagi 2019). In Canada demand for soy oil and protein begun to soar as World War II begun and the country experienced critical shortages of oil and protein. Both imports and domestic soy production (finally) started to take off (Shurtleff and Aoyagi 2019).

Meanwhile, back in the United States, the crushing industry developed swiftly and both oil and meal were used in a wide range of commodity chains. In 1933, only 23% of soybeans consumed in the United States were crushed; by 1935, by contrast, the figure was more than 50% (ASA 1949). Between 1936 and 1942, the United States surpassed both Manchuria and Germany and became the world's leading soybean crusher (Shurtleff and Aoyagi 2016). In 1942, more than 70% of all soybeans were crushed (Kromer and Gilliland 1954, 14), and, after 1945, the amount exceeded 80% (ASA 1949). Although the company Archer Daniel Midlands (ADM, founded in 1902) had tried to develop solvent extraction systems for soybeans at one of its plants, the meal it rendered was 44% protein meal – good for feed – but lower oil content, making it very hard to sell. So, despite some efforts, the continuous screw press model remained the most common way to extract oil from soybeans in the United States until the late 1940s (Kenyon, Kruse, and Clark 1948). ADM, however, spent large amounts of money on research and marketing efforts for the new meal. It soon managed to take full advantage of the phenomenal increase in demand for soybean meal in Europe and Japan. On the eve of the Second World War at the end of 1940, ADM, with it headquarters still in Minneapolis, had six soybean processing plants and by 1942 the company had become a major producer of soy flour. ADM is still one of the handful of the biggest soy trading/processing companies in the world; together with Bunge (founded in 1818), Cargill (founded in 1867) and Dreyfus (founded in 1851). Collectively known as ABCD, these companies engaged in commodity trading and specialized in logistics and finance and by the turn of the century they were the world leaders in their sector (Dalla Costa and Silva 2018; Cargill 2018; 2022; Bunge 2022; LDC 2022). Their combined dominance has allowed them to establish new standards and "rules" for international grain trade. For example, in 1929, Cargill organized its first export department and started selling grains "FOB" (free-on-board), where financing, carrying and shipping charges are a part of the selling price (Cargill 2022). This has remained the standard format of international grain trade ever since – resulting in extremely high entry costs to participation in the trading stages of the soybean chain (see Chapter 5). Cargill is, in fact, the oldest crushing company in the United States,; it owned hundreds of rural elevators and mills. It did not begin crushing soybeans until 1943, however, beginning with the purchase of a soybean plant, elevator and office building in Springfield from the Illinois Soy Products Company (Shurtleff and Aoyagi 2020b, 5). Even if it only began crushing soybeans during World War II, Cargill had one previous experience with soy. In the 1930s, it had been active in hauling grains, including some soybeans, down the Mississippi River through a subsidiary company (Shurtleff and Aoyagi 2020b). In the 1940s, Cargill begun producing barges, and buying ocean-going tankers and established *Port Cargill* on the Minnesota

River (Cargill 2022). Bunge purchased its first sizable grain facility, Midway, in 1935. It was adjacent to a rail terminal in Minneapolis; adding physical facilities to its grain trading capabilities (Bunge 2022). In short, these traders began to branch out into different sectors: finance, industry, international trade and agricultural properties. They bred animals, financed farmers, bought infrastructure (mills, elevators and food processing plants) and industrialized products (Dalla Costa and Silva 2018).

One increasingly important driver for US soy was overseas demand. All of Europe was in dire need of proteins and fats: France, for instance, needed at least 50% more fats to approach pre-war diet levels (Fletcher 1950, 118). During the War, the United States established soy trade routes with Europe, supplying the allied countries with whole soybeans as well as with soy oil, flour and meal. While European agricultural policies had become more protectionist after the Great Depression, Europe's dependence on cheap fat and oils and, to growing extent, feed, meant that soy imports were still allowed relatively free of protection. By the time Japan was defeated in 1945, US soybean production had already replaced Manchuria's earlier output (Wolff 2000, 249). Moreover, the US government started purchasing soy flour for shipment to Britain under the "Food for Freedom" program (Roth, 2018, 121). As the chairman of the Soybean Nutritional Council, Lamar Kishlar wrote in 1944: "[W]ar has given the soybean its big opportunity" (Kishlar 1944, 2). The giant traders ADM, Cargill and Bunge were also important drivers behind the increasing soy trade; establishing sales offices in Europe and investing in export-oriented infrastructure. Overseas demand truly increased farm incomes and drove the expansion of soybeans. Besides the international trade in soy, commodity trade, in general, was spurred by the expanding activities of the Chicago Board of Trade, where trading in soybean futures contracts was established in 1937. After this, four giant trading, marketing and processing firms began to compete aggressively for control over profits from "miracle bean".

Domestic US demand was also stimulated by the war. American-grown soybeans became the nation's agricultural "star performer" (Prodöhl 2013, 463) as consumption of soy oil during the war tripled.Apart from becoming the United States' first soy flour producer in 1949, ADM also expanded operations that year and moved into edible oils, installing an edible oil refinery, which allowed ADM to move beyond the production of crude soy oil into sales of refined oil to margarine and shortening manufacturers. With this decision, ADM could now also supply the food industry with cooking and salad oil and furnish large consumers with bulk edible refined soy oil. Domestic demand for soy meals also rose dramatically in the United States during the war. Farmers raised more livestock and were encouraged to feed each cow and chicken with larger vegetable protein rations to increase milk, egg and meat production (Figure 4.3).

Soy also became an important ingredient in diets for the army both in the United States and in Europe, as well as a more common substitute in civilian wartime diets (Roth 2018). Soon after the war, however, interest in soy for food faded. Throughout the United States and Europe, the dietary preference was for animal protein. However, interest in soy as feed continued growing. The big traders

Figure 4.3 World War II Poster "Your Country Needs Soybeans, for Food, Feeds, Guns. Grow More in '44".

began to move into the business of producing cheap feed from both soybeans and corn. In 1947, for instance, Cargill acquired the company Nutrena Feed Mills, a pioneer in the feed industry and, in 1951 Cargill's Feed Division (Nutrena Mills, Incorporated) merged with the Royal Feed & Milling Co. of Memphis to manufacture feeds for livestock and poultry (Cargill 2022). Meanwhile, as demand for soy meal rose rapidly so did soybean supply lines. Fearing a fall in soy prices and reasoning that the world was hungry and in need of low-cost sources of protein and oils, ASA began to focus on expanding soy exports into new markets (Shurtleff and Aoyagi 2004). Both elements – soy meal for increased meat production and different strategies to keep soy exports up – formed the kernel of the third soy cycle's regime phase, which we move into in the following chapter.

Reflections on the soybean and its functions

As we have seen in this chapter, the rapid acceptance of soy in Europe had to do with a context of scarcity of fats and oils, at the same time as new scientific and technological advances had created novel industrial applications and uses for fats and oils. Innovators had, since the late 19th century, begun to use science-based predictive models to guide their exploration and, when applied with the traditional methods of trial-and-error, the pace of technological development in advanced capitalist nations rose rapidly. Long before the West started to demand cheap soy for oil, however, the supply side of soy had expanded rapidly in Manchuria due to other pressures. Cheap soy from Manchuria would not have been possible without the preceding opening of ports and lands in Manchuria for foreign exploitation and the decision to allow Han settlement by the weakened Qing Government. Moreover, that Manchurian soy could be bought at a relatively cheap price was because both land and Han-peasant labor were cheap and transport costs were falling quickly, spurred by Russian and Japanese investments in railway and port infrastructure. Together these factors resulted in soy being produced in rising quantities at falling costs. As the rest of the world realized that soy could assume a plethora of uses at much cheaper prices than other oils and fats and easily substitute for them, soybeans became an established staple in world trade. Moreover, as demand for soy oil was high, new uses were found for its "byproducts"; most noteably the protein-rich cake/meal used as cheap animal feed. The emerging modern capitalist agrofood system also saw the construction of new mechanisms for standardization, with new tools to measure and grade quality, which further facilitated trade and reduced transaction costs (Baraibar Norberg 2020, 59).

While technological and scientific advances were starting to be translated into higher performance of agriculture and food throughout the world, the increasing international division of labor resulted in growing specialization and concentration. The frontier areas of both the Americas and Eurasia – previously sparsely populated and extensively cultivated – had productive systems typically based on monoculture and overexploitation of the soil imposed upon them (Friedmann and McMichael 1989, 95–96). The specialization of grain and bean-producing regions, which is an important feature of the international division of agricultural labor and the global food economy, and which we tend to take for granted nowadays, is in fact fairly recent. It was not until the late 19th century that specialized zones of mass production for distant markets leapt forward. This specialization of economic systems, moreover, meant a separation of tasks, where participants do fewer tasks or even one single economic activity. Two major agroecological consequences have been a massive loss of biodiversity and broken nutrient cycles. The greatest losses of diversity were (and are) coupled to the destruction of native ecosystems. As a legume, soy still contributed nitrogen to local soils, which enabled intensified cultivation of sorghum and other more soil-depleting crops. However, most Manchurian soy contributed to fertility in intensive farming systems far south in China and across the sea in Japan.

Meanwhile, Western industries modernized, applying the latest insights from science and technology to new agrofood products and processes, and discoveries in the agricultural sciences led to increasing production and yields. In *Feeding the World: An Economic History of Agriculture, 1800–2000*, Giovanni Federico builds on scattered sources and Maddison's estimates of GDP per capita growth to conclude that agricultural performance between 1870 and 1950 was considerably high compared to the stagnation of previous centuries (Maddison 2003; Federico 2010, 18–19). World agricultural output per capita grew around 0.2–0.3% annually between 1870 and 1938, and the share of livestock products in world gross output also grew (Federico 2010, 26–27). While agricultural expansion slowed down somewhat between 1913 and 1938, and international trade retracted (Federico 2010, 18–19), soy output and trade continued to rise.

An important element here was that soy cultivation finally began to strongly take off in the United States after many years of persistent work by scientists and state officials to incorporate thousands of different varieties of soybean seeds from Asia and tediously adapt them to local soils. Moreover, an improved understanding of soy's capacity to fixate nitrogen and the complex interactions between the plant's root nodules with bacteria in the soil led to significant yield improvement.

This entire period of seven decades, from 1870 to post-World War II, was a relatively rapid transition of Western agriculture from organic, circular agroeconomies based on on-farm inputs toward industrial systems with increasing use of external inputs. When soybeans started to expand in the United States (and to a lesser extent in Europe) they were mainly intended to enrich soils, but also to act as a forage crop that could reduce feed costs and increase self-reliance in predominantly mixed family farms. Adoption of new agricultural models, however, gradually increased reliance on external inputs, e.g., feed concentrates and fertilizers. The idea of mining nutrients and hauling them to the farm made its appearance first in the form of guano and later mined nitrates imported from South America. Finally, with the breakthrough of the Haber-Bosch process, coal and petroleum fueled the availability of industrial artificial fertilizers. This process delivered nitrogen in almost unlimited quantities for the production of fertilizers. As we have seen throughout this book, humans have historically cultivated legumes (including soy) to mobilize nitrogen from its non-reactive form or have accessed it through animal manure, guano and nitrate salts, but the development of the Haber-Bosch method in early 1900 circumvented the need for these sources (Mejia 2022, 185–87). Soon, chemical industries would revolutionize per-hectare yields based on petrochemicals (Pomeranz and Topik 2006, 101). During the first decades of the 20th century, synthetic fertilizer was still just a supplement; the major part of the nitrogen supply came from farmyard manure and nitrogen-fixing crops in the mixed systems. But by the end of the period commercial fertilizer use eventually allowed industrial societies to bypass ecological constraints that limited growth (Gorman 2013, 54). In this way, the soybean (and other legumes) which had provided the ecosystem service of nitrogen fixation to maintain soil nutrients for thousands of years was no longer needed. The value of the restorative power of legumes in crop rotations to enrich soils with nitrogen began to be lost in advanced agricultural economies (Gorman 2013, 61). Instead, other functions of the soybean gained value.

Meanwhile, animal farmers began relying on increasing amounts of purchased animal feed inputs instead of producing them on-farm. Farming models based on self-sufficiency, thus began to disappear, particularly in the United States. The separation of crops and animals and the growing drive to export for production broke local nutrient cycles. This meant mined soils on crop farms and nutrient overflows (pollution) on animal farms – starting the so-called metabolic rift (Bellamy Foster 1999). The specialization and separation of farm tasks, moreover, decreased economic diversity and erased traditional biocultural practices as well as food cultures tied to native landscapes. Soy played a very central role in this shift: as we saw, the increased use of soy oil resulted in massive amounts of soy meal around. Much money and effort were put into finding the most profitable way of using this byproduct of oil pressing. Not least the meat industry became very interested in investigating how it could take a more central role as a protein source within the general trend to intensify animal production. Economic actors, operating within the social relations of a capitalist system, always try to maximize earnings by systematically cutting costs to be able to bring more goods to the market at a lower cost.

At the same time, it was clear that meat was considered the most desirable food among European and American populations. The high status of meat had a long history in Europe and was deeply embedded in food cultures, but perhaps in the post-war context (after years of rations and *ersatz* food), meat became a symbol of affluence and welfare even more than before. Politicians were well aware that if voters could afford a diet rich in meat, they would consider themselves rich. Accordingly, to produce meat in the cheapest way possible became a political priority.

Mixed farming, thus, disintegrated as an entirely new regime characterized by soy crops sold as a cheap ingredient for animal production elsewhere was emerging. Thus, by the end of the 1940s, the roots of this new soy model had spread from East to West and deeply into US soils, but taking on entirely new roles and meanings – where the most important was soy facilitating "meatification".

Note

1 1904 is the date of the second edition.

References

ADM. 2022. "Our Company – History – Founders' Story." Archer Daniels Midland. 2022.

Arntzen, Charles J., and Ellen M. Ritter. 1994. *Encyclopedia of Agricultural Science*. San Diego, CA: Academic Press.

ASA. 1949. "Soybean Bluebook." American Soybean Association, Hudson Iowa.

———. 2022. "American Soybean Association – Our Timeline." 2022.

Bacon, L. B., and F. C. Schloemer. 1940. *World Trade in Agricultural Products: Its Growth; Its Crisis; and the New Trade Policies*. Rome: International Institute of Agriculture.

Baraibar Norberg, Matilda. 2020. "Changes and Continuities in Agrofood Relations, 1870–1970s." In *The Political Economy of Agrarian Change in Latin America*, by Baraibar, 57–116. Cham: Palgrave Macmillan, Springer Nature.

Barbier, Edward. B. 2011. *Scarcity and Frontiers: How Economies Have Developed through Natural Resource Exploitation*. Cambridge: Cambridge University Press.

Bedier, Mohammad. 2018. "Chapter 1: Demystification of Mergers and Historical Overview." In *Cross-Border Mergers and Acquisitions the Case of Merger Control v. Merger Deregulation*. Cheltenham, UK: Edward Elgar Publishing.

Bellamy Foster, John. 1999. "Marx's Theory of Metabolic Rift: Classical Foundations for Environmental Sociology." *American Journal of Sociology* 105 (2): 366–405.

Berg, B. Å. 2013. "Margarine – A Scapegoat Milked by Many Maids. A Brief History of Margarine in Sweden with an International Background." In *From Local Champions to Global Players: Essays on the History of the Dairy Sector*, edited by Rytkönen, García Hernández, and Jonsson, 95–110. Stockholm: Acta Universitatis Stocholmiensis.

Bix, Herbert. 1972. "Japanese Imperialism and the Manchurian Economy, 1900–31." *The China Quarterly* 51: 425–43. http://www.jstor.org/stable/652483.

Brother, George H., C. W. Murray, and Francis P. Griffiths. 1951. "Concentrates and Meals of Alfalfa." In *USDA Yearbook of Agriculture 1950–1951*, 345–48. Washington, DC: USDA.

Bunge. 2022. "Who We Are – Our History." https://www.bunge.com/who-we-are/our-history.

Buzan, Barry, and George Lawson. 2013. "The Global Transformation: The Nineteenth Century and the Making of Modern International Relations." *International Studies Quarterly* 57 (3): 620–34.

Cargill. 2018. "Momentum – 2017 Annual Report." Annual Report. Cargill, Incorporated.

———. 2022. "Cargill Company Timeline: 1865–Present." https://www.cargill.com/doc/1432078093613/pdf-cargill-timeline.pdf.

Chang, Kwang-chih. 1963. *The Archaeology of Ancient China*. 1st ed. New Haven, CT: Yale University Press.

Dalla Costa, Armando, and Gustavo Pereira da Silva. 2018. "Bunge and His First Fiftieth Anniversary in Brazil (1905–1955): The Construction of an Economic Group." *Investigaciones de Historia Económica – Economic History Research* 14 (3): 199–209.

Dennys, Nicholas B. 2012 (originally published in 1867). "Newchwang." In *The Treaty Ports of China and Japan: A Complete Guide to the Open Ports of Those Countries, Together with Peking, Yedo, Hongkong and Macao*, edited by King, Dennys and Mayers 538–46. Cambridge Library Collection – East and South-East Asian History. Cambridge: Cambridge University Press.

Du Bois, Christine M. 2018. *The Story of Soy*. London: Reaktion Books.

Du Bois, Christine M., Chee-Beng Tan, and Sidney W. Mintz, eds. 2008. *The World of Soy*. Singapore: NUS Press.

Eaglesham, Allan R. J. 1989. "Global Importance of Rhizobium as an Inoculant." In *Microbial Inoculation of Crop Plants*, edited by Campbell and Macdonald. Special Publications of the Society for General Microbiology 25. McLean, Virginia: IRL Press at Oxford University Press.

Eckstein, Alexander, Kang Chao, and John Chang. 1974. "The Economic Development of Manchuria: The Rise of a Frontier Economy." The Journal of Economic History 34 (1): 239–64.

Federico, Giovanni. 2010. "Feeding the World." In *Feeding the World: An Economic History of Agriculture, 1800-2000*. Princeton: Princeton University Press.

Fletcher, Merna Irene. 1950. "World Soybean Production in 1948." *The Scientific Monthly* 70 (2): 116–21.

Friedmann, Harriet. 1982. "The Political Economy of Food: The Rise and Fall of the Postwar International Food Order." *American Journal of Sociology* 88: S248–86.

Friedmann, Harriet, and Philip McMichael. 1989. "Agriculture and the State System: The Rise and Fall of National Agricultures, 1870 to the Present." *Sociologia Ruralis* 29 (2): 93–117.

Ginsburg, Norton S. 1949. "Manchurian Railway Development." *The Journal of Asian Studies* 8 (4): 398–411.

Gorman, Hugh. 2013. "N and the Rise of Science." In *The Story of N*, 52–63. A Social History of the Nitrogen Cycle and the Challenge of Sustainability. New Brunswick, NJ: Rutgers University Press.

He, Fanneng, Quansheng Ge, Junhu Dai, and Yujuan Rao. 2008. "Forest Change of China in Recent 300 Years." *Journal of Geographical Sciences* 18 (1): 59–72.

Horowitz, Richard S. 2006. "Politics, Power and the Chinese Maritime Customs: The Qing Restoration and the Ascent of Robert Hart." *Modern Asian Studies* 40 (3): 549–81.

Hosie, Alexander. 1904. *Manchuria: Its People, Resources and Recent History*. 2nd ed. New York: Charles Scribner's sons.

Hu, Xue-Mei. 2002. The North-East Soybean Exports and the Development of China's North-East Region (1860–1931). Northern Cultural Relics, Total No. 71 (2002, No.3), pp. 93—99. 胡雪梅，2002。东北大豆出口贸易与近代中国东北开发（1860-1931）。北方文物，71: 3, pp. 93-99.

Hymowitz, Theodore. 1970. "On the Domestication of the Soybean." *Economic Botany* 24 (4): 408–21.

———. 1990. "Soybeans: The Success Story" in *Advances in New Crops*, edited by Janick and Simon, 59–63. Portland, OR: Timber Press.

IIA (*Institut International d'Agricuture* [International Institute of Agriculture]). 1923. [International Institute of Agriculture]). *Oleaginous Products and Vegetable Oils Production and Trade*, edited by Capone and Grinenco; Rome: International Institute of Agriculture. Bureau of Statistics.

———. 1936. *Le Soja dans le Monde*. Rome: *Institut International d'Agriculture*.

———. 1939. *La Production et Le Commerce International Des Huiles et Graisses*. Rome: Imprimerie de la Chambre des Deputes, Charles Colombo.

———. 1940. *Les Grands Produits Agricoles, Compendium International de Statistiques*. Rome: *Institut International d'Agriculture*.

———. 1944. *Les Grands Produits Agricoles, Compendium International de Statistiques*. Rome: *Institut International d'Agriculture*.

Isett, Christopher. 2007. *State. Peasant, and Merchant in Qing Manchuria, 1644–1862*. Stanford, CA: Stanford University Press.

Isett, Christopher, and Stephen Miller. 2016. *The Social History of Agriculture: From the Origins to the Current Crisis*. Lanham, MD: Rowman & Littlefield.

Jonsson, Ulf and Ronny Pettersson. 1991. "Specialisering Och Diversifiering i Europeiskt Jordbruk" in *Diversifiering Och Specialisering i Europeiskt Jordbruk Tiden 1750–1950: Redovisning Av Agrarhistoriskt Seminarium [Uppsala] Den 17 September 1990*. edited by Ljungren and Myrdal, 5–14. Stockholm: Kungl. Skogs--och Lantbruksakademi.

Kenyon, Richard L., N. F. Kruse, and S. P. Clark. 1948. "Solvent Extraction of Oil from Soybeans." *Industrial & Engineering Chemistry* 40 (2): 186–94.

Kishlar, Lamar. 1944. "Our Common Interests." *The Soybean Digest* 4 (5):9. Chicago, IL: The American Soybean Association.

Kromer, George Wayne. 1961. *Trends in Soybean Acreage and Production, 1924–60*. Washington, DC: U.S. Dept. of Agriculture, Economic Research Service.

Kromer, George Wayne, and Celma Bunyan Gilliland. 1954. *Processing the Three Major Oilseeds*. Washington, DC: U.S. Dept. of Agriculture, Agricultural Marketing Service, 56.

Lander, Brian, and Thomas David DuBois. 2022. "A History of Soy in China: From Weedy Bean to Global Commodity." in *The Age of the Soybean: An Environmental History of*

Soy during the Great Acceleration, edited by da Silva and de Majo, 29–47. Winwick, Cambridgeshire: The White Horse Press.

Langthaler, Ernst. 2018. "Friedrich Haberlandt's Failed Vision: Soy in European Food Cultures, 1873–1945." *Environment & Society Portal, Arcadia*, Rachel Carson Center for Environment and Society, 3.

Langthaler, Ernst. 2020. "Broadening and Deepening: Soy Expansions in a World-Historical Perspective." *Historia Ambiental Latinoamericana y Caribeña (HALAC) Revista de La Solcha* 10 (1): 244–77.

Laranjo, Marta, Ana Alexandre, and Solange Oliveira. 2014. "Legume Growth-Promoting Rhizobia: An Overview on the Mesorhizobium Genus." *Special Issue on Plant Growth Promotion* 169 (1): 2–17.

LDC. 2022. "LDC Annual Report 2021." 2022. https://www.ldc.com/annual-report-2021/. (Accessed 2022-10-17).

List, G. R. 2015. "Soybean Lecithin: Food, Industrial Uses, and Other Applications." In *Polar Lipids*, edited by Ahmad and Xu, 1–33. Urbana, IL: AOCS Press.

Lyons, Thomas P. 2003. *China Maritime Customs and China's Trade Statistics, 1859–1948*. Trumansburg, NY: Willow Creek of Trumansburg.

Maddison, Angus. 2003. *The World Economy: Historical Statistics*. Paris: OECD.

Magnan, André. 2012. "Food Regimes." In *The Oxford Handbook of Food History*, edited by Pilcher, 370–88. Oxford: Oxford University Press.

McMichael, Philip. 2013. *Food Regimes and Agrarian Questions. Agrarian Change & Peasant Studies*. Winnipeg, Canada: Fernwood Publishing.

McNeill, J. R. 2001. *Something New under the Sun: An Environmental History of the Twentieth-Century World*. First published as a Norton paperback. New York: W. W. Norton.

Mejia, Enrique Antonio. 2022. "Five Decades of Soybean Agriculture:" In *The Age of the Soybean: An Environmental History of Soy during the Great Acceleration*, edited by da Silva and Majo, 185–204. Winwick, Cambridgeshire: White Horse Press.

Myrdal, Janken. 2020. "Agricultural Treatises in Eurasia before ca. 1500: The Iceberg Method, a Survey, and a Grand Hypothesis." *The Medieval Globe* 6 (2): 1–47.

Noguchi, Masayoshi, and Trevor Boyns. 2013. "The South Manchuria Railway Company and Its Interactions with the Military: An Accounting and Financial History." *The Japanese Accounting Review* 3 (2013): 61–101.

O'Rourke, Kevin and, Jeffrey Williamson. 1999. *Globalization and History: The Evolution of a Nineteenth-Century Atlantic Economy*. Cambridge, MA: The MIT Press.

Peruchi Moretto, Samira, Eunice Nodari, and Rubens Nodari. 2022. "Soybean as a Critical Resource: Domestication, Dissemination and Introductions (Glycine Max)." In *The Age of the Soybean: An Environmental History of the Soyacene during the Great Acceleration*, edited by da Silva and de Majo Chapter 1:19–28. Winwick, Cambridgeshire: White Horse Press.

Piper, Charles Vancouver, and William Joseph Morse. 1916. "The Soy Bean, with Special Reference to Its Utilization for Oil, Cake, and Other Products." U.S. Department of Agriculture Bulletin 439. U.S. Department of Agriculture Bulletin. Washington D.C.: U.S. Department of Agriculture.

———. 1923. *The Soybean*. New York, NY: McGraw-Hill Book Company, Inc.

Pomeranz, Kenneth. 2021. *The Great Divergence: China, Europe, and the Making of the Modern World Economy*. Princeton: Princeton University Press.

Pomeranz, Kenneth, and Steven Topik. 2006. *The World That Trade Created: Society, Culture, and the World Economy, 1400 to the Present*. 2nd ed. Armonk, NY: M.E. Sharpe.

Prodöhl, Ines. 2010. "'A Miracle Bean'. How Soy Conquered the West, 1909–1950." *Bulletin of the German Historical Institute Washington, DC* no. 46: 111–29.

———. 2013. "Versatile and Cheap: A Global History of Soy in the First Half of the Twentieth Century." *Journal of Global History* 8 (3): 461–82.

Qi, Wu. 2020. "The Relation between Japanese and Chinese Merchants in the Perspective of Northeast Soybean Business in Late Qing Dynasty." *Researches in Chinese Economic History* 4: 150–61.

Reardon-Anderson, James. 2005. *Reluctant Pioneers: China's Expansion Northward, 1644–1937.* Stanford, CA: Stanford University Press.

Roberts, John M. 2004. *The New Penguin History of the World.* 4th ed. London: Penguin Books.

Roth, Matthew David. 2018. *Magic Bean: The Rise of Soy in America.* Lawrence, KS: University Press of Kansas.

Ryzhova, Natalia. 2022. "The Failure of the 'Soybean Revolution' in the Ussr." *Ab Imperio* 1: 191–222.

Seth, Michael J. 2010. *A History of Korea: From Antiquity to the Present.* Rowman & Littlefield Publishers.

Shan, Patrick Fuliang. 2014. *Taming China's Wilderness: Immigration, Settlement and the Shaping of the Heilongjiang Frontier, 1900–1931.* Routledge.

Sheikh, S. M., and K. S. Zakiuddin. 2019. "History of Human Powered Oil Expeller: A Literature Review." In *Explorations in the History and Heritage of Machines and Mechanisms*, edited by Zhang and Ceccarelli, 77–88. Cham: Springer International Publishing.

Shurtleff, William, and Akiko Aoyagi. 2004. "American Soybean Association: History of Work with Soy – A Chapter from the Unpublished Manuscript, History of Soybeans and Soyfoods, 1100 B.C. to the 1980s." A Chapter from the Unpublished Manuscript, History of Soybeans and Soyfoods, 1100 B.C. to the 1980s. Lafayette, CA.

———. 2007. "History of Soy in Europe (Incl. Eastern Europe and the USSR (1597–Mid 1980s)." In *History of Soybeans and Soyfoods, 1100 B.C. to the 1980s (Unpublished Manuscript).* Lafayette, CA: Soyinfo Center.

———. 2009. *History of Soybeans and Soyfoods in South America (1882–2009): Extensively Annotated Bibliography and Sourcebook.* Lafayette: Soyinfo Center.

———. 2014. *Early History of Soybeans and Soyfoods Worldwide (1024 BCE to 1899): Extensively Annotated Bibliography and Sourcebook.* Lafayette, CA: Soyinfo Center.

———. 2016. *History of Soybean Crushing, Soy Oil, and Soybean Meal (980–2016): Extensively Annotated Bibliography and Sourcebook.* http://www.soyinfocenter.com/pdf/196/Crus.pdf.

———. 2018. *History of Research on Nitrogen Fixation in Soybeans (1887–2018): Extensively Annotated Bibliography and Sourcebook.* Lafayette, CA: Soyinfo Center. http://www.soyinfocenter.com/pdf/210/NiFi.pdf.

———. 2019. *History of the National Oilseed Processors Association: 1930–2019.* Lafayette, CA: Soyinfo Center.

———. 2020a. *History of ADM (Archer Daniels Midland Co.) and the Andreas Family's Work with Soybeans and Soyfoods (1884–2020): Extensively Annotated Bibliography and Sourcebook.* Lafayette, CA: Soyinfo Center.

———. 2020b. *History of Cargill's Work with Soybeans and Soy Ingredients (1940–2020).* Lafayette, CA: Soyinfo Center.

———. 2021. *History of International Trade in Soybeans, Soy Oil and Soybean Meal, and Trade Policy (1859–2021): Extensively Annotated Bibliography and Sourcebook.*

———. 2022. *History of Soybeans and Soyfoods in Manchuria (1833–2022): Extensively Annotated Bibliography and Sourcebook*. Lafayette, CA: Soyinfo Center.

Thompson, F. M. L. 1968. "The Second Agricultural Revolution, 1815–1880." *The Economic History Review* 21 (1): 62–77.

US Bureau of Foreign Commerce. 1880. *Report Upon the Commercial Relations of the United States with Foreign Countries for the Year 1879*. Vol. 1. Washington, DC: U.S. Government Printing Office.

U.S. Soybean Export Council. 2008. "How the Global Oilseeds and Grain Trade Works." South West Harbour: High Quest Partners and Soyatech; Soyatech.

Veraart, Frank. 2022. "Catalysing Socio-Ecological Change: The Extraction and Processing of Edible Oils, 1910-1940". *Global Environment*, 15 (2): 370–397.

Werner, Dietrich, and William E. Newton. 2005. *Nitrogen Fixation in Agriculture, Forestry, Ecology, and the Environment*. Dordrecht: Springer Science & Business Media. 350 pp.

Williams, Thomas A. 1897. "The Soy Bean as a Forage Crop." *Farmers' Bulletin (USDA)* 58: 1–19.

Wolff, David. 2000. "Bean There: Toward a Soy-Based History of Northeast Asia." *The South Atlantic Quarterly* 99 (1): 241–52.

Zatsepine, Victor. 2014. "Surveying Manchuria: Imperial Russia's Topographers at Work." In *Entangled Histories: The Transcultural Past of Northeast China*, edited by Ben-Canaan, Grüner, and Prodöhl, 177–88. Transcultural Research – Heidelberg Studies on Asia and Europe in a Global Context. Cham: Springer International Publishing.

Zhang, Bo, 2008. Policy Adjustment and Changes of Bean Goods Trade in Northeastern China in the Qing Dynasty. *Journal of Tianjin University* (Social Science), 10, (6): 525—528. 张博，2008。 制度调整与清代东北豆货贸易格局的变迁。天津大学学报（社会科学版）第十卷第六期页 525—528。

5 The regime of the third soybean cycle (1950–today)

> Cargill China earns the U.S. State Department's Award for Corporate Excellence. The State Department was particularly impressed by the farmer training program of Cargill Animal Nutrition. More than 2 million Chinese farmers have gone through the free education program.
>
> (Cargill 2022)

Above quote comes from the home page of Cargill – the largest privately held corporation in the United States (US) (by revenues) and one of the world's largest soybean traders. It tells how this big transnational company was awarded the "Corporate Excellence" prize in 2008 by the US government. The quote is in fact rather illustrative of the current regime phase of the third soybean cycle as it alludes to the proactive role of US-based transnational firms in the Chinese "nutritional transition" (where per capita meat consumption is rapidly rising). It also proudly recognizes a production model based in increasing use of soymeal for animal feed. The quote further reveals that while rising meat consumption in China is often pointed to as the main driver behind contemporary soybean expansion, China's dominance in soybean imports is not simply a spontaneous shift in response to farmers' demands, but in fact is also the result of concerted efforts by large, powerful private corporations acting together with the US government to strategically promote a specific soy-based food system model throughout the world. This model – and how it has been maintained until today – is the core of this chapter.

In the previous chapter, the agrarian and the industrial revolutions coupled with imperialism to accelerate human activities across the globe. We saw how a new model of soybean production – a cash crop, mass produced under monocultural forms (or very simple crop rotations) – spread across the Manchurian frontier. This contrasted sharply with integrated cropping systems in Northern China where soybeans originally played a role for centuries. In fact, historically and globally, crop and animal production had been closely integrated in mixed farming systems. It was not until after World War II that agricultural specialization and separation of crops and animals became significant. However, when soybean farming spread throughout the Midwest, it was among US farmers who were already mechanized and capitalized – using tractors and combines, silos and mills, fertilizers and improved seeds – that a "third way" of industrial soybean

DOI: 10.4324/9780367822866-5

production was born. While China and Manchuria were still important places for soybean production, it was this "US Soy Model" that became the most successful and dynamic. In a few decades, the United States surpassed China (now including Manchuria again) as the world's most important soybean producer. It was the US model of both production and consumption of soybeans that was exported to the world during this regime which started in 1950.

The beginning of this period coincides with what food regime scholars consider the second international food regime (1950s–1973). The United States took the role once dominated by the British, transferring surplus domestic food production to its post-colonial empire, not least through international aid (McMichael 2009, 140–41). Food aid became important in this regime, since food subsidies and agricultural support ensured a political agenda to defend against a communist world order. This meant that selected developing countries adopted national agro-industrialization strategies, land reforms and increased market ties (McMichael 2009, 141). Linking national agricultural sectors to global supply chains made agribusiness more transnational and, coupled with US development aims, led to an agricultural system with a new division of labor (McMichael 2013, 32). Some food regime scholars argue the second food regime fell apart simultaneously with the Bretton-Woods system in 1973 and a third, "corporate" food regime began in the late 1980s. This third regime is characterized by increased neo-liberalism, intensification of transnational links in agribusinesses, the universalization of export agriculture and ever-increasing distances between the food we eat and who produces it and where it is grown (McMichael 2009). This neo-liberal globalization turned previously development-driven state agricultural intervention with wider societal goals into market-oriented processes. Moreover, a deregulation of financial flows worldwide not only resulted in increased trade by corporate entities (McMichael, 2013, 47), but the line between financial institutions and agricultural corporations was blurred, leading to a mimicking of behavior of financial institutions (Burch and Lawrence 2009, 277). Financialization transformed and extended corporate control over the entire food system, including agriculture. Yet, although the state no longer controls markets it is still depended upon to aid the market in times of food crises.

Through the lens of the soybean, however, we argue that the period from 1950 to today is one regime. This entire time period is characterized by the Great Acceleration (Steffen et al 2015) – i.e. intensifying already established legacies. It is not a break – or threshold that has been passed – which is key for whether a system enters into a new regime. The changes we highlight are the same (production and trade) model, but taken to an even higher level of intensification, e.g. with the Gene Revolution the model becomes further entrenched – strengthening and concentrating certain connections in the system, namely that corporations now gather the revenues from more of the activities in the chain. Thus, that the entire regime is characterized by a "great acceleration" of relentless growth, industrialization, financialization and transnationalization while ignoring the social-ecological repercussions of agriculture throughout the entire food system, are in fact important continuities.

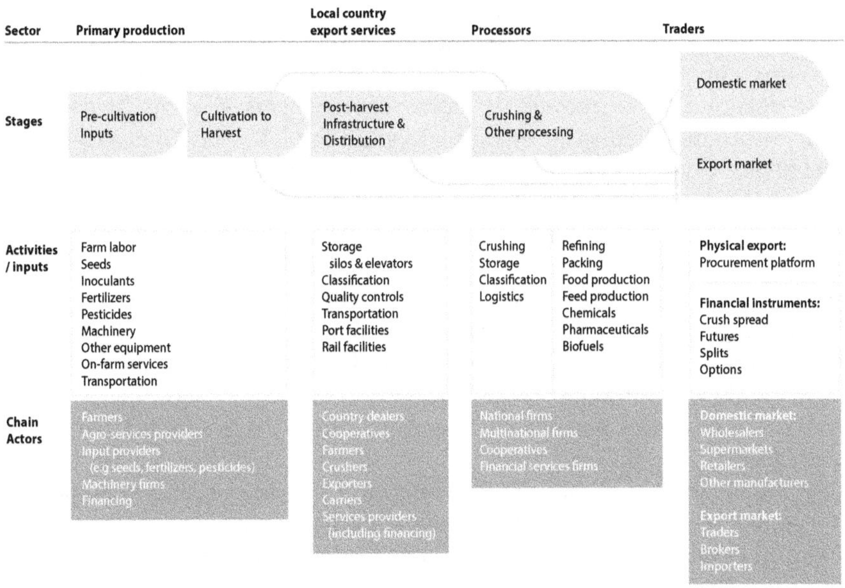

Figure 5.1 The actors and activities in the different stages of the soybean commodity chain today.

After this introduction, we first critically reflect on empirical sources before we delve into regime section. At the heart of this chapter is our usual thematic presentation of the main roles and functions of soybeans in the agrofood system: on-farm cultivation, international trade, animal feed, oil and soyfoods. This exploration begins with the roles played by soybeans on-farm, including production growth, expansion and intensification of croplands, geospatial shifts to the South and the Gene Revolution. We continue with soybeans' roles in international trade, including deregulation, financialization and the vertical and horizontal integration of traders. We then explore the new main function of soymeal as feed in the Soy-Meat Complex and then oil's different uses for food and fuel. We finally discuss soy as a nutritious food before we reflect on the regime as a whole. Hopefully, Figure 5.1 will help to follow the actors and activities in the different stages of the global soybean commodity chain.

A critical note on sources

There is a huge and growing literature dealing with soy during this period. Just like for Chapter 4, the main difficulties in terms of sources are related to selection, rather than access. Most data in this chapter come from the archives and websites of the Food and Agriculture Organization of the United Nations

(FAO) and United States Department of Agriculture (USDA). We began with FAO production and trade yearbooks until 1961 then we used FAOSTAT – a collection of online databases containing more than 1 million time-series records covering international agricultural statistics for over 200 countries, from 1961 until recently. Data are provided by national governments or extrapolated by FAO staff. We triangulated with several sources in the USDA Economics, Statistics and Market Information System (ESMIS) which contains over 2,100 publications from five agencies of the USDA. Statistical data from China was highly unreliable up until the death of Mao Zedong in 1976 and is today still regarded as susceptible to political influence (Crook 1988; Federico 2010). For a duration of 20 years, the Chinese government refrained from publishing any statistical trade data, prompting the FAO to base its estimations on trade statistics collected from China's trading partners (Crook 1988, 1–5). We have adjusted volume and area data from 1950 to 1960 and checked later years in FAOSTAT according to USDA statistician Fredrick Crook (1988). Source information is specified in the respective figures.

We have also made wide use of written material produced by the powerful soy agribusiness actors themselves. This includes corporate annual reports, webpages and even advertisements by key private actors involved in different stages of the soy chain, e.g. seed and agrochemical producers, traders and processors, as well as important soybean producers' organizations. Beyond most countries' legal imperative on companies to write annual reports, corporations are keen to use annual and/or sustainability reports as public relations documents; to construct a particular picture of themselves and communicate with their various publics (Stanton and Stanton 2002). So, it is important to keep in mind that the *raison d'etre* of these reports is to present the organization in a positive light. However, they often include data about companies' global activities and performance which is difficult to otherwise access because of the methodological nationalism that still characterizes most national accounts. Here, as always, we have tried to triangulate information when possible.

The soybean's roles on-farm: snowballing across the US Corn Belt, pioneering into Latin American frontiers and stagnation in China

Global soy production

Global soy production has exploded in the last 70 years. When considering world soybean production 1950–2020, some very clear and distinct features emerge (see Figure 5.2). First of all is relentless production growth. The dramatic and sustained rapid growth in world soybean production during this period is unequaled by any other crop. Although soybeans held significance in 1950, their global production experienced a surge from that point onwards. The world's soybean production was 16 million tons (MMT) in 1950. Subsequently, after 25 and 50 years, in 1975 and 2000, production surged to 64 and 161 MMT, respectively. As of 2020, global soybean

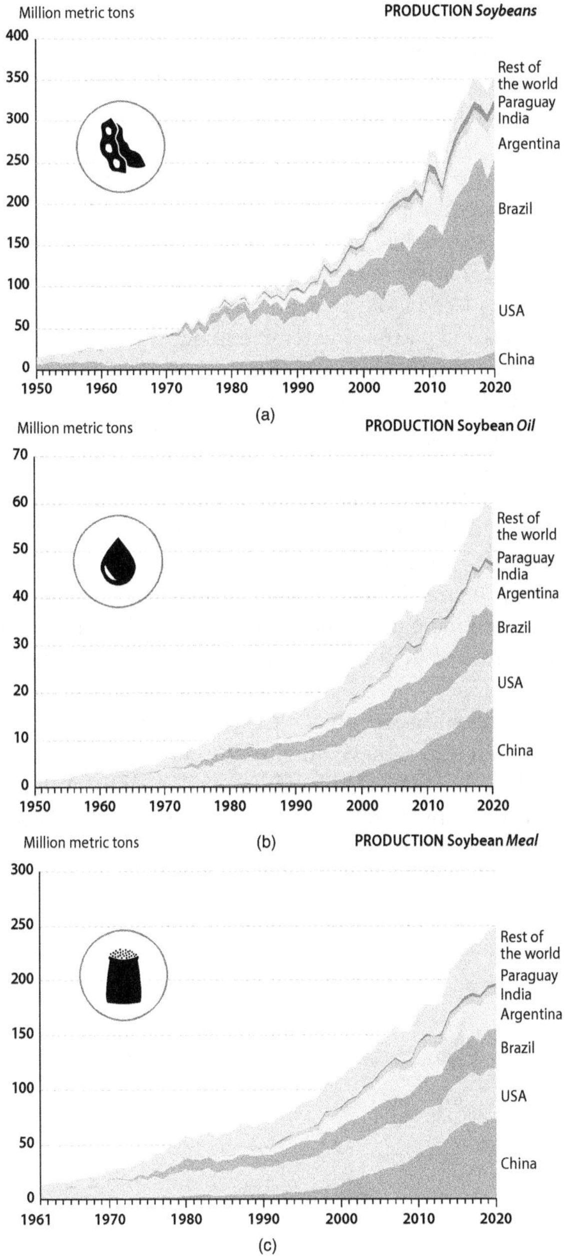

Figure 5.2 Global production by major nations and the rest of the world: (a) soybeans 1950–2020; (b) soy oil 1950–2020; (c) soymeal 1961–2020.

Sources: (A) USDA NASS (1952, 161; 1954, 126; 1956, 127; 1958, 133; 1959, 138; 1960, 138; 1961, 138; 1962, 165), Crook (1988, 126), Barnhart (1954, 223), FAOSTAT (2022); (B) Kromer (1961, 27), Hacklander (1986, 33), USDA FAS (1957, 6; 1959, 148; 1961, 148), USDA NASS (1957, 174), FAOSTAT (2022); and (C) FAOSTAT (2022).

production reached 354 MMT (USDA 1952 production, FAOSTAT 2022). Hence, between the start of this regime in 1950 and 2020, global soybean production witnessed an impressive 22-fold increase, while the world's population "only" grew to three times its size, from 2.5 billion to 7.7 billion (De Maria et al. 2020). Although production of maize, wheat, rice and cotton also rose significantly, no other crop has grown as fast as the soybean and its derivatives (Howard 2016, 92–93).

The most fundamental input necessary to produce millions of tons of soybeans is, of course, land. In absolute terms, soybean acreage grew from 16.5 million hectares (Mha) to 127 Mha between 1950 and 2020, representing an increase of almost eight times (see Figure 5.3). Soybean cultivation initially expanded by replacing other crops and pastures. Later, it extended further into frontier regions, leading to the displacement of natural grasslands, forests, wetlands and savannahs. At the beginning of this regime, in 1950, under 10% of the Earth's surface was croplands (1,220 Mha) – with about 100 Mha of them in oil crops (soy areas were about 16 Mha). By the end of the regime, nearly 1,600 Mha were in cultivation globally and of them 300 Mha were in oil crops – of which over 120 Mha of them were now in soybeans alone. Of the major crop groups, grains are largest today with almost 700 Mha and coarse grains occupy about 340 Mha. Over the regime cycle, grains increased slightly while coarse grains were stable. However, oil crops expanded dramatically (almost 300%) and soybeans' contribution to this growth went from 16% to 40% of areas. (USDA NASS 1952; Ritchie and Roser 2019; FAOSTAT 2022)

In addition to extensification, soy cultivation has contributed to a significant intensification of land use, achieved mainly through the adoption of more technology (improved seeds, pesticides, fertilizers and machines) that enabled a rise in yields and greatly contributed to increasing production. World average soy

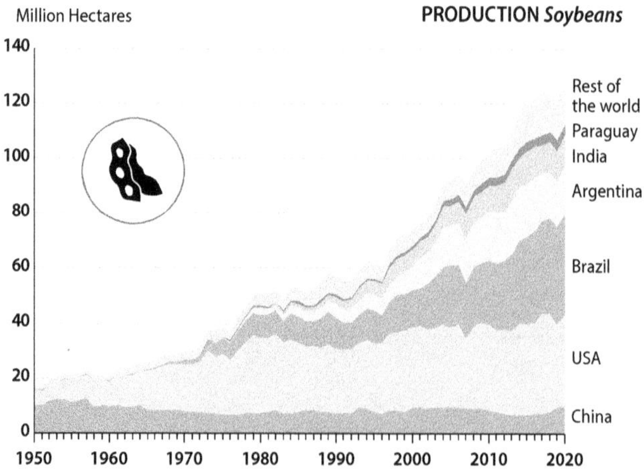

Figure 5.3 Global areas under soybean cultivation 1950–2020 by main producing nations. Source: USDA NASS (1952, 161; 1954, 126; 1956, 127; 1958, 133; 1959, 138; 1960, 138; 1961, 138; 1962, 165), Crook (1988, 126), Barnhart (1954, 223), and FAOSTAT (2022).

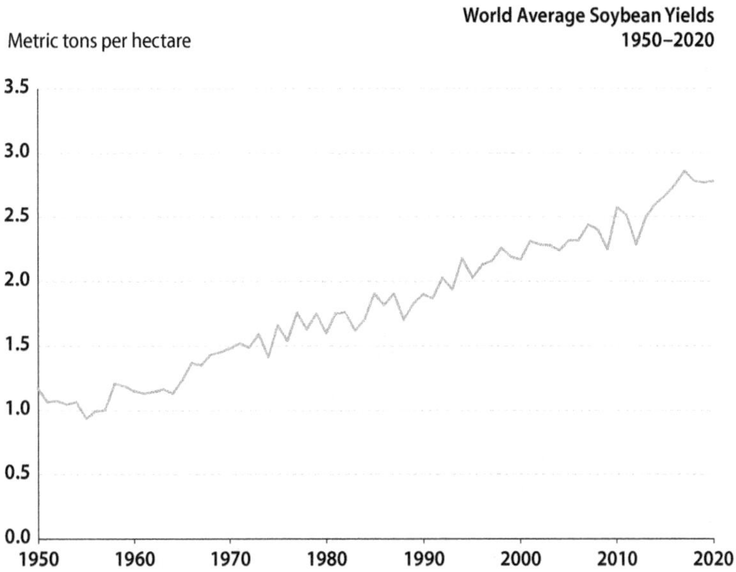

Figure 5.4 World average soybean yields 1950–2020.
Source: 1950–2011 (Brown 2012); 2012–2020 (FAOSTAT 2022).

yield more than doubled between 1950 and 2020 from 1.18 to 2.78 tons (MT)/ha (beneath this average, however, lay important local variation and significant yield gaps) (see Figure 5.4) (Brown 2012; FAOSTAT 2022). Soy croplands expanded 672% and yields rose 134% – enabling production to rise over 2,000% in 70 years.

As we delve into the various on-farm roles of soybeans during this period, we will take a closer look at three distinct geographic regions: the dynamic soy model that originated in the United States, the rapid expansion of soybean cultivation in South America over the past 50 years and China – which still remains one of the most significant soybean producers in the world, albeit stagnant.

As we saw in Chapter 4, the United States' share of world soybean production rose rapidly during and after World War II. By 1960, the United States already represented 67% (15 MMT) and, in 1969, 76% of the world's soy production. Meanwhile, soybean production in war-torn Europe was negligible, despite the lingering dearth of vegetable oils and protein, and despite government encouragement (Fletcher 1950, 118). Europe instead increased imports, as explored in the following trade section. In 1969, farmers in the United States and China grew 76 and 17% of the total world production, respectively. As the United States became the most important country for soybean cultivation, soy became the most important crop for the United States and its cultivation snowballed across the US Corn Belt and then expanded further as the Corn Belt itself included new areas. By 1973, soybeans had surpassed both wheat and corn to become the country's most important cash crop (Mintz, Tan, and Du Bois 2008, 5). In 2020, the United States produced 112.5 MMT of soybeans, much more than the entire world only two decades earlier, but now US volumes were only 32% of global soybean production. Thus, in spite of the continuous growth of US soybean production, rates of growth were even higher for new entrants from South America.

When soybeans first arrived in Latin America in the late 19th century there was agronomic interest in the bean, but it was not until the 1960s and 1970s that the soybean became successful outside of experimental stations. In 1950, Brazil produced 0.036 MMT and by 1969 it produced its first MMT (USDA NASS 1952; FAOSTAT 2022). In just one decade, between 1971 and 1974, Brazil almost quadrupled its production, from 2.1 to 7.9 MMT – and then doubled it again by 1980 producing 15.2 MMT. Brazil was at this time the second largest soybean producer in the world – well ahead of China (7.9 MMT), but behind the United States (48.9 MMT). However, by 2000, Brazil already produced 121.8 MMT and, by 2020, the Latin American giant produced 138 MMT, surpassing the United States as the world's biggest soy producer (FAOSTAT 2022). Public agricultural research and seed adaptation to sub-tropical and tropical climates, as well as significant government promotion through subsidized credits, bank debt write-offs and export subsidies, ignited phenomenal soy production growth (Relly and de Majo 2022). As Brazilian farmers expanded production, it even "spilled over" into neighboring Paraguay. Soy expansion into Eastern Paraguay was also supported by agricultural "modernization" programs, e.g. the wheat program promoted soybeans as a rotation crop (Baraibar Norberg 2020b). Paraguay went from producing only 0.002 MMT in 1961 to 11 MMT in 2020 (FAOSTAT 2022). Soy became the country's number one crop. Just south of these nations, Argentina emerged as the next soy giant. Argentine production increased from just 0.08 MMT in 1961 to 48.8 MMT by 2020 (Ibid). Already in 1992, Argentina (11.3 MMT) passed Chinese production volumes (10.3 MMT) to become the world's third largest soybean producer (Ibid). As in the case of Brazil, government policies promoting soybean production through research and extension played a key role in the rapid initial increase of production. Both public and private work with seed adaptation began in the 1960s and were crucial in making soy production thrive in completely new biomes and climates. After the turn of the millennium, Uruguay and Bolivia also had impressive production growth, going from almost no production in 2000 to more than 3 MMT two decades later (Ibid).

Outside of the Americas, only China and India are major producers. At the onset of this regime phase, China – the traditional heartland of soy – had already lost its position as the world's leading soy production hub. While Chinese farmers all over the country still cultivated soy (Lander and DuBois 2022, 37–38), production was stagnant (Buck 1937, 404–17). Even with Manchuria re-incorporated in 1954 statistics, China "only" produced 44% of the world's soybeans. This can be attributed to various factors, such as the continued use of rudimentary production methods and tools (Yang 1945, 22), the abrupt loss of the Japanese market post-World War II and the gradual global adoption of modern chemical fertilizers that reduced the demand for soybean cake (Lander and DuBois 2022, 41). Last, but not least, the radical shifts under Mao Zedong and the Chinese Communist Party, further contributed to the decline of soybean cultivation in the country. While Mao instigated major land reforms, with five-year plans for agriculture and ambitious plans for overall industrialization – *the Great Leap Forward* – agricultural output stagnated (Federico 2010, 206). Exacerbated by drought, the largest famine in human history became reality when approximately 30 million people starved to death in China between 1959 and 1961 (Smil 2004, 80).

While soybeans were mainly regarded as a crucial industrial crop in much of the Western world, in China, they held official status as a food grain. However, the Communist government clearly gave more preference to other crops, such as wheat and rice (Federico 2010, 209), leading to a decline in soy production between 1957 and 1977, from around 10.1 MMT to 7.3 MMT (Lander and DuBois 2022, 42). It should, nevertheless, be remembered that the statistics from this period are uncertain as the Chinese government did not release any official statistical data between 1956 and 1976 (Crook 1988, 2–3). Following Mao Zedong's death in 1976, there was a renewed emphasis on agriculture and expansion of food production. A series of agricultural policies were implemented, with a focus on achieving national food self-sufficiency and promoting the adoption of technologies to increase production intensity (Surls and Tuan 1981). From 1961 to 1989, gross agricultural output rose by an annual rate of 2.99% driven by both expansion and increased yields (Crook 1988, 21; Federico 2010, 209). The growth rates accelerated towards the end of this period (Federico 2010, 211). Despite this growth, China's share of the world's production of soybeans was on a steep decline, dropping from 44% in 1954 to 33% in 1960 and further to 9% by 1980 (Shurtleff and Aoyagi 2007).

In the last 20 years, China has improved its soybean productivity, adopting "modern" technology in agricultural inputs for domestic grains production. However, its yields remain clearly lower than in Western countries (Werner and Newton 2005, 67). Between 2000 and 2020, soybean acreage experienced a slight decrease from 9.3 to 9.1 Mha, while production increased from 15 to 18 MMT due to higher yields (Werner and Newton 2005, 73; USDA FAS 2022).

To feed one-fifth of the world's population with less than 10% of the world's arable land is a huge challenge. Thus, as we will explore in the following sections, despite being a significant domestic producer, China's food self-sufficiency ratio decreased from 93.6 percent to 65.8 percent between 2000 and 2020, leading it to become the world's largest food importer (Zoe Liu 2023).

In 1961, India produced only 0.005 MMT, but by 2020, it produced 11.2 MMT – more than half the production of China (19.6 MMT) (FAOSTAT 2022). Early attempts to produce soybeans in India were not successful, as we saw in Chapter 4, but after collaboration with the United States during the 1970s resulting in several new varieties, volumes and yields gradually increased (Werner and Newton 2005, 4:45). Today, soy is a major cash crop – often referred to as the "miracle golden bean". Productivity, tenancy forms, farm size and management practices of soy cultivation in India vary significantly (Ibid 2005, 4:59), in stark contrast to soy production throughout the Americas.

While new players have emerged, global soy production throughout this period has remained rather concentrated to a few biomes in a few nations. In 2020, three countries (United States, Brazil and Argentina) produced 80% of the world's soybeans, around 283 MMT and adding China, India and Paraguay – equals almost 92% (Ibid). There are some important emerging soy states, e.g. Ukrainian and Russian production took off in the 2000s with both countries reaching their first MMT in 2009 and 2010, respectively. In 2020, Ukraine produced 2.8 MMT and Russia 4.3 MMT (Ibid). Notably, in Russia, soybean production is largely driven by Chinese investors and farmers in provinces close to the Chinese border. In 2022, soybean production was expected to actually rise substantially in Ukraine as a consequence of the Russian invasion – if corn farmers switch to soybeans since these can be exported by rail to Europe without needing port access (Singh and Kesavan 2022). The only EU country producing over 1 MMT of soybeans is Italy. The continent with the smallest soybean production is Africa with South Africa as the largest producer (1.254 MMT), followed by Nigeria (0.6 MMT) in 2020 – both former British colonies (FAOSTAT 2022).

The land expansion and intensification contributing to soy's successful growth also brought high social-ecological risks and costs. Of course, these varied greatly depending on what land use or land cover preceded soybean cultivation and what management practices and social policies were pursued. While the whole period is characterized by increasing industrialization, monoculturalization and substitution of labor with capital, we have divided our discussions of the regime's soy production into two subsections; before and after the introduction of biotechnology in the mid-1990s. However, while the global rise in production has been impressive, global scale data conceals a wide range of differentiated patterns and realities at finer scales. We will expand on key features such as the development of the "US Soy Model", its expansion to South America, the impact of the so-called Gene Revolution and the commodity chain drivers in the pre-cultivation stages.

Creating an efficient commodity chain: the "US Soy Model"

Soybean production in the American Midwest witnessed a significant surge during World War II and the post-war era and this upward trend has persisted ever since, albeit with intermittent fluctuations in production, as is typical in the agricultural sector. When soy production took off in the US Corn Belt, between 1950 and 1970, soybeans mainly expanded over existing agricultural lands. In this way, soy benefited from the large infrastructure of equipment, storage and rail, barge and truck transportation that were already installed in the area. Since soy was mainly incorporated into crop rotations alongside other grains, its relation with other crops was generally complementary; although soy quickly became the major cash crop of these rotations (Roth 2018, 168).

The price dynamics of agricultural commodities were significantly influenced by government policies aimed at enhancing productivity and regulating markets through supply control measures. Beginning from the late 1950s, the federal government implemented programs that imposed acreage restrictions as a means of curtailing surpluses. However, soybeans and sorghum were notably exempted from these programs, which allowed them to expand rapidly. Additionally, the government provided support for domestic production through various agricultural policies, typically encapsulated within the Farm Bill (Thoyer 1998), e.g. subsidies in the enactment of PL480 Agriculture and Food Act of 1954.

The agricultural practices in the Corn-Hog Belt of the Midwest, where soy had gained prominence, followed the principles of the *Second Agrarian Revolution* until the mid-1960s. This involved crop rotations and mixed farming, with on-farm production of animal fodder, seeds and other inputs, and a limited use of chemical inputs (Jonsson and Pettersson 1991; Myrdal 2022, 35). This farming model was also prevalent in Continental Europe and the breadbasket regions of South America, including Southern Brazil and the Argentine Pampas. So, soy was most commonly grown for hay, until the notable shift towards crop specialization in the US between 1965 and 1975. This specialization was primarily driven by the desire to capture economies of scale and reduce high production costs, as reflected in the 1973 Farm Bill. As a result, areas previously designated for hay, grazing and green manure diminished (Kromer 1961, 25). The two-year rotation of corn-soybean became the dominant way of growing soybeans, but soy could also be rotated with wheat or sorghum (Schnitkey, G 2013; Werner and Newton 2005). At the same time, industrial inputs grew in importance and farming systems became increasingly both intensive and specialized. Building on decades of previous research, plant breeders developed higher yielding soybean varieties (Roth 2018, 125, 162–65; Relly and de Majo 2022, 144) and varieties with higher oil content (Wilcox 1970). The new seed varieties, however, required more capital-intensive practices in order to yield better. Thus, the production of soybeans became more reliant on off-farm inputs like commercial seeds (instead of saved seeds), inoculants (either bacteria-laden dirt in fields or coating seeds with bacterial cultures), chemical control of weeds (herbicides) and pests (pesticides), synthetic fertilizers, irrigation and machinery. Commercial farmers willingly adopted the capital-intensive innovations and machines, like combines, which reduced labor and increased yields. This model fit well in the Midwest where many farmers could afford the innovations and labor supply was low. Accordingly, the highly modern and economically effective US Soybean Model was established. By 1965, soy yields in the United States were already almost double those of China, Argentina and Brazil (De Maria et al. 2020). This industrialization of agriculture was not a soybean-specific phenomenon, all commercial agriculture moved into the highly specialized, high input-output model. Yet, the "stars" of the Green Revolution with the most spectacular rises in yields per hectare were high-yielding hybrids of corn, wheat, sorghum and rice – not soybeans. Meanwhile, livestock production also specialized – reducing on-farm fodder crops and increasing amounts of feed supplements, e.g. soybeans. And this was "mass production for mass consumption" (Friedmann and McMichael 1989, 108); Not only did agriculture industrialize, but also food itself increasingly shifted from final use to inputs into manufactured products (Ibid, 103).

The soy technology package, however, was even more mechanized and labor-sparing than for, e.g. peanuts or cotton; cotton required around 18 times more labor per acreage than soybeans (Roth 2018, 169). Thus, as soybeans expanded south and partly displaced cotton in states like Louisiana, demand for labor decreased. In short, the soy model gave a competitive edge to the large farms – with large economies of scale – and put small farmers out of business. The success of soy over other crops accordingly spurred the removal of tenant farmers and laborers, particularly in the South, while farmland was increasingly concentrated in the hands of a small number of large, powerful growers (Roth 2018). The peak number of farms in the United States was 6.8 million farms in 1935, and this number has steadily decreased since then with an estimated 2 million farms in 2018 (Spangler, Burchfield, and Schumacher 2020). In 2018, the top 7.5% of the farms controlled 40.8% of the land (Ibid).

Specialization in soy and ensuing intensification also brought concerns over land degradation, widespread erosion, biodiversity loss and increased water use. The reduction of the number of crops in rotations, particularly monocultures, generally leads to reductions in soil carbon and nitrogen (McDaniel et al 2014) – and the increasing use of synthetic nitrogen fertilizers decreases soil organic matter (Mulvaney et al 2009). In addition, soy not only displaced other crops, but increasingly also grasslands (Spangler, Burchfield and Schumacher 2020). Conversion of grassland to cropland is a major loss of biodiversity and accruement of a significant carbon debt (Wright and Wimberly 2013). None of these concerns even slowed the spread of soy. So, while the total acreage of crops planted in the United States have actually remained fairly constant over the past 50 years, with most of the production increase being a result of land-use intensification, the cultivation of soy witnessed a surge in acreage due to the introduction of regionally adapted varieties, which facilitated its expansion into the US South. In fact, the combined area of harvested soybean and corn crops witnessed a remarkable surge from 1963 to 2017, increasing by 76%, which is equivalent to 29.9 Mha (Spangler, Burchfield and Schumacher 2020). In 2019, soy and corn accounted for 56.6% (67.2 Mha) of all harvested cropland acres in the United States (Ibid).

The US Soy Model goes South

As seen in Chapter 4, field trials with soybeans started in South America in the late 19th century, and experimental stations, agronomists and researchers worked hard on local seed adaptation was done throughout the 1930s and 1940s. Soybeans first really started to spread among family farmers in the Southern State of Brazil (Rio Grande do Sul), mainly as a rotation crop to boost pastures or as forage. In the 1950s, soybeans began expanding into more tropical regions, where they were used as green manure in coffee plantations or as feed for pigs (Peruchi Moretto, Nodari, and Nodari 2022, 25). Agribusiness actors also played a key role promoting soybeans in South America, e.g. the big trader and food processor company Bunge y Born encouraged soybean cultivation in Brazil by distributing free seeds and providing small financial advances and technical assistance to farmers as well as inaugurating, in 1969, the first soy crushing plant in Latin America, S. A. Moinhos Rio Grandenses (Dalla Costa and Silva 2018; Bunge 2022). Cargill – another of the handful of giant traders and processors – had also been present in South America since the 1930s and in Brazil since 1965 (Cargill 2022). Export-oriented infrastructure was thus already in place in many areas, particularly in fertile areas close to ports (Baraibar Norberg 2020b, 64–65). Soy took off as a cash crop for export in the 1970s in Brazil, encouraged by both governmental and private seed adaptation developments (Peruchi Moretto, Nodari, and Nodari 2022, 22–26).

Latin America had been an important global agrofood provider since colonial times, but despite this agriculture was less modernized than in the United States and agricultural productivity was significantly lower (Bulmer-Thomas 2003). The low yields in South America were often understood to be a consequence of the resilient *latifundio-minifundio* structure whereby large landowners could compensate low productivity with vast amounts of land, while peasants lacked the capital and technology to invest in the land (Baraibar Norberg 2020b, 66–67, 80). The use of external inputs and mechanization was relatively low as the Green Revolution technologies had never been fully adopted. Agriculture, in this way, on average, used significantly lower levels of agrochemicals than the United States or Europe. Soy first took off among the small middle segment of commercialized farmers in the traditional breadbasket regions of the Argentine Pampas and the Brazilian South, where it first followed a path similar to the United States, largely replacing other crops – e.g. sunflower and corn in Argentina and beans, corn and cotton in Brazil. As farmers increasingly focused solely on crop systems, led by soy, the traditional mixed farming systems that involved pasture rotations declined. The trend towards soy specialization was especially evident in areas with relatively high land prices, where it led to land-use intensification through increased use of fertilizers, agrochemicals and irrigation. Moreover, the fact that numerous soy farmers leased their land (they were not the owners), meant they felt they had no alternative but to specialize in the activities that generated the highest annual economic margins in order to cover the steep rise in land lease prices (Baraibar Norberg, 2020). Soy cultivation in these areas, thus, led to notable social-ecological simplification, raising many concerns: the displacement of local food crops (such as black beans in Brazil), soil compaction and erosion, concentration of land ownership and displacement of family farmers (Baraibar 2008; Shurtleff and Aoyagi 2009, 229).

Besides supplanting other crops, soy in South America also took a completely different route than in the US, expanding beyond the traditional breadbasket and mixed farming regions, resulting in significant conversion of grasslands to croplands, as well as forests to croplands. Brazil had been investing in developing adapted cultivars that could thrive in tropical climates for, and research began to bear fruit in the 1980s. As a result, the soybean frontier shifted from being mainly concentrated in the South, close to ports, to the vast central Western states where land was "plentiful" and "cheap" (Shurtleff and Aoyagi 2009, 449, 501). This movement was further facilitated by infrastructure improvements that cut transport and storage costs, thus allowing millions of new hectares to be more readily incorporated into the soy complex. As noted by Masuda and Goldsmith (2009), the most efficient way to increase soybean output was by replacing native vegetation, such as grasslands, wetlands and forests. As soy farmers expanded into new areas beyond traditional agricultural regions, the Atlantic rainforest became an early casualty. This forest once covered extensive land areas, stretching from as far north as the Brazilian state of *Rio Grande do Norte* to as far south as *Rio Grande do Sul* and also extending into the interior of the continent, encompassing Northeastern Argentina and Central Paraguay (Nunes et al. 2022). Numerous Brazilian soy farmers ventured across the border into Paraguay, where land prices were still lower than Brazil (Baraibar Norberg 2020d, 241–47; Baraibar Norberg 2020b, 93–95). This "movement" along with other commercial crops and infrastructure projects was encouraged and endorsed by public policies in both Brazil and Paraguay, with forest areas viewed as untapped resources in need of "development" (Baraibar Norberg 2020e, 178; 2022). However, the destruction of the highly diverse flora and fauna was not the only consequence, as indigenous communities and small farmers were also widely displaced, leading to the creation of soy "deserts," land conflicts and occupations.

Brazilian soy farmers also expanded into densely populated rural areas in Paraguay, where small-scale agricultural activities, particularly cotton farming, formed the basis of livelihoods for the majority of the population. These farmers, known as "*brasiguayos*," often employed dubious methods and took advantage of corrupt land title programs under the Stroessner regime (Baraibar Norberg 2020b, 93–95). In contrast to the soybean's expansion in the US Corn Belt, where labor was typically scarce, the labor-saving soy model in these areas created a growing surplus of labor and soaring urbanization (Baraibar Norberg 2020d, 226).

Soy areas expanded even further as neo-liberal reforms (under the "Washington Consensus"') swept the region in the 1990s and the former state-interventionist and Import Substitute Industrialization policies were strongly discredited (Kay 1989, 11, 39–45). By debt-driven necessity or consent, all Latin American governments adopted reforms for, e.g. deregulation of markets, privatization of infrastructure and promotion of foreign investments (Baraibar Norberg 2020e, 165–67). These reforms created increasingly beneficial conditions for the export-oriented soy businesses to grow and consolidate in the region. These regulative shifts in Latin America did not operate in a vacuum, but interacted with regulative shifts in other countries and the international arena, particularly, calls for multilateral liberalization by the United States and WTO, as well as diversification strategies adopted by the increasingly transnational agribusiness firms. Transnational traders also played a vital role here, e.g. the Brazilian subsidiary of the Bunge alone acquired 150 operational units: including factories, mills, silos, distribution centers and ports (Dalla Costa and Silva 2018).

During the latter decades of the 20th century, then, soy cultivation experienced rapid growth in South America. However, it was not until just a few years before the turn of the millennium, when genetically engineered (GE) soy arrived, that the region became the world's biggest producer of soy.

The Gene Revolution: the US Soy Model goes corporate

The first commercially successful genetically engineered (GE) soybean was the herbicide tolerant (HT) soybean GTS 40-3-2, sold under the brand name *Roundup Ready* (RR) in the United States in 1995. It was designed and patented by Monsanto. Monsanto was an agrochemical company founded in 1901, which has manufactured many controversial products, such as the insecticide DDT, PCBs for may uses and the exfoliant *Agent Orange*. Its premium product since the 1970s was a full-spectrum weed killer (herbicide) glyphosate – sold as *Roundup*. As Monsanto diversified into plant biotechnology it developed a plant trait that provided crops with tolerance to its herbicide.[1] The idea was thus to have a cheap weed killer to "clear the land" of all other plants (i.e. weeds) except the desired crop (Baraibar Norberg 2020a, 134–36). With this technological package, the agrochemical, biotech and seed stages became tightly integrated. The GTS 40-3-2 trait was approved for food, feed and planting first in the United States and only a few months later in Argentina (Trigo and Cap 2003). This technology was then rapidly diffused through Monsanto's strategy of mergers and acquisitions of seed companies as well as widespread licensing (Moss 2009). While the adoption rate of RR soybeans was fast in both countries, it was even faster in Argentina – from 0% in 1996 to well over 90% of all soybean production by 2003 (Ibid). Monsanto's RR soy then quickly spread from Argentina to Brazil and Paraguay, where it became widely diffused despite that Brazil had not yet established a regulatory mechanism for approval of GE (until 2005), and Paraguay did not officially approve RR soybeans until 2004. By 2010, more then 95% of all soybeans in the United States, Argentina, Brazil, Paraguay and Uruguay contained GE technology (ISAAA 2019).

The adoption of GE is considered the fastest among crop technologies worldwide, with a total of 2.7 billion hectares of biotech crops sown from 1996 to 2019. In 2019 alone, the area covered by biotech crops reached over 190 Mha. Almost 99% of global biotech crop area is dedicated to the cultivation of the "big four" crops (soybeans, maize, cotton and canola). Soybeans, in particular, are the most successful GE crop, covering 91.9 Mha or more than 48% of the global biotech area (see Figure 5.5). (ISAAA 2019)

Although the biotech companies themselves often emphasize the importance of food crops like rice, squash, papaya, eggplant and potato for food security, these crops only account for a small percentage of the total global GE crop area.

Unlike the Americas where GE soybeans are widely adopted, the world's fourth and fifth largest soy producers – China and India – only cultivate non-GE soybeans (see Figure 5.6). However, many countries that have not approved production of GE soy, still allow its import, as will be further discussed in the next section. Although only nine countries have approved the cultivation of GE soy (United States, Brazil, Argentina, Paraguay, Uruguay, Bolivia, Canada, Japan and South Africa), more than 75% of global soybean production is GE soybeans. This is primarily due to the fact that the United States, Brazil and Argentina are responsible for 80% of global soy production and heavily rely on

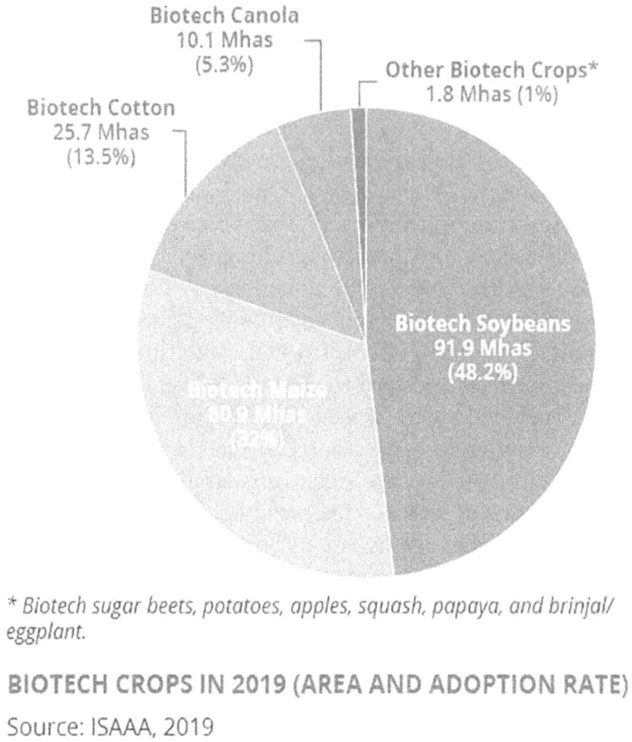

Figure 5.5 Global biotech crop areas. There were over 190 Mha of biotech soybeans, maize (corn), cotton, canola (rapeseed) and other crops in 2019. More than 90% of global biotech crop areas are cultivated by one of the "big four" crops: soybeans, maize, cotton and canola. So, while biotech promoters often mention rice, squash, papaya, eggplant, potato and other food crops are often mentioned, these are only a small fraction of total areas.

Source: ISAAA (2019; 2022).

GE soybeans. Therefore, given the significance of GE soy in the Americas, it is important to examine it in greater detail.

What contributed to the success of GE soy in the Americas? One factor is that GE soybeans – engineered to be used with glyphosate – provided farmers with a cost-effective means of weed control without the need for tillage.[2] The use of no-till and glyphosate helped save both labor and fuel costs. Furthermore, GE technology was paired with early-maturing soybean varieties, enabling two harvests in areas where only one was possible previously, leading to significant land-use intensification (including the increase in the use of agrochemicals and mechanization) but also significantly increasing production levels on the same land area. The adoption of the technological package consisting of HT soybeans, glyphosate and no-till farming led to higher economic margins for soybean

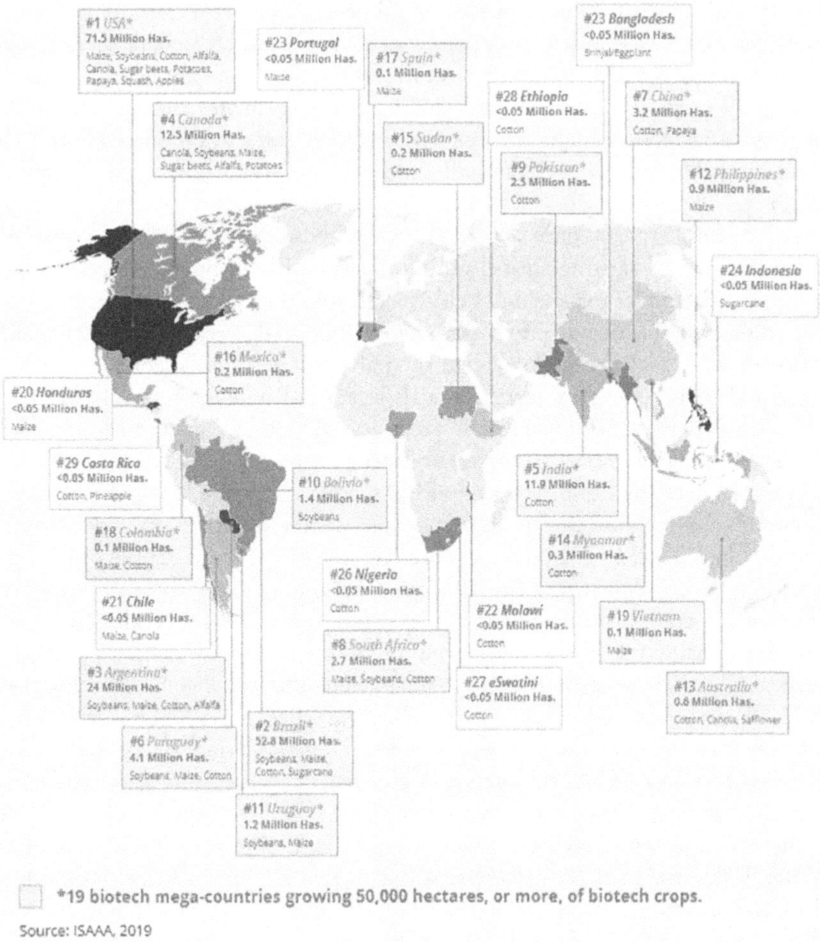

Figure 5.6 Biotech producing countries in 2019. The International Service for the Acquisition of Agri-biotech Applications' map of 29 countries growing biotech crops - 19 countries grow more than 50,000 ha (ISAAA 2019).

production compared to other land uses. This drove up land prices, which in turn created even further specialization in soybeans. This production package also rendered the accumulated knowledge of local peasants outdated and irrelevant.

The introduction of GE-soy brought about a belated "agrarian modernization" that changed Latin America, just as industrialization had previously transformed agriculture in Europe and the United States (Baraibar Norberg 2020b, 77–80).

The adoption of this capital-intensive and labor-saving model led to the concentration of land and the displacement of small-scale farmers and rural workers. Although land concentration had been a long-term trend in the region since the late 19th century, the speed and scale of change brought about by soy farmers were unprecedented, with soy farmers having the largest average holdings among all crops (Baraibar Norberg 2020c, 7–10). Estimates indicate that today large-scale farmers account for over 80% of total soybean production, with the remaining 20% in the hands of small-scale farmers (Voora, Larrea, and Bermudez 2020).

GE technology not only continued to increase growth in production and intensification of existing agricultural production, but in South America it also facilitated the direct expansion of cultivation into natural areas that were previously untouched – new frontiers. While conventional breeding and local adaptation of seeds had already expanded the palette of geographic conditions where soy could thrive, the new technological package (RR soybeans + glyphosate + no-tillage) allowed soy production in areas not previously considered apt for cultivation (soils with low organic matter and structurally fragile). Consequently, with GE, the pace of soy expansion in South America accelerated, deforesting native biomes and expanding over indigenous reservations and traditional farms (Relly and de Majo 2022, 145). In fact, soy became one of the main drivers of land-use change in South America during this regime phase.

GE soy deeply transformed the region. More than 33 Mha of additional land (roughly equal to the size of Vietnam) were incorporated into soybean production between 2000 and 2020 – a land-use change called *sojización* (Baraibar Norberg 2022, 91). Important driving actors were big firms that specialized in soybean production and adopted strategies of geographical diversification. By expanding into new territories and attracting external investment, these firms spread risks and increased profits, as investors financed production in exchange for a share of the profits after harvest (Guibert, Grosso, Arbeletche, Bellini 2011; Baraibar Norberg 2020e, 174). Around 10 Mha of expansion took place in Argentina, particularly in the Chaco, in Northeastern *Santiago del Estero*, driving massive deforestation and social conflicts (De Maria et al. 2020; *Dirección de Estimaciones Agrícolas* 2020; 2022). It was also big Argentinean firms that incorporated Bolivia and Uruguay into the soy complex, leasing and buying land for GE soy cultivation. This expansion was facilitated by liberalized land markets and fueled by new capital as many of these soy firms became joint-stock companies, receiving infusions of capital from foreign investors and pension funds (Baraibar 2014). In less than a decade soy expanded over 1 Mha in tiny Uruguay as well as Bolivia, becoming by far the most cultivated crop in both countries. In Uruguay, soy mainly replaced mixed farming systems and pastures and led the way for a general land use intensification and "displacement" of family farmers (Baraibar 2014). In Bolivia, soy expanded into two distinct forest areas: the *Chiquitanos* and *Chaco* (Volante et al. 2015). While soy areas in Paraguay tripled from 1.2 Mha in 2000 to 3.4 Mha in 2017; mostly in the eastern region – replacing cotton and the remains of the Atlantic Rainforest. Yet it was in Brazil that soy expanded the most in absolute terms, from 13.4 to 34.2 Mha between 2001 and 2019 (Song et al. 2021). With the most rapid expansion in the Brazilian Amazon and *Cerrado* biomes (Ibid). Currently, about 60% of all soybeans produced in Brazil are in the former *Cerrado* savanna (Reuters 2021).

Besides deforestation and resultant biodiversity loss, soybean production in Brazil is responsible for more than one third of all pesticide use in the country (Peruchi Moretto, Nodari, and Nodari 2022, 26). In Brazil, Argentina, Bolivia and Paraguay, soy is linked to land grabbing and irregular land transfers that have caused social polarization, displacement and urbanization. In Paraguay this pattern has led to violent conflicts. Resistance to the widespread use of agrochemicals emerged in several rural communities in Brazil and Argentina. At the same time, the massive use of fertilizers led to pollution both as the potent greenhouse gas nitrous oxide and eutrophication of waters. We will discuss the emerging resistance movements and counterpoints to the present GE soy model in Chapter 6.

Simultaneously, the expansive "success" of glyphosate-tolerant soy in monocultural forms and over huge areas, created an intense agroecological response when glyphosate resistance in weeds such as amaranthus, lolium, conyza and malva emerged. To combat this, Monsanto and the handful of competitors developed new varieties of soy tolerant to other weed killers – or to both glyphosate and some other herbicide (ISAAA 2019; ISAAA 2022; MN Department of Agriculture 2022).

Besides herbicide tolerant (HT) traits, there are insect resistant (IR) soy traits (mostly for moths and butterflies), as well as "quality enhancing" traits (mostly for higher levels of oleic acid). It is increasingly common with stacked traits, for example, a combination of HT (Glufosinate, 2,4-D, dicamba and/or glyphosate) with IR (ISAAA 2022). In Latin America, a few years before the patent on RR soybeans expired in 2016, Monsanto successfully launched Intacta Roundup Ready 2 Pro soybeans (MON87701 x MON89788 – often referred to as RR2 soybeans). This second-generation RR soybean contains stacked traits of glyphosate HT and IR (Baraibar Norberg 2020a, 136–37; ISAAA 2022). However, just as in the case of herbicide resistance in the weed community, the widespread use of RR2 (and other Lepidopteran insect resistant strains) has resulted in the emergence of resistance among insect communities in Brazil (Vergara-Camus and Kay 2017). The rising resistance to specific pesticides can be seen as a variation of the classical "technological treadmill", which has created a significant challenge to the soy model, as is further discussed in Chapter 6.

Chain drivers in pre-cultivation: global concentration through vertical and horizontal integration

The way the "technological treadmill" is articulated in the soy complex illustrates the fact that the Gene Revolution, in stark contrast to the Green Revolution, is characterized by a relative absence of public R&D (Schenkelaars et al. 2011).

Instead, it is driven by a concentrated group of giant private firms controlling many activities in the chain (Clapp 2016, 78–79).

The strong patent protection on traits, seed genome and other immaterial innovations, allow owners of GE technology far-reaching control over access to and use of their protected products for 20 years, including the overruling of farmers' rights to save their own seeds (Howard 2016, 3:108–10). For example, Monsanto aggressively and successfully filed lawsuits against farmers who used the technology in saved soy seeds, without paying for the patent. By the early 2000s, Monsanto managed to achieve a close to monopoly position in the soybean GE market (Moss 2010) and farmers' traditional practice of saving seeds dropped.[3] Critics argue that GE, in this way, has led to a dramatic losses in farmers' knowledge concerning seed diversity and their ability to adapt to changing climate and social conditions (Guilherme Fowler de Ávila Monteiro 2009; Hubbard 2022).

Presently, GE traits are completely dominated by "the big four" – Bayer, BASF, DuPont Pioneer and Dow AgroSciences, based either in the United States or Europe (Waltz 2015; ISGA 2017; ISAAA 2022); although Argentina recently emerged as a challenger, pioneering with stacked Glyphosate and drought tolerance traits (HB4 x GTS 40-3-2 and DBN 9004). The *alma mater* of biotech soybeans, Monsanto, managed to remain the world's biggest supplier of GE seeds until 2018 when it was acquired by the German agrochemical and biotech company Bayer for USD 62.5B – the largest cash deal on record.[4] However, to satisfy the requirements of the US Justice Department for approval of the deal, Bayer sold off some of its soybean businesses (LibertyLink soy and glufosinate weed killer) to German BASF in an USD 8B deal. Bayer is now the largest seed and agrochemical company in the world, with more than a quarter of the global market share in seeds and pesticides (DeCarlo 2018). The decision to drop the brand name Monsanto was probably due to the company being one of the most condemned companies in the world. The Bayer-Monsanto deal has rightfully received a lot of attention, and while it represents the most spectacular case, it should be seen in the light of a wider movement of big mergers and acquisitions in the global biotech, agrochemical and seeds industries – and the resulting extreme concentration. Biotechnology is also increasingly vertically integrated, where the previously separate input markets of biotech, seeds and pesticides are now under the umbrella of one technological package, leading to extreme degrees of concentration (Moss 2009; Bonny 2017; OECD 2018; Deconinck 2019.

This horizontally and vertically integrated business model has meant that the spread of GE soy traits included the increased use of specific seeds and agrochemicals (in one single package), to be combined with increasingly specialized (and big) machinery and fertilizers to complete the substitution of labor and natural processes with capital and technology (Baraibar Norberg 2022). By 2015, the "Big 6" (Bayer, BASF, Dow, DuPont, Monsanto and Syngenta) controlled around 75% of the agrochemical market and 61% of the seed market (DeCarlo 2018). Just two years later, the "Big 6" had been reduced to the "Big 4" (Bayer; DowDupont, ChemChina-Syngenta, BASF).[5] Over this short time span, Dow and DuPont (the

4th and 5th largest biotechnology and seed companies in the world, respectively) merged into DowDuPont and the Chinese state-owned agrochemical company ChemChina acquired the seed and agrochemical giant Syngenta for USD 43B. This marked a new era with China becoming a global giant in the field of agrochemicals and seeds. In 2021, China grew further and completed the merger of ChemChina and Sinochem Group, forming the world's largest agrochemical producer, operating in more than 150 countries, along with 220,000 staff (Simung 2021).

International soybean trade: whole soybeans, meal and oil

The political economy of soy trade

By the 1940s, the United States had an excess of soybeans and many other crops compared to the domestic demand, which was a result of both the Depression-era stimulus policies and the high European demand during the World Wars. To cope with the excess export capacity, an extensive institutional framework was established, aimed at promoting agricultural exports and enhancing the competitiveness of US agriculture in the global marketplace. For example, in 1953, the USDA established the Foreign Agricultural Service (FAS) for export promotion. Additionally, in 1954, Public Law 480 (PL 480) was enacted and later amended with the Food for Peace Act of 1966 to address the surplus of crops, including soybeans, by allowing the US government to purchase them and sell them at low prices to poor and indebted nations as a form of development aid. The American Soybean Association (ASA) played a significant role in shaping these policies, which not only helped to maintain prices for American farmers but also became a crucial instrument of security policy to manage the "Third World" and prevent the spread of communism. By using surplus soybeans as aid, they were given an additional role: food aid for political sway (Lappé 1980).

It is crucial to bear in mind that during this period, agricultural trade was characterized by significant national protectionism, with most countries implementing some form of restriction, such as strict quantitative quotas, variable levies or licensing requirements. The protectionist policies that emerged after the Great Depression continued throughout the Bretton Woods era, which lasted until 1973. In fact, agriculture was excluded from the first four rounds of the General Agreement on Tariffs and Trade (GATT) and agrofood markets remained heavily regulated and restricted until the establishment of the World Trade Organization (WTO) in 1995. (WTO 1995; 1996). Unlike most other agricultural commodities, however, soybeans were traded with fewer restrictions during this period. The Dillon Round of GATT in 1960-61 provided tariff-free access to the European market for US soybeans and other feed crops, but at the cost of accepting the high levels of European protection on wheat (Secretary of State for Economic Affairs 1970). This "exceptionalism" of soybeans was mainly supported by European meat and dairy farmers who increasingly replaced their locally grown fodder with low-cost soy from the US Midwest. Outside of Europe, Japan was an important buyer of US soy. In tandem with Japan's postwar political and economic dependence on the United States, the country became the single largest customer of US soybeans and soybean products.

In 1960, even though the United States accounted for "only" 57% of global soy production, it contributed almost 90% of all soy exports, including beans, oil and meal (Cartter and Hartwig 1962). Thus, in the postwar international food order, the United States mantled a hegemonic role and used its position to find new markets overseas for its "surplus" products and where soy received exceptional treatment and was traded freely.

As mentioned earlier, the second largest soy producer at this time was China (including Manchuria again), but the cradle of the soybean consumed most of its production domestically. Chinese soybean exports averaged about 0.9 MMT a year 1955–58, representing only about 11% of its production. In the lean years after Mao´s Great Leap Forward (1958–62), soybean production and exports fell sharply (FAO 1958; 1959; 1960). In 1974, China became a net importer of soybeans for the first time in history, with 0.28 MMT imported (FAOSTAT 2022). This was due in part to the cost savings of importing foreign soybeans to supply southern consumption, rather than transporting food inland across China, and instead selling northern-grown soybeans to nearby Japan.

Meanwhile, the USDA launched a significant "export push" after the first oil crisis in 1973, with soybean exports accounting for approximately USD 3B, or about 5% of all exports. Increasing soybean exports was viewed as crucial for maintaining the US trade balance in the face of skyrocketing petroleum prices. Unexpected market events enabled soybeans to further enhance US trade revenues, particularly the Russian "Grain Robbery" of 1972 – when the Soviet Union quietly purchased over one-fourth of US wheat harvests to increase their own livestock production. In combination with grain production shortages in 1973, wheat prices tripled and corn and soy prices doubled (Peters, Langley, and Westcott 2009). Further, the simultaneous rupture of the Bretton Woods system and establishment of a floating exchange rate regime enhanced access to international markets, further accelerating soy prices and trade. Worried about the domestic meat industry in need of cheap feed, US President Nixon imposed an export embargo in 1973 to keep soybean prices low. It quickly cut domestic soy feed prices in half, but also irked importers and the embargo was quickly reversed and the "free trade" discourse returned. The size of the US domestic economy usually gave it considerable leverage to induce other states to enter into negotiations for market access and generally liberalize trade (Silver and Arrighi 2003). However, the experience of Nixon's embargo encouraged Japan to choose the free market to diversify its own soy suppliers and it began direct cooperation with Brazil, financing research on seed adaptation to secure ties (Du Bois 2018, 18). Soy continued to boom in the United States, and by 1979, soybeans and soybean products were agricultural export superstars, netting USD 7.5B. Soy was now the top US agricultural export earner and the second overall behind aircraft (Shurtleff and Aoyagi 2004).

> ASA played a very active role during the recently concluded Multilateral Trade Negotiations (MTN). Its officers and staff met with key persons in government and industry to preserve soybean and soybean product trade benefits. Zero bindings on soybeans and soybean meal were maintained in the European Community and zero bindings on soybeans were obtained from Japan. ASA also actively supported and helped develop legislation designed to expand U.S. agricultural exports. The Agricultural Trade Expansion Act of 1978 as approved will create new trade offices, upgrade the status of many agricultural attaches and provide for CCC credit for the People's Republic of China. ASA has successfully fought efforts to set higher soybean loan levels, establish target prices, restrict soybean acreage, set up government monitoring boards, allow soybeans on setaside acres and many other issues that restrict free trade for soybeans. [...] Through its market development, research, government relations and information activities, ASA is providing programs which have made soybeans the number one U.S. cash crop and number one U.S. export. These program efforts continue to have the single goal of building and maintaining profitability in soybean production. The future looks extremely bright. Soybeans are truly the 'gold that grows.'
>
> (Smith 1979 in Shurtleff and Aoyagi 2021, 1089)

Again, it was a combination of the US state and powerful actors like the American Soybean Association (ASA) who worked proactively and long term to continuously expand soy exports. Keith Smith, the Director of Research at ASA, made it clear in the above 1979 excerpt (from the book *50 Years with Soybeans* by the National Soybean Crop Improvement Council) how extremely proactive the ASA was in efforts to increase US soy exports and that it did not work alone. The soy export strategy was backed by the state in several ways, not least in the Agricultural Trade Expansion Act and the CCC (the Commodity Credit Corporation, a wholly owned government corporation which funded farm income and price supports and conservation). Following diplomatic recognition of China in 1978, ASA traveled to China to nurture potentially huge markets in hog and poultry industries growing to feed over 1 billion people. ASA opened an office in Beijing in 1982 and engaged in programs to stimulate Chinese imports of US soybeans and to spread the model of industrial meat production with feed concentrates including soy (Du Bois 2018, 12).

In the 1970s, many big agribusiness companies with US headquarters, particularly traders, began to outgrow the state-centered national model and also lobbied hard for international deregulation of agricultural markets (McMichael 2009). While they had benefited from domestic agricultural policies in the post-war period, particularly US subsidies to farmers, they now shifted focus to a business model that sourced agricultural commodities wherever they were cheapest. As a part of their globalization strategies, traders started acquiring export-related infrastructure, such as silos, elevators, port terminals, transportation and large processing plants, in addition to purchasing and trading grains. As mentioned in the previous section, Latin America adopted liberalization policies that included export promotion and measures to attract foreign direct investment, often through the provision of available land (Baraibar Norberg 2020e, 165–69, 193–96).

The resulting spectacular growth of soybean exports from Latin America – particularly Brazil – brought increased competition for US farmers and downward price pressure. Accordingly, US producers and their powerful organizations asked for government support, arguing, among other things, that they were disadvantaged by the strong US dollar which made it cheaper for the large traders, such as Cargill, Bunge and ADM, to ship by sea from Brazil or Argentina than to use the railway from Iowa (Shurtleff and Aoyagi 2009, 242, 350, 351). Since farmers had the government's ear, the United States increasingly combined policies of farm support and protectionism at home, while insisting on free trade internationally (Silver and Arrighi 2003). Backed by more government support, US soy production and exports continued to increase – but soy exports from Latin America also rose – with transnational traders as both promoters and facilitators.

The establishment of the WTO in 1995 and its "Agreement on Agriculture" marked a significant shift in the political and economic governance of food production, security and trade. This change moved away from the "national developmentalist" model adopted to varying degrees worldwide after World War II to a global agrofood system that is increasingly liberalized, deregulated and financialized (Stiglitz and Foundation CIDOB and Policy Dialogue 2004). Although countries in the South removed trade barriers on agricultural products, states in the North devised new, more subtle methods to continue supporting domestic farmers and protect their markets (Clapp 2006, 567; 2016).

China long represented an important exception to liberalization with its high self-sufficiency targets in its national food security plans (He 2016). Outside of quota restrictions, China's import tariff on soy was 180% (da Silva et al. 2017, 5). However, after several poor soy harvests in the mid-1990s, the country decided to increase its soy imports. Within only a few years, import volumes skyrocketed – increasing 86 times between 1995 and 1999 (4.3 MMT) (Ibid). After becoming a member of WTO in 2001, restrictions were further relaxed, tariffs were lowered and soy imports rose further (Zhang and Liu 2014; MacDonald et al. 2015, 275; Clever and Xinping 2016; Hairong, Yiyuan, and Bun 2016, 374). These regulative shifts interacted with population and economic growth leading to an unprecedented rise in soy imports. In 2011, China's soybean consumption reached 70 MMT, but its domestic production only amounted to 14 MMT. As a result, the country had to import 56 MMT of soybeans from foreign fields, primarily in the Americas. However, less than a decade later, in 2020, China's soybean imports had risen to 103 MMT, representing 62% of all imported soybeans worldwide (167 MMT) (FAOSTAT 2022). While no other country compares to the outstanding position of China as the main buyer of soybeans, several other nations in Asia, Russia and the Middle East have also become increasingly dependent on foreign agroecosystems for cheap protein to feed their animals.

While most agricultural products in the world are still consumed locally, every other soybean produced in 2019 was actually traded on the international market. Soybeans and soy products are among the most traded agricultural commodities in the world. While almost all soybeans end up crushed for oil and meal

– oscillating between 80 and 96% since the 1970s (Shurtleff and Aoyagi 2009, 263) – there is significant trade with whole beans, since crushing is often done in the country of import, as we shall see.

Trade with whole soybeans

Figure 5.7a shows that exports of whole beans grew steadily from 1950 until the acceleration in the late 1990s when Latin America significantly augmented the market. In comparison, average world soybean consumption per capita

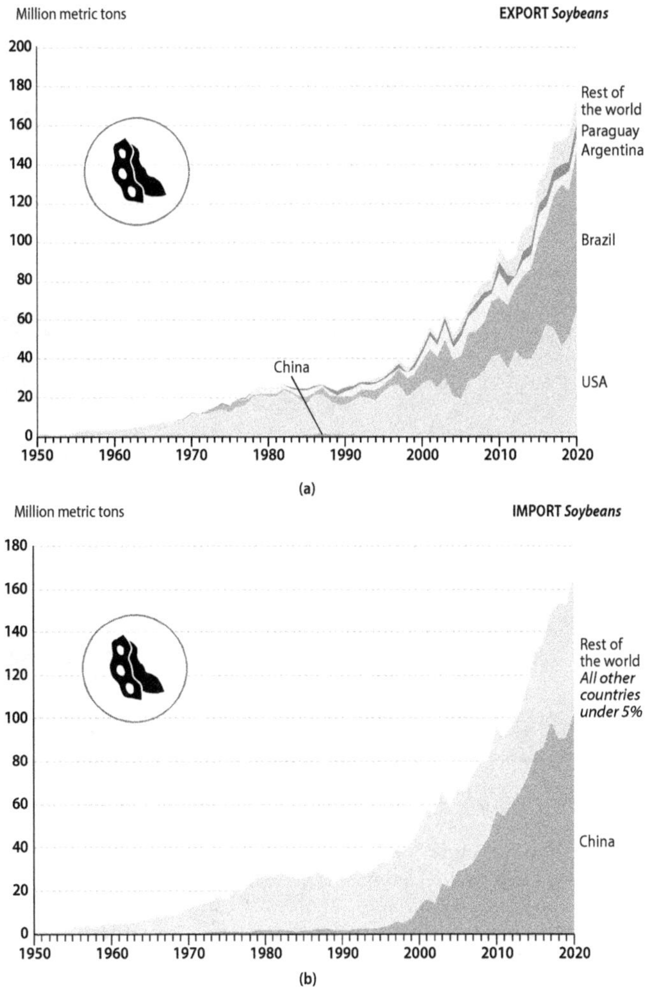

Figure 5.7 World trade volumes of whole soybeans 1950–2020 by top countries: (a) exports and (b) imports.

Source: FAO (1951, 68; 1952, 98; 1953, 144; 1954, 132; 1955, 109; 1957, 170–71; 1958, 171–72; 1959, 182–83; 1960, 155–56; 1961, 196–97; 1962, 196–97; 1963, 288–89), Comtrade UNSD (2022), and FAOSTAT (2022).

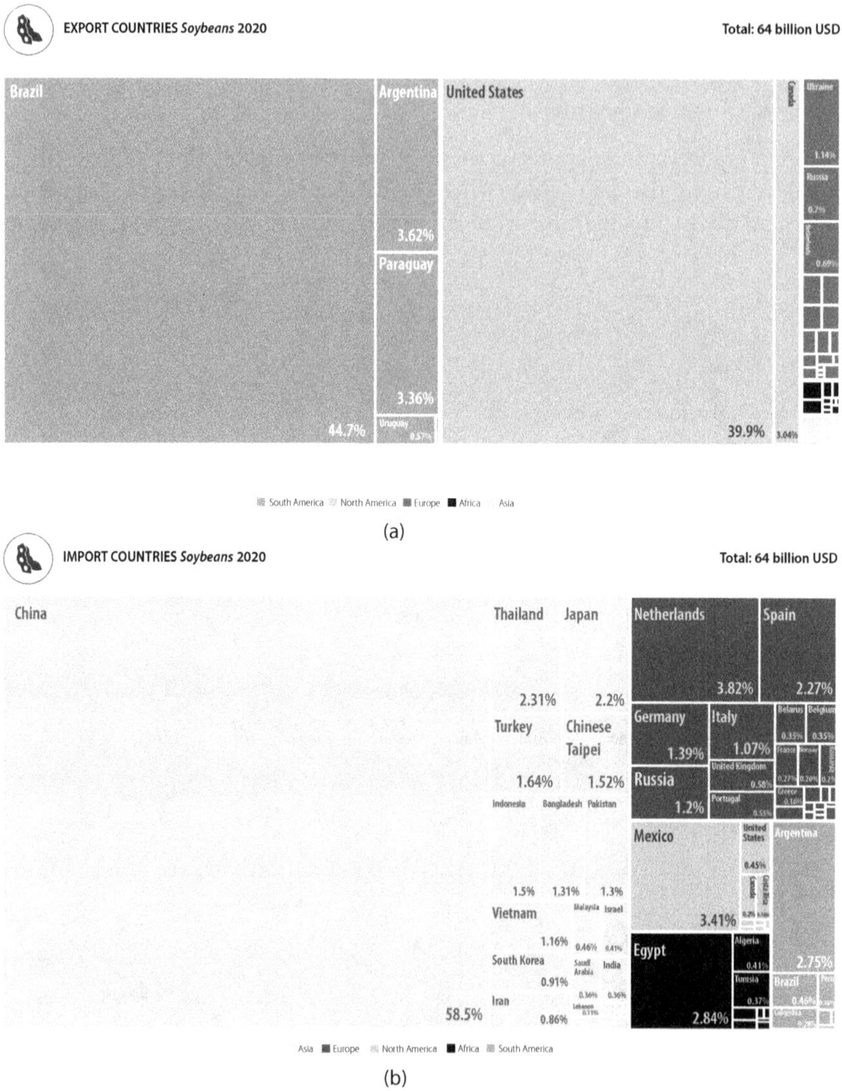

Figure 5.8 Top export and import nations of whole soybeans in 2020 by value. (a) Brazil and the United States clearly dominate bean exports with 85% of total export values. (b) China dominates imports with 60% of world bean imports. The next largest importer is Netherlands with 4% – all other nations import less than 4% of total imports.

Source: OEC (2022).

was 19.5 kilos (kg) in 1990 and grew over 40% to 28 kg, in 2000, and by 2020 consumption was 37 kg – almost doubling in 30 years (USDA PSD 2022). It is also clear from Figure 5.7a that the United States clearly dominated exports throughout most of the regime. However, Brazil has recently equaled and just surpassed US volumes. In 2020, whole soybeans were the world's 29th most traded product, with a total trade value of USD 64B (billion) representing 0.38% of total world trade Figure 5.8. In 2020, Brazil was the world's largest exporter of whole soybeans (USD 28.6B), closely followed by the United States (25.6B) (Figure 5.8a). While the most important exporters of whole soybeans have been countries in the Americas throughout this regime (Figure 5.7a), imports have, as already discussed, been extremely concentrated to China since the mid-2000s (volumes in Figure 5.7b) and (USD 37.4B in 2020 in Figure 5.8b). The EU has been the second largest buyer for many years with the Netherlands alone buying for 2.44B in 2020, followed by Mexico (2.19B) (Figure 5.8b). Today Thailand, Egypt and Japan are also important buyers, but the exact order between them fluctuates (Figure 5.8b). Note that Argentina is not only an important exporter of whole soybeans (Figures 5.7a and 5.8a), but also importer in 2020 (1.76B in Figure 5.8b). This is a consequence of Argentina's large crushing capacity.

It is hard to overstate the importance of China in the international soy market. First of all, its rapidly increasing demand has kept soy prices up despite of increases in supply and driven the fast soybean expansion. China's appetite for whole soybeans is the main driver of the dramatic soybean expansion in Latin America during recent decades (Hairong, Yiyuan, and Bun 2016; CAS 2018). The level of specialization in soybeans in the export structure is very high for Paraguay, Brazil, Argentina and Uruguay, where the Chinese-driven soy commodity boom enabled economic growth, high export earnings and, in countries that managed to tax exports (Argentina), it enabled socio-distributive policies. As China became South America's most important trading partner, however, there has been a (re)primarization of the export baskets in the region (Baraibar Norberg 2020d, 211; De Maria et al. 2020). This was partly due to China's overall development strategy of importing primary commodities, while "adding value" at home. Yet, these commodities – including soybeans – are natural resource intensive (e.g. measured in GHG, land or water) and also prone to social conflicts (Ray and Gallagher 2016).

Trade with soy oil and meal

The soybean oil market is much smaller than that of whole beans and meal, but this is hardly surprising since one unit of soybeans produces only about 18% oil and 80% soybean meal. Export trade in oil was dominated by the United States until the 1970s when world volumes boomed and other exporters took shares of the growing market. By the late 1990s Argentina dominated the market and continues today (Figure 5.9a). No one country dominated oil imports until the mid-2010s when India emerged as a major buyer of over a quarter of volumes (Figure 5.9b). In 2020, trade with soy oil had a total value of USD 10.4B (Figure 5.10) – around one sixth the value of whole soybeans (Figure 5.8) and less than half of soybean meal (Figure 5.12) (OEC 2022). The top five exporters of soybean oil in 2020 were (in USD): Argentina (3.9B), the United States (1.03B), Brazil (767M), the Netherlands (545M) and Russia (444M) (Figure 5.10a). The import market of

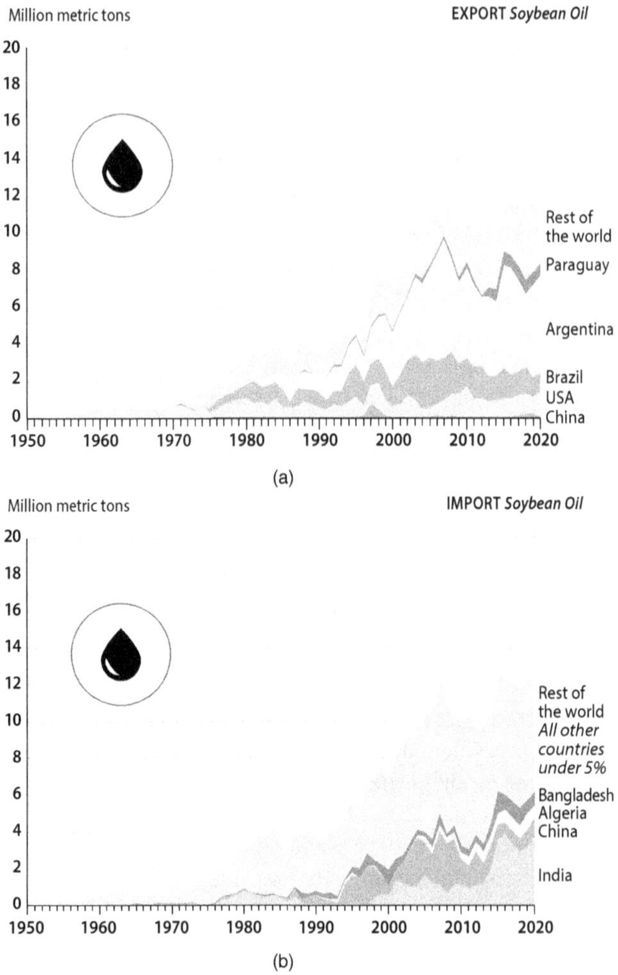

Figure 5.9 World trade volumes of soybean oil 1950–2020 by top countries: (a) exports and (b) imports.

Source: FAO (1951, 91–92; 1952, 119; 1953, 136; 1954, 157; 1955, 133; 1957, 221; 1958, 212; 1959, 223; 1960, 196–97; 1961, 236–37; 1962, 236–37; 1963, 334) and FAOSTAT (2022).

soy oil is less concentrated than other soy markets, but Asia claims half of all oil imports, with India importing almost 30% and China 8% (Figure 5.10b). In 2018, the average tariff for Soybean oil was 12.6%, but the variance between countries is huge. Many countries have high import tariffs on oil and meal in order to incentivize domestic crushing (OEC 2022).

In Figure 5.11a, we see that soybean meal export volumes were dominated by the United States until the 1980s and since then US export volumes have not grown as fast as other nations. Brazil who entered the market in earnest with

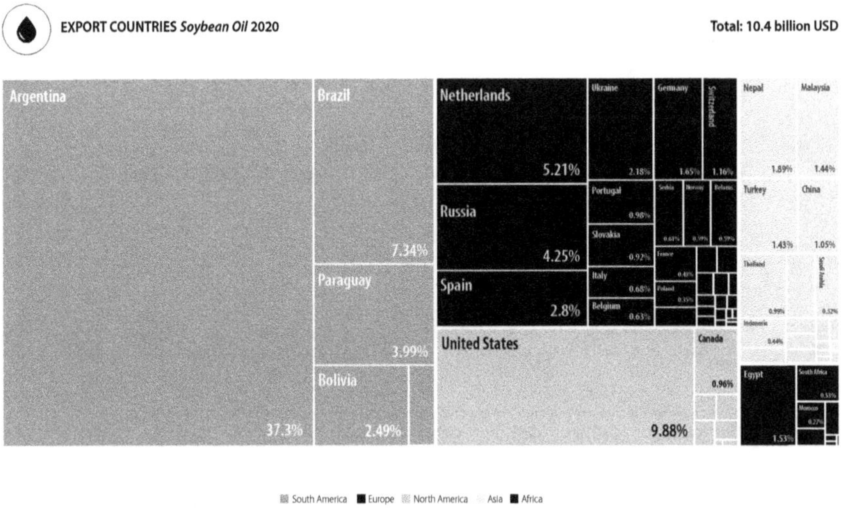

(a)

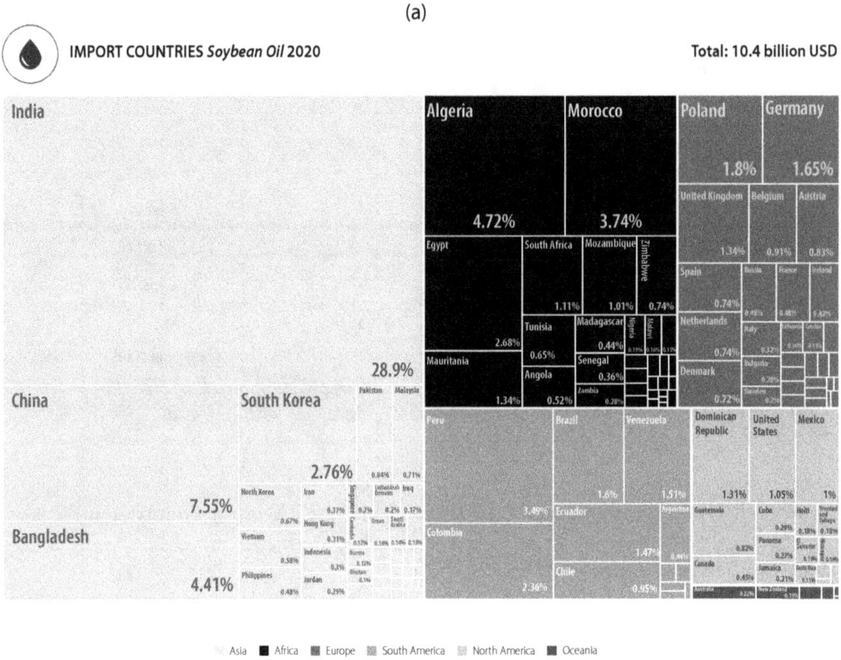

(b)

Figure 5.10 Top export and import nations of soybean oil in 2020 by value: (a) Argentina exports 37% of all oil and together with Brazil and the US exports more than half of all oil, and (b) Asia claims half of all oil imports, with India importing almost 30% and China 8%.

Source: OEC (2022).

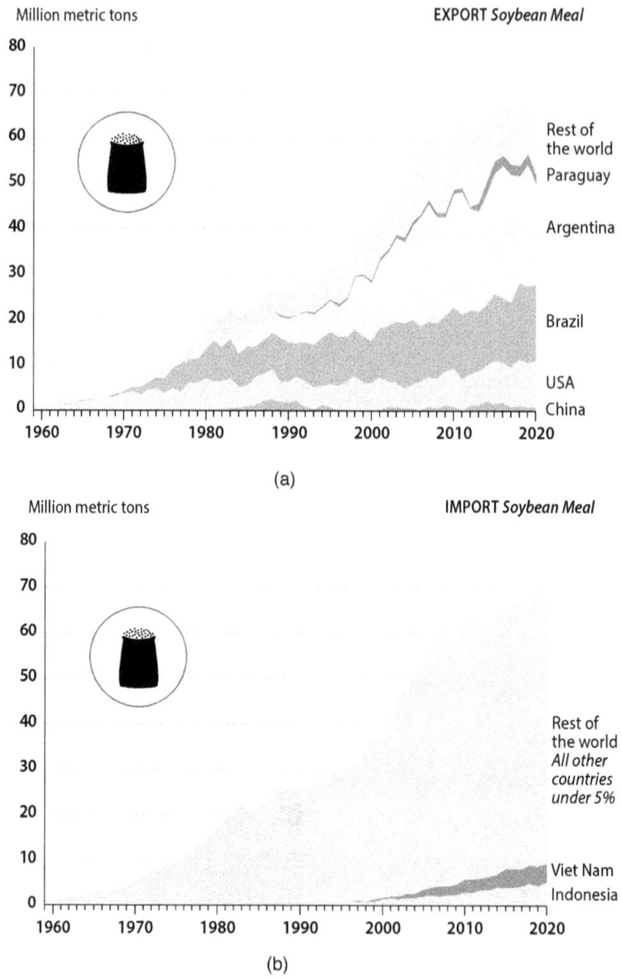

Figure 5.11 World trade volumes of soybean meal 1959–2020 by top countries: (a) exports, (b) imports.

Source: FAO (1964, 270) and FAOSTAT (2022).

volumes in the 1970s had equaled the United States by the 1980s and has since approximately doubled volumes. While Argentina already exported in the 1980s, they really increased rapidly in the late 1990s and dominate today. In contrast, soybean meal imports are not dominated by any one buyer (Figure 5.11b). Today, Indonesia and Vietnam are the two largest buyers, but only with a combined volume of less than 15%.

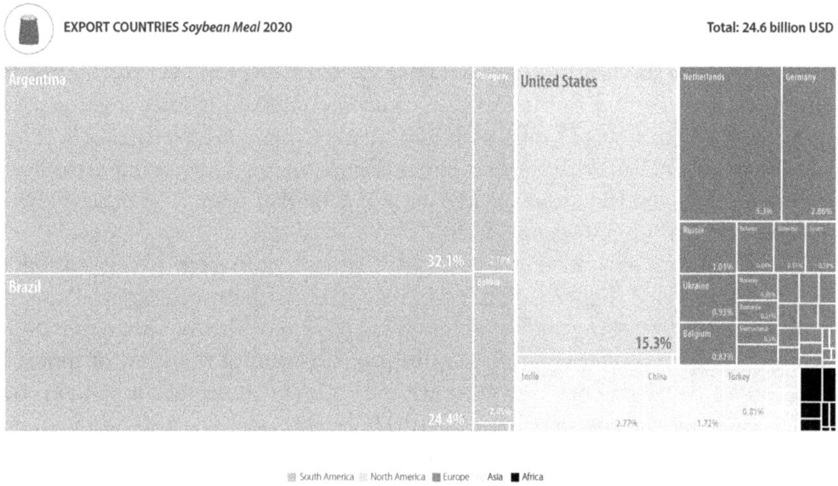

(a)

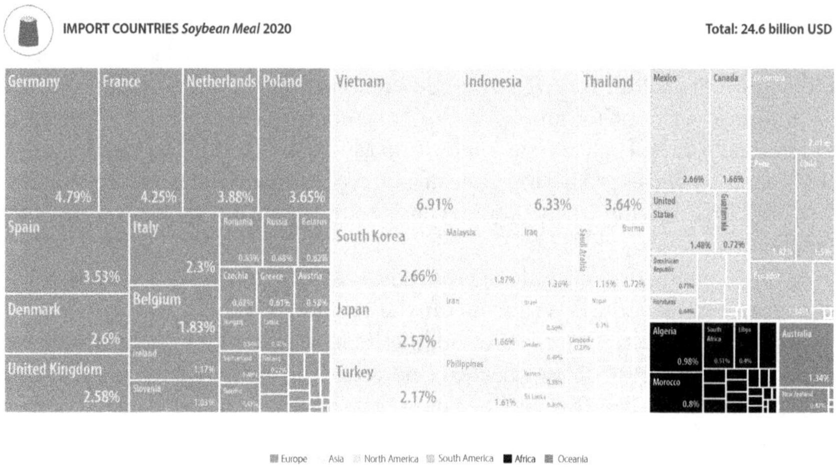

(b)

Figure 5.12 Top export and import nations of soybean meal in 2020 by value. (a) Argentina alone exports one-third of all meal (32%) and Brazil one-quarter (24%). Together with the United States (15%) these 3 nations provide almost 75% of all exports, (b) importers of meal are many and only Indonesia and Vietnam buy more than 5% of total meal imports.

Source: OEC (2022).

In 2020, Soybean meal was traded for USD 24.6B (Figure 5.12). In 2018, the average tariff for Soybean meal was 3%, which is relatively low, but it conceals large variations with several countries adopting high import tariffs, around 40%, to support domestic crushing (OEC 2022). In 2020, the top five exporters were Argentina (7.89B) and Brazil (5.99B) – with a combined market share of over 56%, the United States (3.76B) with 15% market share and Netherlands (1.3B) and Germany (703M) with 8% combined (Figure 5.12a). The top importers were Vietnam (1.7B) and Indonesia (1.56B) with 13% market share – no other nations imported more than 5% (Figure 5.12b).

When soybeans expanded in Brazil and Argentina during the 1970s and 1980s, governments levied higher export taxes on soybeans than meal and oil to promote domestic processing and avoid becoming mere providers of raw commodities (leaning on Latin American structuralist-inspired ideas of the need of industrialization for development). Consequently, the ABCD companies invested heavily in crushing capacity and both countries became global giants of meal and oil supply. As the meal industry matured, Brazil increased meal export taxes to capture more valued added by developing their own soy-based pork and chicken industries (Shurtleff and Aoyagi 2009, 173, 234). These measures were partly successful as an increasing share of soybean meal stayed in the country, expanding and intensifying animal farming (Peine 2013; Turzi 2016; Paolera, Amorocho, and Musacchio 2018). In Argentina, differentiated export taxes made soybean oil and meal account for around 80% of the country's soy exports (Castellano and Goizueta 2011; Paolera, Amorocho, and Musacchio 2018). Targeted blends of biodiesel and policies promoting intensification of the meat industry have resulted in more beans staying in the country and being crushed into oil and meal for export.

As the combination of trade figures indicates, China mainly demands whole soybeans. Significantly lower tariffs on the import of whole soybeans than on meal and oil protected more value-adding domestic activities, but still allowed imports (Ward 2018). Tariffs on meat and other high value-added products are even greater, thus China has supplied most of its poultry and pork demand domestically (USDA FAS 2022). This differentiated tariff system has had an important impact on international price relations between beans, meal and oil and created tensions with Argentina and Brazil (the world's largest and second largest soybean meal exporters). In 2019, however, following decades of negotiations, Argentina won long-sought Chinese approval to sell soymeal. According to several analysts, one reason for China's radical change of direction was deteriorating US trade relations during the Trump Administration. To avoid US soybeans, China increased its reliance on Latin America.[6]

At the same time, it is worth mentioning that exporting meal and oil instead of whole beans does not provide nearly as much value-added as to export soy-fed meat or cheese as both the United States and Brazil do. By using a large share of their domestic soy production to feed hogs, chickens and dairy cows and manufacture margarine and biodiesel, more technology, labor and capital are incorporated in the complex, enhancing backward and forward linkages (Lapitz,

Evia, and Gudynas 2004, 33). In this way, although Argentina has had explicit policies to use soy as a springboard for moving up the value ladder, it has so far only managed to climb the few first rungs (Baraibar Norberg 2020d). Argentina is not alone among the natural resource-rich countries of the Global South failing to "upgrade" its export structure. One reason is that within the international trade system, soy and other primary products are traded "freely", i.e. face almost no restrictions, compared to higher value agricultural products where most countries have high import tariffs outside of restricted quotas, e.g. on meat (Lapitz, Evia and Gudynas 2004, 69–70; Baraibar Norberg 2020a, 126–28). In recent years, however, China has lowered beef import tariffs and raised import quotas, leading to an increase in the number of soy-fed animals being raised overseas. In this way, many Latin American countries now sell soy both in the forms of "meat to be" (i.e. feed) and meat.

As we saw in Figures 5.11 and 5.12 of soymeal trade, many European countries are buyers and, as a bloc, the EU is the world's biggest soybean meal importer, reflecting the fact that soymeal is the primary source of protein for EU animal feed – providing more than 60% of vegetable protein inputs (European Parliament 2015). There are, however, hardly any imports of GE soybeans for food on the EU market – probably due to the Union's compulsory labeling of food containing GE. Consumers' aversion to GE food does not, however, translate into the soybean meal market, since animals can be fed with GM soybeans without meat, eggs or milk being labeled as GE-derived products. While imported GE soy is allowed in the EU, the approval process is longer and stricter, including a complex procedure of risk assessment. For example, Monsanto applied in 2010 for approval of the stacked traits (MON 87769 × MON 89788), and it took 12 years before the European Commission granted approval.[7] The biotech industry and the International Soybean Growers Alliance (ISGA) have criticized EU's long approval process as "unsound non-tariff barriers to trade" (ISGA 2017). In fact, Europe's livestock and feed manufacturing industries have also lobbied for EU to authorize new GE soybean events for import more swiftly, since they depend heavily on them. NGOs, the media and the European Parliament, however, have expressed criticism about the Union being too GE-friendly.

Chain drivers in commercialization: vertical and horizontal integration of traders

As we have seen, large quantities of soybeans, oil and meal are traded on world markets. While official production and trade statistics remain organized according to countries, the main economic agents behind rising soybean trade were (and are) big transnational corporations (Murphy, Burch, and Clapp 2012; Clapp 2015; Gordon et al. 2017; Baraibar Norberg 2020a) – just like as is the case for the large producers of inputs in the pre-cultivation stage of the soybean commodity chain. The handful of the biggest traders – Archer Daniels Midland (ADM), Bunge, Cargill and Dreyfus (LDC) – collectively known as the ABCDs because of the coincidence of their initials – have been in the international arena since the late 19th century (see Chapter 4). The ABCD corporations have controlled the majority of global grain and oilseed trade flows since the early 1900s. As large agrochemical-biotech companies hold great power through their technologies, the giant traders govern of their parts of the chain by controlling and coordinating market information. Information and the mechanisms for delivering it are the unifying force that holds together the structure of business in increasingly distant and complex commodity chains (Gereffi 1999). Traders link together key activities across several stages in the commodity chain both upstream and downstream from trading (please review Figure 5.1 again). They have vertically integrated to include these additional activities: local export infrastructure (e.g. storage and ports); financial trading as well as physical trading; processing (e.g. crushing and mills); and intermediate consumption (e.g. own animal production facilities and biofuel processing). The traders have also horizontally integrated within stages by buying up local companies to consolidate power and take advantage of global economies of scale and diversify risks. Starting upstream here are some examples of their strategies.

As we saw in Chapter 4, investment in local export infrastructure started early amongst the traders – in the 1940s Cargill and Bunge owned physical facilities a river port and rail terminal in the United States, respectively. However, there was a wave of mergers and acquisitions in the 1970s, when traders began investing heavily in local export infrastructure across the world. Moreover, in 2009, Cargill became the first Western company to wholly own a port in China when Grain & Oilseed Supply Chain (GOSC), China & Korea completed negotiations to acquire *Yangjiang* Port (ADM 2022; Bunge 2022; Cargill 2022; LDC 2022).

Although traders began in physical trading, they are deeply involved in financial trading as well. In the early 2000s, the four ABCD mega-firms controlled around 75% of soy exports (de LT Oliveira and Hecht 2017) and around 70% of the world trade of corn, wheat and soy taken together (Clapp 2016, 105). Since then, a new giant has been added to the "big four" China National Cereals, Oils and Foodstuffs Corporation (COFCO) (Wesz Jr 2016). COFCO is one among many illustrative indicators of China's resurgence as a great food power and changed global strategy. This state-owned trading company is rapidly increasing its market share. COFCO restructured in the 1990s into an internationally operating entity, but it is only in recent years that it started to invest heavily abroad. The company has a global coverage of grain and oil producing areas and owns a sophisticated global production and procurement platform and trade network. In a few years, it went from nothing to earning more than 50% of its operating income from overseas business.[8] In 2018, COFCO passed Cargill in Paraguay as the most important soybean trader, exporting 8.27 MMT of soybeans, which represents 15% of total soybean exports.[9] It has moved significantly into infrastructure, such as storage, processing, transportation and port facilities, creating a whole value chain that stretches around the world (Gaudreau 2015). In 2014, COFCO bought two "classic" agribusiness firms, Nidera and Noble Agri, and then in 2018 it sold the seed business part of Nidera to Syngenta, which was bought by another Chinese state-owned enterprise, ChemChina – the previously mentioned upcoming giant in the international agrochemical and seed business.

Traders are also increasingly involved in financial activities. This trend was strengthened with deregulation in the 1990s, when rules surrounding position limits and financial intermediaries were relaxed (Clapp 2014, 797–99). Moreover, dispute settlement mechanisms protecting private firms vis-à-vis states in third countries have further spurred trade with both physical commodities and with futures contracts – but without engaging in a physical transaction (Chang 2009, 480; Wolford et al. 2013, 2; Clapp 2016, 64–67). Financial investors, such as pension funds and other institutional investors, have substantially increased their participation in commodity futures markets since the early 2000s, leading to a large inflow of investment capital into the agrofood system (Irwin, Sanders, and Yan 2022). The turnover of commodity futures contracts traded in organized exchanges has increased exponentially since 2005; from around 800 million contracts per year in 2005 to 8,000 million contracts in 2019 and hitting new records in 2021, surpassing 29.28 billion contracts according to the Futures Industry Association (FIA).[10] For many traders, financial instruments offer higher profits than the physical activities (Howard 2016, 74). Soybeans, meal and oil are among the top most traded items on future and option markets. Traders make money by buying soybeans at one discount rate and selling them at another. The availability of soybean, soybean meal and soybean oil futures allow processors/traders to profit on changing price relations between soybeans and processed oil and meal. The "crush spread" is the term used to denote the difference between the value of soybeans and its byproducts. Financialization of soy trade has, in this way, expanded in terms of both scope and depth. The big transnational traders use their information advantage to take positions in the commodity market, in particular, in commodity index funds such as trade with futures contracts based on speculation of price movements. As these companies are involved in many stages, they can profit from taking positions on expected price swings in all scenarios (Clapp 2014, 797–99; 2016, 18, 133–35; Gonzaga Belluzo 2015). As Reuters recently concluded about the war in Ukraine: "Supply chain middlemen like ADM and Bunge thrive when crises such as droughts or war trigger shortages in parts of the world".[11] Traders now not only control and coordinate market information – but also access to finance.

The Dalian Commodity Exchange traded the most soybean meal futures contracts in 2018, with over 238 million trades (see Figure 5.13). As significant amounts of financial capital entered the soybean market, price became slightly detached from shifts in supply and demand, although there is no conclusive evidence of longer-term systemic effects on volatility (OECD/FAO 2011, 18; USDA 2022). The political economist and critical agrofood scholar, Jennifer Clapp argues that the financialization converts agrarian products into highly complex agricultural commodity derivatives traded on the futures exchange markets, which implies an increased distancing of food from its biophysical environment of production and obscures the economic relations at different stages along the

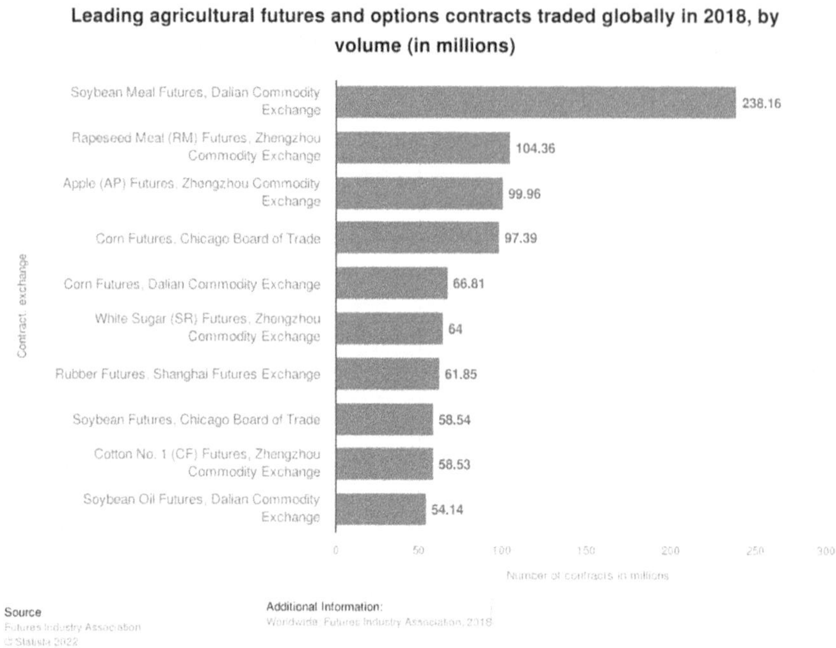

Figure 5.13 Leading agricultural futures and options contracts traded worldwide in 2018 by volume.
Source: Statista (2022).

commodity/value chains (Clapp 2014; 2015; 2016). Financialization of soybeans added to the increasing "abstraction" of food, disembedding it from where it was produced and strengthening the influence of financial actors over others in the food system. The economic sociologist Peter Gibbon suggests that rather than entirely buyer-driven, agrofood chains are strongly trader-driven, since the large trading houses (ADM, Cargill, Dreyfus and Bunge) often play the lead (nodal) firm role (Gibbon 2001).[12]

The crushing industry was actually the first activity that traders expanded into (see Chapter 4) – it was downstream to trading if trade was in whole soybeans, but later become upstream when sales were of oil and meal. Mid 1980s, in the United States alone Bunge had 15 crushing plants as well as five oil refineries and plants producing soy flour, soy flakes and textured soy protein (Shurtleff and Aoyagi 2020a). In addition to processing, traders have increasingly moved further downstream into a whole range of market segments that can be considered intermediate consumption from food oils, into animal nutrition, biofuels and even meat packaging making the chain into a more complex web of activities (ADM 2022; Bunge 2022; Cargill 2022; LDC 2022).

Traders have also horizontally integrated as they expanded internationally through strategic acquisition of local firms in grain storage, transportation, port facilities and export and trading operations across the Americas, Europe and Asia (Kneen 2002; Lieber 2002; Traders of Dreyfus 2008; Dalla Costa and Silva 2018; Shurtleff and Aoyagi 2020a; 2020b). Already in 1973, the soy commodity chain was dominated by a handful of mega-actors, with Cargill as the largest US soy processor with an 18% market share, followed by Archer Daniels Midland (ADM) with 17% (Shurtleff and Aoyagi 2020b). ADM and Cargill continued alternating between first and second place. By 1984, Bunge had completed enough acquisitions to add six overseas crushing plants to those in the United States and make it the largest soybean crusher in the world.

As shown in Chapter 4, powerful agribusiness firms, such as Cargill and ADM, played important roles in growing the roots of the soy success in the United States in the present regime. The outstanding flexibility and geographical adaptability of the soybean, together with globalization strategies of these agribusiness firms coupled well with the decreasing costs and increased capacities for storage and container shipping to transform soy into the ultimate commodity; a standardized product that could be sourced from anywhere (i.e. nowhere) for the handful of transnational grain traders to make profits on arbitrage. As the quote that began this chapter indicates, Cargill, the largest privately held corporation in the United States, has been actively working as far downstream in the chain as foreign meat producers to create overseas demand for soy production – thereby influencing the choices of producers far upstream for a particular crop, soybeans, in feed concentrates for animal feed in China. Traders such as Cargill have been some of the biggest winners of rising international agrofood trade. Since *Forbes* begun tracking the largest private companies in the United States in the mid-1980s, Cargill has held the number one position several times – including the last two years, with revenues of USD 165B in 2022 (Forbes 2022).

Soymeal – the ultimate commodity to feed the "US Soy-Meat Complex"

After oil is removed from soybeans, the remaining defatted flakes are toasted, dried and ground to be used as a protein source for livestock, pets, human food and industrial uses. There are different estimates, but according to most sources almost 80% of the world's soybean crop is processed into meal (and 7% of beans are fed whole directly). Virtually all soymeal is used for animal feed – mostly for poultry (49%), then hogs (26%), 7% for aquaculture, only 3% for beef and dairy production and the rest feeds other livestock and pets (Ritchie and Roser 2021). Soymeal consists of 40% protein, enabling animals to grow and fatten rapidly. In combination with its low price, soybean meal became the most common animal feed input the world (De Maria et al. 2020). Soy has in this way played a pivotal role in making chicken, pork and beef, as well as milk, eggs and aquaculture products cheaper – allowing global meat consumption to almost quadruple over the past 50 years.

The contemporary dependence on soymeal in animal production represents a clear break with traditional mixed farming systems typically relatively self-sufficient in feed (mostly on-farm legumes and grains), that dominated both in Europe and in the United States up until World War II, as shown Chapter 4. But, with increasing amounts of cheap byproducts from crushing soybeans for oil, in combination with many years of both public and private research and experimentation in animal nutrition to maximize animal growth at the lowest cost, soymeal became an increasingly central ingredient in the intensification of broiler and hog production in the United States (Turzi 2016, 7–8). While consumption of chicken was relatively low in the late 1940s, it became the most common item on the US meat plate by the mid-1990s, and with it, soymeal demand rose. Although maize was also an important component in feed, soymeal was the preferred source of protein for all types of poultry. Accordingly, big processors and traders such as Cargill and ADM, with access to cheap soy and corn, entered the broiler industry in the 1960s through a long list of acquisitions of companies producing a broad line of animal feeds and other agricultural and nutritional products (Cargill 2018; 2022). The hog industry followed a similar path, becoming increasingly industrialized, large-scale and specialized. The feeder-to-finish operations are the largest consumers of soymeal. The top three agribusiness corporations – Bunge, ADM and Cargill – already controlled most domestic soybean crushing, and since the 1970s and 1980s, they also became dominant players in the market segments of animal feed in the United States (Howard 2016, 74; Shurtleff and Aoyagi 2019, 5).

Soybean farmers were concerned about oversupply despite the increasing demand for soymeal, as the supply of soybeans had also increased (Du Bois 2018, 98–99). ASA and the US Government found that by increasing overseas demand for US crops, price levels could stay up, which in turn made it possible for the government to pay less in price supports and other subsidies to farmers. Different mechanisms, including the Marshall Plan and other types of aid, were used to persuade foreign countries to buy US soy surpluses and ASA "experts" travelled around the world teaching producers how to use soymeal in animal feeds (Ibid). The ABCD companies were also involved in this development, e.g. Cargill entered the European feed market already in 1964 with the purchase of the Hens Voeders Co (Cargill 2022). Another important facilitator of the use of US soymeal in Europe was the aforementioned implicit agreement with Europe to allow free trade with soybeans in exchange for US acceptance of European protection of their wheat. Thus, Western Europe soon became increasingly reliant on soymeal, gradually abandoning traditional mixed farming systems for poultry and pork production (DeCarlo 2018). European farmers who specialized in intensive systems (e.g. pork production in Denmark, Holland and Southern Europe) wanted cheap soymeal, since feed costs typically account for 50–80% of total operating costs in intensive animal systems (Davis and D'Odorico 2015; MacDonald et al. 2015; le Polain de Waroux et al. 2017). A seeming threat to intensive animal farming – the outbreak of bovine spongiform encephalopathy (BSE) or "mad cow disease" in the 1980s – led to regulations that prohibited the use of bones or other "animal-derivatives" in feeds – further boosting use of soymeal.

Thus the intensive "US Soy-Meat Complex" spread around the world, particularly in the industries of poultry and pork, where animals are raised indoors under factory-like conditions and given protein-rich cakes to fatten quickly (Lapitz, Evia, and Gudynas 2004, 57). As this feed model gained ground, with high-protein inputs at its core, so did the global sourcing of specialized feed-stuffs for local livestock industries (McMichael 1997, 647). The ABCD companies played an important role here as a proactive force behind this global development by moving into animal nutrition. An illustrative example is how the ABCD companies acquired major poultry and pork producer companies throughout the world, not least in Brazil and Asia. Already by 2004, Cargill had 163 animal nutrition plants in 22 countries and developed a "modern" large-scale hog and chicken industry based on soymeal (Cargill 2022). This development was also promoted by public policies to "add more value" to agricultural production and support domestic chicken and pork export industries. While Brazil is still far from consuming as much soymeal as the United States or Europe, its domestic use of soymeal has increased rapidly since the 1990s, and it is increasingly exporting broilers and pork instead of just soy (Shurtleff and Aoyagi 2009, 318). Other Latin American countries have followed similar paths, but have been less successful in "adding value" to soy through pork and chicken industries and instead buy these items from Brazil (OEC 2022).

While hog and poultry production were "modernized" early – i.e. factory farms with feed concentrates – US cattle remained mainly free-ranging until the 1960s. Feeding cattle additional rations, rather than letting them graze freely, increased productivity substantially. In the most intensive systems – so-called feedlots or Confined Animal Feeding Operations (CAFOs) – cattle are fed exclusively on "total mixed rations" – a controlled mix of forages, grains, proteins, vitamins and minerals to boost production. This "factory-farm" model – made possible by abundant grain and soy harvests – allowed livestock farmers to feed large numbers of cattle faster than ever before (fewer days for animals to reach slaughter weight) and in one place (cutting transport costs). Between the early 1960s and 1990s, beef cattle in US feedlots grew immensely (Baraibar Norberg 2020a, 122–23). Consequently, big traders/crushers increasingly also moved into this business, e.g. in 1974, Cargill purchased its first cattle feedlot – Caprock Industries (Cargill 2022). In Europe, industrial feedlots for beef production are still not widely adopted, but dairy cows are increasingly fed soybean inputs. Soybean meal is a staple in the diet of high-producing dairy cows, increasing milk yield and milk protein content. Thus, the use of feed rations in Europe has increased outside of "factory farming". While cattle are still mostly fed on grasses and pastures, they are given increasing amounts of rations in the final "fattening" or "finishing" stage – to gain weight rapidly before slaughter (USDA 2016). It is also common to supplement with soybean meal in grazing, which results in higher forage intake and nutrient digestibility. The percentage of each ingredient in these mixes depends on several factors, not least price relations.

In Latin American beef-producing countries, production remained mainly extensive and grass-fed for some time. Even though several countries were important exporters of both soy and meat, these businesses were typically not as deeply intertwined as they were in the United States and use of domestic soymeal was low. Thus, the feedlot model of animal production initially had an indirect impact in Latin America – through the enormous surge in global demand for feed, which resulted in huge expanses of savannas and forests being plowed up to plant soybeans. The lion's share of soybeans was produced by fewer, specialized, large crop producers, while cattle and sheep farmers were mostly small family farmers. There were also some big "*latifundistas*" which typically held pasturelands (Baraibar Norberg 2020). However, ranching activities have become more and more entwined with soy as *sojización* involves similar self-reinforcing feedbacks, e.g. the expansion of soy led to increased land prices which created pressures for intensification and as more soybeans are readily available close by (i.e. without having to pay international shipping costs or import taxes) animal production becomes relatively cheaper and more competitive, especially in areas that used to be far from suppliers. Thus, in the wake of rising land prices, ranchers are increasingly "finishing" with an increasing amount of soymeal in their rations.

Moreover, since the 1990s, the expansion of industrial aquaculture has further fueled demand for soymeal as a source of protein to replace fishmeal which has a more limited supply related to overfishing and due to large fluctuations in the key supply nations of Peru and Chile due to El Niño (Deutsch et al. 2007). Today, every other bite of fish comes from aquaculture and soybeans played a key role in this production expansion (Troell et al. 2014). Thus, an even wider array of animal products worldwide have become, and to an increasing degree, "soy products". In short, soybeans became the single most important ingredient in concentrated protein fodder and arguably the most essential ingredient in the global food system (Turzi 2016, 2–5). In 2014, soybean meal production reached 243 MMT and accounted for 62.5% of oil meals (Soybean Meal Info Center, 2018). It represents two-thirds of the total world output of protein feedstuffs, including all other major oil meals and fish meal.

The most important driver is increasing global meat demand. In the West, the new meat-based diet developed after World War II was first and foremost based in consumption of white meat, particularly poultry. Spearheaded by the United States, poultry production further accelerated in the early 1990s as demand from developing countries grew. The rapid rise in the consumption of meat and dairy products was particularly striking in East and Southeast Asia where China was (again) by far the most spectacular case – where per capita consumption increased 15-fold since 1961 (Ritchie and Roser 2021). China became the world's largest meat producer. This is often understood to be linked to the nutritional transition, or "Bennett's Law", positing that when poor people get more money, they tend to first go through an expansion phase (more of the same food), then a substitution phase (more energy-rich foods such as meat and those with a high concentration of vegetable oils and sugar). The production of high-energy food, in turn, requires more resources, for example, instead of grain being directly consumed by humans, it is used as animal feed for livestock production (Godfray et al. 2010, 2771; Godfray 2015). While this might explain part of this shift, there is more to it. As we have mentioned several times, transnational corporations, farmers' associations and public officials from the United States have been traveling to China for decades with the explicit aim of convincing them to "intensify" meat production using of protein-rich soyfeed. The opening quote to this chapter, when Cargill China earned the US State Department's Award for Corporate Excellence for training more than two million Chinese farmers, is a clear illustration that the "nutritional transition" in China is not a mere spontaneous shift in response like "Bennett's Law", but also the result of strong actors that have been strategically pushing for this shift to occur. Of course, population growth, urbanization and increasing purchasing power in Asia have also effected dietary shifts. Globally, meat production has more than quadrupled to 340 MMT in 2018 (Ritchie and Roser 2021) and to feed these livestock we presently use 35% of all arable land to produce mainly soybeans and maize. According to UNEP (United Nations Environment Programme), the calories that are lost by feeding cereals to animals, instead of eating them directly as food, could feed an extra 3.5 billion people.

Soybean oil for food and fuel: valuable coproducts

While soymeal is the main driver of soy, the same crushing process also generates oil – central to several other value chains. Although oil was initially mainly appreciated for industrial uses, technological advances managed to successfully remove the "beany" flavor in the 1970s and after that, soy oil began to be used widely in commercial frying and baking (Du Bois 2018, 16) and became mainly destined for food. By 1975, soy provided 61% of US food use of fats and oils (animal and vegetable combined) (Hacklander 1986). The rise of soy oil as the most important edible fat in the United States (and second in the world after palm oil) was thus in large part a story about advancements in science and technology (Mintz, Tan, and Du Bois 2008, p. 5). Eventhough the most common end products became those for cooking and margarine, the versatile oil had multiple uses.

As soybean acreage replaced other oilseeds, there was substantial and rising soy oil availability which spurred new uses. Outside of the mainstream uses (i.e. edible oils, feed inputs and exports) there is still a lot of experimentation with soybean derivatives, and it plays a key role in several niches of the economy. Two examples are lecithin and soy ink. Developed in Germany in 1922, lecithin is one of the most versatile and valuable byproducts of soy oil (List 2015, 1). While lecithin has been an important food additive for a century, it gained recent importance as a food supplement ingredient. Soy lecithin is actually one of the most ubiquitous additives in our food system.[13] In foods it provides about a dozen functions, e.g. an emulsifier, wetting agent and an anti-spattering and stabilizing agent (Bradley 1951). Breads, chocolate, ice-cream, infant formulas, cakes and peanut butter all contain lecithin, but only in small amounts (Du Bois 2018, 17). It is also used in the pharmaceutical, cosmetic, feed and paint industries, including textiles, coatings, rubber, resins and soaps (List 2015). While lecithin can be produced from many sources (e.g. egg yolks, cottonseed or sunflower oil), soy lecithin is historically the most widely used. However, in the EU, lecithin from GE-free canola and sunflower oil became more important due to EU labeling requirements for GE foods. In the United States, virtually all food-grade lecithin is derived from soy oil. Another example is soy ink for packaging and newspapers (Du Bois 2018, 20). Successfully tested first in 1987, by *The Gazette* in Iowa, it now dons the pages of more than 90% of US daily papers and is more environmentally friendly than traditional petroleum-based ink.[14]

But the latest big increase in uses for industrial applications happened in the late 1990s when biodiesel became the second largest user of soy oil in the United States. Biodiesel is mainly produced from oilseeds such as soybean, corn and canola and can partially replace diesel and jet fuel. Once again, this demand did not simply arise. US soybean farmers, worried by the amount of excess soy oil on the market needed to find more buyers.[15] To create a new market, they created the United Soybean Board and the National Biodiesel Board and successfully lobbied for the use of soybean oil in biodiesel.[16] Biodiesel is estimated to have increased economic margins for farmers by 13%.[17] And a welcome "side-effect" was lower prices for soymeal, so, livestock and poultry producers also reaped the benefits when biodiesel increased demand for soy oil by 300% in a decade. The promotion of biodiesel in the United States centered on arguments of job creation for "rural America" and energy security (sovereignty), at least as much as arguments to reduce greenhouse gas emissions. About 100 biodiesel production plants in the United States produce biodiesel mainly from soybean oil. In 2019, around 6.5 billion liters of biodiesel were produced in the United States from ~5 billion kg of fats and oils of which more than half was soybean oil (Khanal and Shah 2021, 907–13).

The United States is not the only country that has promoted this new market outlet. In the early 2000s most soybean-producing countries in South America had adopted ambitious biofuel targets (both ethanol and biodiesel). In Argentina, Paraguay and Uruguay public policies were often presented as efforts to "add value" to the soybean chain – to use the "boom" as a springboard for development (Baraibar Norberg 2020b). The Brazilian state had been promoting both ethanol and biodiesel since the 1960s and by 2010 had targeted blends of 20% for biodiesel and 25% for ethanol (Texo, Bentancur, and Duque 2009, 25). While other oilseed crops have higher oil content than soybeans, most biodiesel from the Americas are sourced from soy due to existing large soybean acreage. Besides high domestic biofuel use, an important portion of South American biodiesel production is exported to Europe. Allowing Europe to fulfill their commitments to reduce fossil fuel using Latin American agroecosystems (Davis and D'Odorico 2015).

While the byproducts of soy oil are certainly not as economically valuable as soymeal, these residuals are highly desirable and one could argue integral to the economic success of the entire global commodity chain. With this expansion of uses they may be closer to coproducts in the sense that they are no longer incidental, but are expected and depended upon as sources of revenue and as ingredients in many products globally, if we consider values generated in the processing phase of the "soybean complex". Thus, the versatile bean has embedded itself into the global market and value chains of other products and made itself ubiquitous and indispensable, creating more incentives to maintain high production levels.

Soy as nutritious food

Only about 7% of all soy produced globally is directly consumed by humans, typically as tofu, or soy milk and edamame, or 20% if soybean oil is included (see Figure 5.14). Direct consumption of oil has been relatively steady since the 1960s, increasing only 4–9 MMT between 1960 and 2019 (Ritchie and Roser 2021). Soybeans are about 35–40% protein and, on a protein cost per kilogram basis, the soybean has become the cheapest source of protein (Judd 1970), thus its popularity for animal's diets. But, as we know, soybeans are also incredibly nutritious for humans. Heated soybeans provide high quality protein with all the essential amino acids necessary for a complete human diet. Thus, they are a fully nutritious alternative to eggs and meat and are a much less expensive source than animal products both in market prices and natural resources, e.g. soybeans eaten directly require less land, water and produce fewer GHG emissions (Deutsch et al. 2010; Deutsch et al. 2013; Ran et al. 2013); notwithstanding the ethical benefits of not keeping sentient animals confined and then slaughtering them (Du Bois 2018, 10). In spite

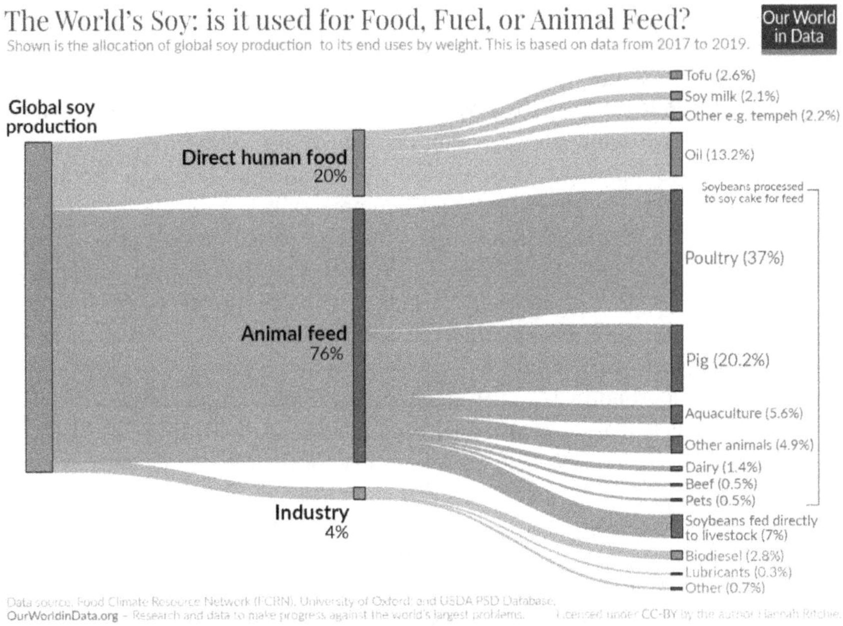

Figure 5.14 The three main uses of soy in 2018 were for human food (7% directly and 13% as food oil), industry (4%) and the rest is animal feed (76%).

Source: Ritchie and Roser (2021).

of these qualities, soybeans have yet to become a major source of food protein in the West.

The nutritious values of soy means that soyfoods have been a counterpoint in the Western food system, advocated by vegetarians and nutritionists for more than a century (Chen 1962; Keys and Keys 1967). Although consumer interest and use remained small, scientific and engineering efforts to improve taste and usage qualities persisted, e.g. with significant research on soy flour. While soy has increasingly entered our diets, it has been as an invisible ingredient. Soy flour contains more than 50% protein and is used in many commercial bakeries. Soy hulls contain a high level of digestible fiber and are further processed into fiber bran breads, cereals and snacks. Oil became a main ingredient in margarine and in ready-made frozen foods in the 1960s and developments in the utilization of soybean protein concentrates, isolates and textured protein have continued to advance significantly. Yet, the wave of interest in soyfoods during the World Wars and in the late 1960s to feed world hunger did not affect mainstream consumption patterns. But this alternative movement has definitely formed a counterpoint to corporate food, fast food and Cheap Meat.

While soyfoods have never became widely popular in the West, Asian communities within the United States, remain faithful to their food traditions, thus soybeans provided another alternative counterpoint: soy sauce, tofu and miso were commonly sold in small businesses owned by Japanese immigrants (Roth 2018, 144). From the 1970s onwards, an increasing part of the non-Asian population also started to consume traditionally oriental soyfood such as tofu and miso (e.g. in Chinese and Japanese restaurants). Since 2000, a new wave of soy food interest has swept the West. Soy-based foods, such as tofu, soy milk and soy sauce have again gained popularity outside Asia. Soy milk was adapted to Western tastes – flavored, sweetened and/or fortified for better nutrition (Liu 2008, 443–46) and is now sold large-scale, commercially.

Accordingly, the big agribusiness firms are also in this segment. Bunge has promoted new soyfoods in Brazil since in 1976 through the subsidiary companies SANBRA and SAMRIG (Bunge 2022) and participated directly in the *Programa Nacional de Alimentação Escolar* (PNAE), a national school meal and nutrition education program. Further, in 2006, Cargill acquired Degussa's food ingredients operations, "strengthening Cargill's global portfolio of texturant ingredients and systems, emulsifiers, flavoring and health promoting ingredients" (Cargill 2022). But, as shown in Figure 5.14 only a small portion of the rapidly increasing global soybean harvest is actually used for foods.

As we saw in previous chapters, soyfoods developed and spread throughout East Asia and by the 1600s soy were a cornerstone in Asian cuisine. One strand of foods is based on fermented soy, such as fermented tofu (*sufu*), soy sauce, fermented soypaste (*miso*), *natto, tempeh* and fermented soy milk (soy yogurt). Another strand is based on non-fermented soy, such as soy sprouts, soy nuts, soy milk film (*yuba*), tofu, soy milk, soy pulp (*okara*), green vegetable soybeans (*edamame*) and toasted soy flour. These soyfoods were traditionally made from whole soybeans, but with today's processing technologies, soy sauce, soymilk and tofu can be made from defatted soymeal or its derivative products, such as soy protein isolate (Liu 2008, 441).

In the cradle of soy, China, the bean remained a fundamental ingredient in the mainly vegetarian diet for millennia. Today, China is the world's largest buyer of soybeans, but for its domestic consumption of soy for direct human consumption (e.g. protein products, soy sauce and food oil) it mainly uses domestically grown soybeans (around 13 MMT) which are not genetically modified. However, as explored in detail in the previous subsections, increased appetites and purchasing power for meat, dairy and eggs, created an exponential growth in soybean demand for feed.

Reflections on the soybean and its functions

The third soy regime represents a complex and contradictory set of relations between production and consumption. The beginning of this period is often referred to as "embedded liberalism" or "national developmentalism", rooted in strong state protection. The world food economy was organized under US hegemony and the Bretton Woods system (Friedmann and McMichael 1989, 103). During the early stages of the regime, industrialized economies sought to strengthen domestic agriculture through policies like the Farm Bill in the United States and the Common Agricultural Policy in Europe. These policies aimed to increase supply through protectionist measures and the adoption of industrialized and modernized agricultural practices that facilitated mass crop production (Ibid, 108). Agriculture became increasingly specialized and mechanized, leading to larger and more concentrated operations. Food became increasingly standardized, processed and "cheap". As the next logical step in this process, specialized animal farms emerged, located separately from crop areas and operating on a large scale. These animal farms are often referred to as "factory farms" because they function like industrial facilities, with animals raised in confined spaces and fed concentrated diets that promote rapid growth and high productivity, e.g. in milk, meat or egg production. Soy is a common ingredient in these diets. A positive feedback between increasing consumption of animal products and economies of scale in the "Soy Model", thus, led to intensification of livestock production, which in turn meant a spiraling increase in demand for the protein-rich soybean for feed concentrates to enable "cheap" meat; what is known as the "Soy-Meat Complex." Thus, in soy's third regime the function of soy as fertilizer on-farm seemingly ceased to be important. At the same time, soy oil, which first mainly was used for industrial applications, became a key ingredient in food oils and demand was steady over the period – but quickly became a byproduct of the main use in animal feeds.

The evolution of the "US Soy Model" can also be attributed to the efforts of various powerful actors. For instance, in the US government, the USDA and land grant colleges worked tirelessly to adapt varieties suitable for different regions. Producer organizations like the ASA lobbied successfully for policy regulations in line with their interest. The US Foreign Agricultural Service worked hard for overseas trade benefits to increase exports, while crushers invested heavily in increasing production, processing and infrastructure capacity. Moreover, as the US food supply grew, the government began redirecting surplus food through aid programs to post-colonial states. These programs aimed to maintain prices for American farmers and increase US influence in the "third world." However, they also contributed to the spread of industrial agriculture and the displacement of traditional farming practices in recipient countries (McMichael 2009, 141).

The state thus played a dominant role in the soy regime at this time, but powerful agribusiness actors were growing and strengthening their position in the soy commodity chain, becoming increasingly transnational and vertically integrated. In fact, by the early 1970s, big agrofood corporations had outgrown the state-centered model (McMichael 2009), and the once-dominant approach of "national developmentalism" was increasingly under attack as neo-liberalism gained popularity. The collapse of the Bretton Woods system in 1971, after the United States unilaterally abandoned the fixed exchange rate system, triggered a range of regulatory changes in trade, finance and intellectual property that further accelerated the production, trade and consumption of soybeans. Together, these efforts paved the way for the transnational expansion and dominance of the soy regime.

During the 1970s and 1980s, China still provided an important exception to the liberalization trends, with very high self-sufficiency targets and small import quotas. However, China faced challenges in fulfilling its national food security strategy due to rapidly increasing per capita meat consumption, from 16 kg/capita in 1980 to 50 kg/capita in 2020 (FAOSTAT 2022). This was driven by rising incomes and urbanization, but it was also influenced by decades of lobbying by transnational companies and the USDA promoting the use of soy as animal feed. By the mid-1990s, China decided to relax self-sufficiency targets on soy and increase imports. This led to a gradual expansion of the global expansion of the Soy-Meat complex, with China developing its own version centered on pork. As a result, China became the world's largest consumer of soybeans once again, but this time by relying heavily on Western agroecosystems for feed crops. In contrast, Chinese domestic soy production is GE-free and is mostly used for soyfoods, not feed. As demand for soy soared in China, soy production expanded further south of the US, incorporating existing cultivations and new frontiers in South America, such as the Brazilian Amazon, Argentina's Pampas and Brazil and Paraguay's Atlantic Forest. This expansion was further accelerated by the liberalization of international agrofood markets, which was partly institutionalized in the establishment of the WTO. Soybeans fit seamlessly into the global vision of agriculture, which prioritizes increased production of cheap food by incorporating the cheapest lands and replacing labor with green revolution technologies such as mechanization, high-yielding seeds, chemical fertilizers and pesticides.

The "Gene Revolution" in the mid-1990s furthered this trend by allowing soy production in South America to expand into new frontiers, such as the Gran Chaco region (Paraguay, Argentina and Bolivia), the Cerrado (Brazil) and the Uruguayan grasslands. This new capital-intensive agricultural package did not require farmers' existing or traditional knowledge or workers' labor. Instead, it was based on the US Soy Model, where farmers needed to "get big" to take advantage of economies of scale and pay off input costs, investments and leases. As a result, soy production not only expanded but also intensified, with specialization and agrochemicals, and concentration began to increase. In production areas, small farmers were often displaced, whereas the remaining farmers became fewer and larger. Unlike the United States, however, South America exported most of its soy and used less of it for domestic meat production, with the bovine sector still largely based on pastures. Additionally, almost all GE soy technology is developed in the North and the patents are held there as well. Nevertheless, state-owned Chinese companies have been rapidly increasing their market shares in biotechnology, seeds and agrochemicals, as well as in trading, in recent years. In general, GE soy production across the Americas has brought extreme concentration in both the pre-cultivation and commercialization phases, with vertically and horizontally integrated big firms becoming the norm. Thus, through their size and their involvement in various stages of the commodity chain – they are uniquely positioned to coordinate entire commodity chains.

The dominance of corporations in the soy commodity chain has significant influence over all stages of production and reproduction. The commodity chain has expanded by creating new uses for soy, resulting in the creation of new markets for this highly versatile crop. Some argue that soy's primary use has been as a cash crop. Although it was briefly promoted as foreign aid, it's unclear whether the aim was to alleviate food scarcity, or to promote the benefits of capitalism over communism, or just to expand trade. In contrast, the use of soybeans as food has remained limited – popular primarily among Asians, both at home and abroad.

Soybean production has had a profound impact on the environment and social relations throughout our global agrofood system. Despite feeding more people and animals and occupying more land, the system has become increasingly disconnected from feedbacks from ecosystems, farmers and consumers, instead becoming tightly linked to biotechnology, commodity trading and the global financial system (Crona et al. 2015). This transformation, known as the Great Acceleration, has altered key ecological processes to the extent that some argue we have entered a new epoch, the Anthropocene (Steffen et al. 2015). Soybeans play a significant role in driving these alterations and pushing us toward local, regional and planetary boundaries for biodiversity loss, nitrogen outputs, land use and GHG emissions (Deutsch et al. 2013). The soybean is a major force behind an unsustainable agrofood system based on "Cheap Meat" that may not withstand the challenges of an uncertain future. In the next chapter, we will explore this question in more detail, drawing on what we have learned from our world history of the soybean.

Notes

1 While Monsanto's patent on glyphosate expired in 2000 (further lowering the cost of glyphosate), Monsanto managed to maintain a large market share (Howard 2016, 109).

2 No till means that seeds are sown directly on the residue of the previous crop. This practice has many benefits, e.g. protects the soil against erosion, maintains soil cover, structure, carbon, stability and moisture content, which in turn helps, e.g., in the regulation of soil temperature and carbon emissions.
3 In the US, the amount of saved seeds went from 63% in the mid-1990s to 10% in 2001 (Howard 2016, 108–10).
4 Cf. Reuters, "Battle of the beans", www.reuters.com/article/us-usa-pesticides-soybeans-insight/battle-of-the-beans-monsanto-faces-a-fight-for-soy-market-idUSKBN1FD0G2 American Antitrust Institute (AAI), "White Paper Monsanto – Bayer" www.antitrustinstitute.org/sites/default/files/White%20Paper_Monsanto%20Bayer_7.26.17_0.pdf (Accessed 2022–10–28).
5 Cf. https://global.handelsblatt.com/companies/basf-winner-bayer-monsanto-merger-901801; https://www.boell.de/en/2017/10/31/monsanto-and-co-from-seven-to-four-growing-by-shrinking (Accessed 2020–07–10).
6 Cf. https://www.reuters.com/article/us-argentina-soyproducts-china-idUSKCN1VW21E (Accessed 2022–07–30).
7 Cf. Document 32022D0798, 2022/798 of the European Parliament and of the Council. https://eur-lex.europa.eu/legal-content/EN/TXT/?uri=CELEX:32022D0798 (Accessed 2022–10–28).
8 COFCO's website: http://www.cofco.com/en/AboutCOFCO/
9 http://www.adndigital.com.py/empresa-china-lidera-ranking-las-exportadoras-soja/
10 Cf. https://www.fia.org/resources/global-futures-and-options-trading-hits-another-record-2021 (Accessed 2022–07–19).
11 Cf. https://www.reuters.com/business/adm-bunge-expected-post-strong-results-ukraine-war-ignites-demand-2022-04-25/ (Accessed 2022–06–13).
12 If the power in the chain is captured by leading firms who own capital-intensive industries (with a high degree of asset specificity) and often involve several stages of vertically organized suppliers – the chain is producer-driven. If the chain is governed by leading firms who coordinate far-reaching subcontracting networks – it is buyer-driven. (Bair 2009, 19–21).
13 Soybeans and Soy Lecithin: https://farrp.unl.edu/soy-lecithin. (Accessed 2022–07–15).
14 ASA, Soy Ink Seal, https://soygrowers.com/soy-ink-seal/. (Accessed 2022–07–15).
15 https://www.biodiesel.org/what-is-biodiesel/why-biodiesel. (Accessed 2022–08–07).
16 https://farm-energy.extension.org/soybeans-for-biodiesel-production/. (Accessed 2022–08–07).
17 https://www.biodiesel.org/what-is-biodiesel/why-biodiesel/benefits-to-soybean-farmers. (Accessed 2022–08–07).

References

ADM. 2022. "Our Company – History – Founders' Story." Archer Daniels Midland. 2022. https://www.adm.com/en-us/about-adm/our-company/history/#founders-story.

Bair, Jennifer. 2009. "Global Commodity Chains: Genealogy and Review." In *Frontiers of Commodity Chain Research*, edited by Bair, 1–281. Stanford: Stanford University Press.

Baraibar, Matilda. 2008. "Hur Uppnås Landsbygdsutveckling? Världsbankens Och MSTs Strategier, Samt Dess Möjliga Konsekvenser i Rio Grande Do Sul." *Stockholm Papers in Latin American Studies* 7.

———. 2014. "Green Deserts or New Opportunities?: Competing and Complementary Views on the Soybean Expansion in Uruguay, 2002–2013." Monograph, Stockholm: Stockholm University.

Baraibar Norberg, Matilda. 2020a. "Agrofood Globalization: The Global Soybean and Beef Commodity Chains." In *The Political Economy of Agrarian Change in Latin America*, 117–63. Cham: Palgrave Macmillan, Springer Nature.

———. 2020b. "Changes and Continuities in Agrofood Relations, 1870–1970s." In *The Political Economy of Agrarian Change in Latin America*, 57–116. Cham: Palgrave Macmillan, Springer Nature.

———. 2020c. "Introduction to the Political Economy of Agrarian Change in Latin America." In *The Political Economy of Agrarian Change in Latin America*, 1–56. Cham: Palgrave Macmillan, Springer Nature.

———. 2020d. "Regulative Shifts and Agrarian Change of the Twenty-First Century." In *The Political Economy of Agrarian Change in Latin America*, 209–326. Cham: Palgrave Macmillan. Springer Nature.

———. 2020e. "Regulative Shifts Paving the Way for Agrarian Change." In *The Political Economy of Agrarian Change in Latin America*, 165–208. Cham: Palgrave Macmillan, Springer Nature.

———. 2022. "Sojización as a New First Movement: A Polanyian Analysis of the South American Soybean 'boom." In *The Age of the Soybean: An Environmental History of the Soyacene during the Great Acceleration*, edited by da Silva and de Majo, 91–114. Winwick, Cambridgeshire: The White Horse Press.

Barnhart, Floyd. 1954. *Soybeans*. Caruthersville, MO: Floyd Barnhart.

Bonny, Sylvie. 2017. "Corporate Concentration and Technological Change in the Global Seed Industry." *Sustainability* 9 (9): 1632.

Bradley, T. F. 1951. "Non-Edible Soybean Oil Products." In *Soybeans and Soybean Products* Vol. II, edited by Markley, 853–90. New York: Interscience Publishers, Inc.

Brown, Lester R. 2012. *Full Planet, Empty Plates: The New Geopolitics of Food Scarcity*. 1st ed. New York: W.W. Norton & Company.

Buck, John Lossing. 1937. *Land Utilization in China: A Study of 16,786 Farms in 168 Localities, and 38,256 Farm Families in Twenty-Two Provinces in China, 1929–1933*. Nanjing: University of Nanking.

Bulmer-Thomas, Victor. 2003. *The Economic History of Latin America since Independence*. New York, NY: Cambridge University Press.

Bunge. 2022. "Who We Are – Our History." https://www.bunge.com/who-we-are/our-history.

Burch, David, and Geoffrey Lawrence. 2009. "Towards a Third Food Regime: Behind the Transformation." *Agriculture and Human Values* 26 (4): 267.

Cargill. 2018. "Momentum – 2017 Annual Report." Annual Report. Cargill, Incorporated.

———. 2022. "Cargill Company Timeline: 1865–Present.

Cartter, Jackson L., and Edgar E. Hartwig. 1962. "The Management of Soybeans." *Advances in Agronomy* 14: 359–412.

CAS. 2018. "Anuario de Comercio Exterior de Base Agraria de Los Países CAS 2013–2017." Translated by Grupo Técnico 2 (GT2) – Sistema de Información de Mercados. Consejo Agropecuario del Sur.

Castellano, Andrés and Goizueta, Mercedes. 2011. "*Agregado de Valor En La Cadena de La Soja: Alternativa de Upgrading Para Productores Primarios*." INTA.

Chang, H. J. 2009. *Bad Samaritans: The Myth of Free Trade and the Secret History of Capitalism*. Bloomsbury Press.

Chen, Philip S. 1962. *Soybeans for Health, Longevity, and Economy*. 2rd ed. South Lancaster, Mass: Chemical Elements.

Chourmert, Johanna, and Pascale Phélinas. 2016. "Is GM Soybean Cultivation in Argentina Sustainable." *Ètudes et Documents* 14: 19.

Clapp, Jennifer. 2006. "WTO Agriculture Negotiations: Implications for the Global South." *Third World Quarterly* 27 (4): 563–77.

———. 2014. "Financialization, Distance and Global Food Politics." *The Journal of Peasant Studies* 41 (5): 797–814.

———. 2015. "ABCD and beyond: From Grain Merchants to Agricultural Value Chain Managers." *Canadian Food Studies/La Revue Canadienne Des Études Sur l'alimentation* 2 (2): 126–35.

———. 2016. *Food*. 2nd ed. Cambridge: Polity Press.

Clever, Jennifer, and Wu Xinping. 2016. "China: Oilseeds and Products Update." CH16038. GAIN Report. USDA.

Comtrade UNSD. 2022. "United Nations Statistics Division International Trade Statistics." Database. UN Comtrade Database. 2022. https://comtrade.un.org/data/.

Crona, Beatrice I., Tim M. Daw, Wilf Swartz, Albert V. Norström, Magnus Nyström, Matilda Thyresson, Carl Folke, et al. 2015. "Masked, Diluted and Drowned Out: How Global Seafood Trade Weakens Signals from Marine Ecosystems." *Fish and Fisheries*. https://doi.org/10.1111/faf.12109.

Crook, Frederick W. 1988. *Agricultural Statistics of the People's Republic of China, 1949–86*. USDA US Department of Agriculture, Economic Research Service.

Dalla Costa, Armando, and Gustavo Pereira da Silva. 2018. "Bunge and His First Fiftieth Anniversary in Brazil (1905–1955): The Construction of an Economic Group." *Investigaciones de Historia Económica – Economic History Research* 14 (3): 199–209.

da Silva, Ramon, Mateus Batistella, Yue Dou, Emilio Moran, Sara Torres, and Jianguo Liu. 2017. "The Sino-Brazilian Telecoupled Soybean System and Cascading Effects for the Exporting Country." *Land* 6 (January): 53.

Davis, Kyle F, and Paolo D'Odorico. 2015. "Livestock Intensification and the Influence of Dietary Change: A Calorie-Based Assessment of Competition for Crop Production." *Science of the Total Environment*, 538: 817–23.

DeCarlo, Samantha. 2018. "Executive Briefings on Trade April 2018." *United States International Trade Commission*, USITC Office of Industries.

De Maria, Robinson, Kangile, Kadigi, Dreoni, Couto, Howai, Peci and Fiennes. 2020. "Global Soybean Trade: The Geopolitics of a Bean." Trade Hub. UK Research and Innovation Global Challenges Research Fund (UKRI GCRF) Trade, Development and the Environment Hub.

Deconinck, Koen. 2019. "New Evidence on Concentration in Seed Markets." *Global Food Security* 23: 135–38.

Deutsch, Lisa, Robert Dyball, and Will Steffen. 2013. "Feeding Cities: Food Security and Ecosystem Support in an Urbanizing World." In *Urbanization, Biodiversity and Ecosystem Services: Challenges and Opportunities: A Global Assessment*, edited by Elmqvist, Fragkias, Goodness, Güneralp, Marcotullio, McDonald, Parnell, et al., 505–37. Dordrecht: Springer Netherlands.

Deutsch, Lisa, Malin Falkenmark, Line Gordon, Johan Rockström, and Carl Folke. 2010. "Water-Mediated Ecological Consequences of Intensification and Expansion of Livestock Production." In *Livestock in a Changing Landscape. Volume 1. Drivers, Consequences and Responses*, edited by Steinfeld, Mooney, Schneider, and Neville, 97–110. Washington, DC: Island Press.

Deutsch, Lisa, Sara Gräslund, Carl Folke, Max Troell, Miriam Huitric, Nils Kautsky, and Louis Lebel. 2007. "Feeding Aquaculture Growth through Globalization: Exploitation of Marine Ecosystems for Fishmeal." *Global Environmental Change* 17 (2): 238–49.

"Dirección de Estimaciones Agrícolas." 2022. Serie de Estadística Agrícola Por Cultivo, Campaña, Provincia y Departamento de La República Argentina. Buenos Aires: Dirección Nacional de Agricultura, Dirección de Estimaciones Agrícolas.

Du Bois, Christine M. 2018. *The Story of Soy*. London: Reaktion Books.

Egas, Juan José, and Carmine Paolo De Salvo. 2018. "Agricultural Support Policies in Latin America and the Caribbean: 2018 Review." Agricultural Policy Reports 2018. Agricultural Policy Reports. Inter-American Development Bank. https://publications.iadb.org/publications/english/document/Agricultural-Support-Policies-in-Latin-America-and-the-Caribbean-2018-Review.pdf.

European Parliament. 2015. "Imports of GM Food and Feed Right of Member States to Opt Out." Briefing EU Legislation in Progress. European Union.

FAO. 1951. "1950 Yearbook of Food and Agricultural Statistics. Part 2. Trade Vol 4." Rome: Food and Agriculture Organization of the United Nations.

———. 1952. "1951 Yearbook of Food and Agricultural Statistics. Part 2. Trade Vol 5." Rome.

———. 1953. "1952 Yearbook of Food and Agricultural Statistics. Part 2. Trade Vol 6." Rome.

———. 1954. "1953 Yearbook of Food and Agricultural Statistics. Part 2. Trade Vol 7." Rome.

———. 1955. "1954 Yearbook of Food and Agricultural Statistics. Part 2. Trade Vol 8." Rome.

———. 1957. "1956 Yearbook of Food and Agricultural Statistics. Part 2. Trade Vol 10." Rome.

———. 1958. "1957 Yearbook of Food and Agricultural Statistics. Part 2. Trade Vol 11." Rome.

———. 1959. "1958 Yearbook of Food and Agricultural Statistics. Part 2. Trade Vol 12." Rome.

———. 1960. "1959 Yearbook of Food and Agricultural Statistics. Part 2. Trade Vol 13." Rome.

———. 1961. "1960 Yearbook of Food and Agricultural Statistics. Part 2. Trade Vol 14." Rome.

———. 1962. "1961 Yearbook of Food and Agricultural Statistics. Part 2. Trade Vol 15." Rome.

———. 1963. "1962 Yearbook of Food and Agricultural Statistics. Part 2. Trade Vol 16." Rome.

———. 1964. "1963 Yearbook of Food and Agricultural Statistics. Part 2. Trade Vol 18." Rome.

FAOSTAT. 2022. "FAOSTAT Statistical Database." Rome: Food and Agriculture Organization of the United Nations FAO. http://faostat.fao.org/.

Federico, Giovanni. 2010. *Feeding the World*. Princeton, NJ: Princeton University Press.

Fletcher, Merna Irene. 1950. "World Soybean Production in 1948." *The Scientific Monthly* 70 (2): 116–21.

Forbes. 2022. "America's Largest Private Companies: The Full List." 2022. https://www.forbes.com/lists/largest-private-companies/?sh=71c7cd7bbac4.

Friedmann, Harriet, and Philip McMichael. 1989. "Agriculture and the State System: The Rise and Fall of National Agricultures, 1870 to the Present." *Sociologia Ruralis* 29 (2): 93–117.

Gaudreau, Matthew. 2015. "Seeds, Grain Trade, and Power Off-Land: Chinese Agribusiness in Global Agrarian Change." *BRICS Initiatives for Critical Agrarian Studies Conference Paper* No. 4: 1–13.

Gereffi, Gary. 1999. "A Commodity Chains Framework for Analyzing Global Industries." *Institute of Development Studies* 8 (12): 1–9.

Gibbon, Peter. 2001. "Upgrading Primary Production: A Global Commodity Chain Approach." *World Development* 29 (2): 345–63.

Godfray, H. Charles J. 2015. "The Debate over Sustainable Intensification." *Food Security* 7 (2): 199–208. https://doi.org/10.1007/s12571-015-0424-2.

Godfray, H. Charles J., Ian R. Crute, Lawrence Haddad, David Lawrence, James F. Muir, Nicholas Nisbett, Jules Pretty, Sherman Robinson, Camilla Toulmin, and Rosalind Whiteley. 2010. "The Future of the Global Food System." *Philosophical Transactions of the Royal Society B: Biological Sciences* 365 (1554): 2769–77. https://doi.org/10.1098/rstb.2010.0180.

Gonzaga Belluzo, Luis. 2015. "La Reciente Internacionalización Del Régimen Del Capital." In *Neoestructuralismo y Corrientes Heterodoxas En América Latina y El Caribe a Inicios Del Siglo XXI.*, edited by Bárcena and Prado, 112–25. Santiago de Chile: CEPAL.

Gordon, Line J., Victoria Bignet, Beatrice Crona, Patrik J. G. Henriksson, Tracy Van Holt, Malin Jonell, Therese Lindahl, Max Troell, Stefan Barthel, S, Lisa Deutsch, et al. 2017. "Rewiring Food Systems to Enhance Human Health and Biosphere Stewardship." *Environmental Research Letters* 12 (10): 100201.

Guibert, Martine, Grosso, Susana, Arbeletche, Pedro and Bellini, María Eva. 2011. "*De Argentina a Uruguay: Espacios y Actores En Una Nueva Lógica de Producción Agrícola.*" PAMPA, *Revista Interuniversitaria de Estudios Territoriales* 7 (Suplemento especial tematica: Impactos territoriales asociados a la reconfiguración del sistema productivo primario) 7:13–38.

Guilherme Fowler de Ávila Monteiro. 2009. "Property Rights and Strategy: Market Structure and Competition in Brazilian Soybean Seed Industry." Presented at the Annual meeting of the International Society for New Institutional Economics, Berkeley, CA, June 20.

Hacklander, Duane. 1986. "United States Department of Agriculture Economic Research Service the Decade Ahead for US Soybeans." ERS-622. United States Department of Agriculture National Agricultural Statistics Service.

Hairong, Yan, Chen Yiyuan, and Ku Hok Bun. 2016. "China's Soybean Crisis: The Logic of Modernization and Its Discontents." *The Journal of Peasant Studies* 43 (2): 373–95.

Howard, Philip. 2016. *Concentration and Power in the Food System: Who Controls What We Eat?* Vol. 3. Contemporary Food Studies: Economy. Culture and Politics. Bloomsbury.

Hubbard, Kiki. 2022. "A Short History of Intellectual Property Rights on Seed and What Farmers Should Know." *The Natural Farmer.*

Irwin, Scott H., Dwight R. Sanders, and Lei Yan. 2022. "The Order Flow Cost of Index Rolling in Commodity Futures Markets." *Applied Economic Perspectives and Policy.* https://doi.org/10.1002/aepp.13297.

ISAAA. 2019. "International Service for the Acquisition of Agri-Biotech Applications Brief 55 – Global Status of Commercialized Biotech/GM Crops in 2019: Biotech Crops Drive Socio-Economic Development and Sustainable Environment in the New Frontier." *ISAAA Brief No. 55.* ISAAA Briefs. Ithaca, NY, USA: ISAAA.

———. 2022. "GM Approval Database." International Service for the Acquisition of Agri-Biotech Applications. https://www.isaaa.org/gmapprovaldatabase/.

ISGA. 2017. "Biotech Soy Production in the Americas and the EU - Today and Tomorrow." The International Soybean Growers Alliance edited by Wellington Andrade, Coexistence in International Trade, Amsterdam: Wagenagen University.

Jonsson, Ulf, and Ronny Petersson. 1991. "*Specialisering Och Diversifiering i Europeiskt Jordbruk.*" In *Diversifiering Och Specialisering i Europeiskt Jordbruk Tiden 1750–1950*, edited by Ljungren and Myrdal, 5–14. Stockholm: Kungl. Skogs-och Lantbruksakad.

Judd, R. W. 1970. "Cost of Protein." *Soybean News* 21 (2): 6.

Kay, Cristóbal. 1989. *Latin American Theories of Development and Underdevelopment*. 1st ed. Routledge Library Editions, Development. London: Routledge.

Keys, Margaret, and Ancel Keys. 1967. *The Benevolent Bean*. Garden City, NY: Doubleday & Company, Inc.

Khanal, Asmita, and Ajay Shah. 2021. "Oilseeds to Biodiesel and Renewable Jet Fuel: An Overview of Feedstock Production, Logistics, and Conversion." *Biofuels, Bioproducts and Biorefining* 15 (February). https://doi.org/10.1002/bbb.2198.

Kneen, B. 2002. *Invisible Giant. Cargill and Its Transnational Strategies*. Null.

Kromer, George Wayne. 1961. *Trends in Soybean Acreage and Production, 1924–60*. Washington, DC: U.S. Dept. of Agriculture, Economic Research Service.

Lander, Brian, and Thomas David DuBois. 2022. "A History of Soy in China: From Weedy Bean to Global Commodity." In *The Age of the Soybean: An Environmental History of Soy during the Great Acceleration*, edited by de Majo and da Silva, 29–47. Winwick, Cambridgeshire: The White Horse Press.

Lapitz, Rocío, Gerardo Evia, and Eduardo Gudynas. 2004. *Soja y Carne En El Mercosur: Comercio, Ambiente y Desarrollo Agropecuario*. Coscoroba Ediciones Montevideo.

Lappé, Frances Moore. 1980. *Aid as Obstacle: Twenty Questions about Our Foreign Aid and the Hungry*. San Francisco, CA: Institute for Food and Development Policy.

LDC. 2022. "LDC Annual Report 2021." https://www.ldc.com/annual-report-2021/.

Lieber, James B. 2002. *Rats in the Grain: The Dirty Tricks and Trials of Archer Daniels Midland, the Supermarket to the World*. Hachette: UK.

List, Gary. 2015. "Soybean Lecithin: Food, Industrial Uses, and Other Applications." In *Polar Lipids*, edited by Ahmad and Xu, 1–33. Elsevier.

Liu, KeShun. 2008. "14 - Food Use of Whole Soybeans." In *Soybeans*, edited by Johnson, White, and Galloway, 441–81. Urbana, IL: AOCS Press.

LT Oliveira, Gustavo de, and Susanna B. Hecht. 2017. *Soy, Globalization, and Environmental Politics in South America*. Routledge.

MacDonald, Graham K., Kate A. Brauman, Shipeng Sun, Kimberly M. Carlson, Emily S. Cassidy, James S. Gerber, and Paul C. West. 2015. "Rethinking Agricultural Trade Relationships in an Era of Globalization." *BioScience* 65 (3): 14.

Masuda, Tadayoshi, and Peter D. Goldsmith. 2009. "World Soybean Production: Area Harvested, Yield, and Long-Term Projections." *International Food and Agribusiness Management Review* 12 (4): 143–62.

McDaniel, Marshall, Lisa K. Tiemann and Stuart Grandy. (2014). Does agricultural crop diversity enhance soil microbial biomass and organic matter dynamics? A meta-analysis. *Ecological Applications* 24: 560–570.

McMichael, Philip. 1997. "Rethinking Globalization: The Agrarian Question Revisited." *Review of International Political Economy* 4 (4): 630–62.

———. 2009. "A Food Regime Genealogy." *The Journal of Peasant Studies* 36 (1): 139–69.

———. 2013. *Food Regimes and Agrarian Questions*. Agrarian Change & Peasant Studies. Winnipeg, Canada: Fernwood Publishing.

Mintz, Sidney W., Chee-Beng Tan, and Christine M. Du Bois. 2008. "Introduction: The Significance of Soy." In *The World of Soy*, edited by Du Bois, Tan, and Mintz, 1–23. Singapore: NUS Press.

MN Department of Agriculture. 2022. "Glufosinate Herbicide." https://www.mda.state.mn.us/glufosinate-herbicide.

Moss, Diana. 2010. "Transgenic Seed: The High Technology Test of Antitrust?" The Antitrust Chronicle.

Moss, Diana L. 2009. *Transgenic Seed Platforms: Competition Between a Rock and a Hard Place?* Washington, DC: The American Antitrust Institute.

Mulvaney, Richard L., Khan, S. A., and Ellsworth, Timothy R. (2009). Synthetic nitrogen fertilizers deplete soil nitrogen: a global dilemma for sustainable cereal production. *Journal of Environmental Quality* 38: 2295–2314.

Murphy, Sophia, David Burch, and Jennifer Clapp. 2012. *Cereal Secrets: The World's Largest Grain Traders and Global Agriculture*. Oxfam Research Report. https://www.oxfam.org/en/research/cereal-secrets-worlds-largest-grain-traders-and-global-agriculture.

Myrdal, Janken. 2022. *The Agricultural History of East and Southeast Asia*. Global Agricultural Systems. Uppsala: Department of Urban and Rural Development, Swedish University of Agricultural Sciences.

Nunes, Cássio Alencar, Erika Berenguer, Filipe França, Joice Ferreira, Alexander C. Lees, Julio Louzada, Emma J. Sayer, et al. 2022. "Linking Land-Use and Land-Cover Transitions to Their Ecological Impact in the Amazon." *Proceedings of the National Academy of Sciences* 119 (27): e2202310119.

OEC. 2022. "The Observatory of Economic Complexity Soybeans." MIT. atlas.media.mit.edu.

OECD. 2018. *Concentration in Seed Markets: Potential Effects and Policy Responses*. Paris: Organisation for Economic Co-operation and Development.

OECD/FAO. 2011. "OECD-FAO Agricultural Outlook 2011–2020." Paris: OECD/FAO.

Paolera, Gerardo della, Xavier H. Duran Amorocho, and Aldo Musacchio. 2018. "The Industrialization of South America Revisited: Evidence from Argentina, Brazil, Chile and Colombia, 1890–2010." National Bureau of Economic Research.

Peine, Emelie K. 2013. "Trading on Pork and Beans: Agribusiness and the Construction of the Brazil-China-Soy-Pork Commodity Complex." In *The Ethics and Economics of Agrifood Competition*, edited by James 193–210. Dordrecht: Springer.

Peruchi Moretto, Samira, Eunice Nodari, and Rubens Nodari. 2022. "Soybean as a Critical Resource: Domestication, Dissemination and Introductions (Glycine Max)." In *The Age of the Soybean: An Environmental History of the Soyacene during the Great Acceleration*, edited by de Majo and da Silva, 19–28. Winwick, Cambridgeshire: White Horse Press.

Peters, May, Suchada Langley, and Paul Westcott. 2009. "Agricultural Commodity Price Spikes in the 1970s and 1990s: Valuable Lessons for Today." Technical Bulletin 966. USDA Economic Research Service./.

Polain de Waroux, Yann le, Rachael D. Garrett, Jordan Graesser, Christoph Nolte, Christopher White, and Eric F. Lambin. 2017. "The Restructuring of South American Soy and Beef Production and Trade Under Changing Environmental Regulations." *World Development*, 121:188–202.

Ran, Ylva, Lisa Deutsch, Mats Lannerstad, and Jens Heinke. 2013. "Rapidly Intensified Beef Production in Uruguay: Impacts on Water-Related Ecosystem Services." *Aquatic Procedia* 1: 77–87.

Ray, Rebecca, and Kevin P. Gallagher. 2016. "China in Latin America: Environment and Development Dimensions." *Revista Tempo Do Mundo (RTM)* 2 (2): 131–154.

Relly, Eduardo, and Claudio de Majo. 2022. "A Soyacene of Fire:" In *The Age of the Soybean: An Environmental History of Soy during the Great Acceleration*, edited by de Majo and da Silva, 143–63. Winwick, Cambridgeshire: White Horse Press.

Reuters. 2021. "Brazil: Big Food Companies Write Statement to Archer Daniels Midland, Bunge, Louis Dreyfus, Cargill, COFCO International and Glencore Asking Them to Stop Sourcing Soy from Deforested Areas in the Cerrado."

Ritchie, Hannah, and Max Roser. 2019. "Land Use." Our World in Data. https://ourworldindata.org/land-use.

———. 2021. "Forests and Deforestation." Our World in Data. https://ourworldindata.org/soy.

Roth, Matthew David. 2018. *Magic Bean: The Rise of Soy in America*. Lawrence, KS: University Press of Kansas.

Schenkelaars, Piet, Huib de Vriend, and Nicholas Kalaitzandonakes. 2011. "Drivers of Consolidation in the Seed Industry and Its Consequences for Innovation | Forum Grüne Vernunft." Commissie Genetische Modificatie (COGEM) Dutch Commission for Genetic Modification.

Schnitkey, Gary 2013. "Concentration of Corn and Soybean Production in the U.S." *Farmdoc Daily* (3) 130. https://farmdocdaily.illinois.edu/2013/07/concentration-corn-soybean-production.html

Secretary of State for Economic Affairs. 1970. "Memorandum from the Deputy Assistant Secretary of State for European Affairs (Springsteen) to the Deputy Under Secretary of State for Economic Affairs (Samuels)." *Foreign Economic Policy*, 1969–1972.

Shurtleff, William, and Akiko Aoyagi. 2004. "History of World Soybean Production and Trade – Part 2. A Chapter from the Unpublished Manuscript, History of Soybeans and Soyfoods: 1100 B.C. to the 1980s." Lafayette, CA: Soyinfo Center.

———. 2007. "History of Soybean Production and Trade in China (1949–1980s) – Part 1. A Chapter from the Unpublished Manuscript, History of Soybeans and Soyfoods, 1100 B.C. to the 1980s." Lafayette, CA: Soyinfo Center

———. 2009. *History of Soybeans and Soyfoods in South America (1882–2009): Extensively Annotated Bibliography and Sourcebook*. Lafayette, CA: Soyinfo Center.

———2019. *History of the national oilseed processors association: 1930-2019*. Lafayette, CA: Soyinfo Center.

———. 2020a. *History of ADM (Archer Daniels Midland Co.) and the Andreas Family's Work with Soybeans and Soyfoods (1884–2020): Extensively Annotated Bibliography and Sourcebook*. Lafayette, CA: Soyinfo Center.

———. 2020b. *History of Cargill's Work with Soybeans and Soy Ingredients (1940–2020)*. Lafayette, CA: Soyinfo Center.

———. 2021. *History of International Trade in Soybeans, Soy Oil and Soybean Meal, and Trade Policy (1859–2021): Extensively Annotated Bibliography and Sourcebook*. Lafayette, CA: Soyinfo Center.

Silver, Beverly J., and Giovanni Arrighi. 2003. "Polanyi's 'Double Movement': The Belle Èpoques of British and U.S. Hegemony Compared." *Politics & Society* 31 (2): 30.

Simung, Wu. 2021. "China Merges Sinochem, ChemChina to Create World's Biggest Chemicals Firm." *Yicai Global*. https://www.yicaiglobal.com/news/sinochem-chemchina-merge-to-form-world-biggest-chemical-giant-.

Singh, Shikha, and Mugunthan Kesavan. 2022. "Ukraine Farmers Switching to Beans from Corn on Export Headwinds: Agribusiness Club." S&P Global Commodity Insights. May 24, 2022. https://www.spglobal.com/commodityinsights/en/market-insights/latest-news/agriculture/052422-interview-ukraine-farmers-switching-to-beans-from-corn-on-export-headwinds-agribusiness-club.

Smil, Vaclav. 2004. *China's Past, China's Future: Energy, Food, Environment*. New York: Routledge Curzon.

Song, Xiao-Peng, Matthew C. Hansen, Peter Potapov, Bernard Adusei, Jeffrey Pickering, Marcos Adami, Andre Lima, et al. 2021. "Massive Soybean Expansion in South America since 2000 and Implications for Conservation." *Nature Sustainability* 4 (9): 784–92.

Soybean Meal Info Center. 2018. INFO Source Newsletter. https://www.soymeal.org/soy-meal-articles/world-soybean-meal-production-august-2018/.

Spangler, Burchfield, and Schumacher. 2020. "Past and Current Dynamics of U.S. Agricultural Land Use and Policy." *Frontiers in Sustainable Food Systems* 4 (98): 1–21.

Stanton, Patricia, and John Stanton. 2002. "Corporate Annual Reports: Research Perspectives Used." *Accounting, Auditing & Accountability Journal* 15 (4): 478–500.

Statista. 2022. "Leading Agricultural Futures Options." www.statista.com OR https://www.statista.com/statistics/380413/leading-agricultural-futures-options-by-volume/

Steffen, Will, Wendy Broadgate, Lisa Deutsch, Owen Gaffney, and Cornelia Ludwig. 2015. "The Trajectory of the Anthropocene: The Great Acceleration." *The Anthropocene Review* 2(1): 81–98.

Stiglitz, Joseph, and Foundation CIDOB and Policy Dialogue. 2004. "The Post Washington Consensus Consensus." Presented at the From the Washington Consensus towards a New Global Governance, Barcelona.

Surls, PM and F.C. Tuan. 1981. *China's Agriculture in the Eighties* Washington, DC: USDA ERS (Economic Research Service).

Texo, Juan, Ignacio Bentancur, and Juan Duque. 2009. "Perspectivas Generales de Desarrollo de La Indústria de Los Biocombustibles En El Uruguay." Montevideo: Universidad de la República, Udelar.

Thoyer, S. 1998. "*L'economie Politique Des Soutiens à L'agriculture Dans Les Pays Développés*." In *Agriculture et Alimentation En Quête de Nouvelles Légitimités*, edited by Gérard Miclet, and Sirieix Thoyer, 269–301. Paris: Economica.

Traders of Dreyfus. Letter to Commerce and Logistics. 2008. "Louis Dreyfus Commodities, LDC," February 19, 2008.

Trigo, Eduardo and Eugenio Cap. 2003. "The Impact of the Introduction of Transgenic Crops in Argentinean Agriculture." *AgBioForum* 6 (3): 87–94.

Troell, Max, Rosamond L. Naylor, Marc Metian, Malcolm Beveridge, Peter H. Tyedmers, Carl Folke, Kenneth J. Arrow, et al. 2014. "Does Aquaculture Add Resilience to the Global Food System?" *Proceedings of the National Academy of Sciences* 111 (37): 13257–63.

Turzi, Mariano. 2016. *The Political Economy of Agricultural Booms: Managing Soybean Production in Argentina, Brazil, and Paraguay*. Cham: Palgrave Macmillan, Springer Nature.

U.S. Soybean Export Council. 2008. "How the Global Oilseeds and Grain Trade Works." South West Harbour: High Quest Partners and Soyatech; Soyatech.

USDA. 2016. *Overview of the United States Cattle Industry*. U.S. Agricultural Trade. Washington, DC: United States Department of Agriculture.

———. 2022. *United States Department of Agriculture Oilseeds: World Market and Trade.* World Agricultural Reports. Washington, DC: United States Department of Agriculture Foreign Agricultural Services.

USDA FAS. 1957. *Foreign Crops and Markets World Summaries.* Washington, DC: United States Department of Agriculture, Foreign Agricultural Service.

———. 1959. *Foreign Crops and Markets World Summaries.* Washington, DC: United States Department of Agriculture, Foreign Agricultural Service.

———. 1961. *Foreign Crops and Markets World Summaries.* Washington, DC: United States Department of Agriculture, Foreign Agricultural Service.

———. 2022. "Oilseeds and Products Update Country: China – People's Republic of Post: Beijin." Oilseeds and Products CH2022-0075. USDA FAS.

USDA NASS. 1952. *Annual Statistics 1951.* US Agricultural Trade. Washington, DC: United States Department of Agriculture National Agricultural Statistics Service.

———. 1954. *Annual Statistics 1953.* US Agricultural Trade. Washington, DC: United States Department of Agriculture National Agricultural Statistics Service.

———. 1956. *Annual Statistics 1955.* US Agricultural Trade. Washington, DC: United States Department of Agriculture National Agricultural Statistics Service.

———. 1957. *Annual Statistics 1956.* US Agricultural Trade. Washington, DC: United States Department of Agriculture National Agricultural Statistics Service.

———. 1958. *Annual Statistics 1957.* US Agricultural Trade. Washington, DC: United States Department of Agriculture National Agricultural Statistics Service.

———. 1959. *Annual Statistics 1958.* US Agricultural Trade. Washington, DC: United States Department of Agriculture National Agricultural Statistics Service.

———. 1960. *Annual Statistics 1959.* US Agricultural Trade. Washington, DC: United States Department of Agriculture National Agricultural Statistics Service.

———. 1961. *Annual Statistics 1960.* US Agricultural Trade. Washington, DC: United States Department of Agriculture National Agricultural Statistics Service.

———. 1962. *Annual Statistics 1961.* US Agricultural Trade. Washington, DC: United States Department of Agriculture National Agricultural Statistics Service.

USDA PSD. 2022. "United States Department of Agriculture Production, Supply and Distribution Online." PSD Online.

Vergara-Camus, Leandro, and Cristobal Kay. 2017. "New Agrarian Democracies: The Pink Tide's Lost Opportunity." *Socialist Register* 54: 224–43.

Vitousek, Peter M., Harold A. Mooney, Jane Lubchenco, and Jerry M. Melillo. 1997. "Human Domination of Earth's Ecosystems." *Science* 277 (5325): 494–99.

Volante, J., J. Mosciaro, M. Morales Poclava, L. Vale, Silvana Castrillo, Jorge Sawchik, Guadalupe Tiscornia, et al. 2015. "*Expansión Agrícola En Argentina, Bolivia, Paraguay, Uruguay y Chile Entre 2000–2010: Caracterización Espacial Mediante Series Temporales de Índices de Vegetación.*" *RIA. Revista de Investigaciones Agropecuarias* 41 (2): 179–91.

Voora, Vivek, Cristina Larrea, and Steffany Bermudez. 2020. "Global Market Report: Soybeans." Sustainable Commodities Marketplace Series 2019. International Institute for Sustainable Development; State of Sustainability Initiatives.

Waltz, Emily. 2015. "First Stress-Tolerant Soybean Gets Go-Ahead in Argentina." *Nature Biotechnology* 33: 682.

Ward, Michael. 2018. "Grain and Feed Update, Peoples Republic of China, Drought, Floods, and Storms Buffet China's Grain Market." GAIN Report. USDA – Foreign Agricultural Service.

Werner, Dietrich, and William E. Newton. 2005. *Nitrogen Fixation in Agriculture, Forestry, Ecology, and the Environment.* Dordrecht: Springer Science & Business Media.

Wesz Jr, Valdemar João. 2016. "Strategies and Hybrid Dynamics of Soy Transnational Companies in the Southern Cone." *The Journal of Peasant Studies* 43 (2): 286–312.

Wilcox, J. F. 1970. "Breeding Soybeans for Protein and Oil Quantity and Quality." In *Soybean Breeding Conference*, edited by Fehr 103–09. Ames, IA: Department of Agronomy, Iowa State University.

Wolford, W., S. M. Borras Jr, R. Hall, and I. Scoones. 2013. *Governing Global Land Deals. The Role of the State in the Rush for Land*. First Edition. Malaysia: Wiley Blackwell.

Wright, Christopher K., and Michael C. Wimberly. 2013. "Recent Land Use Change in the Western Corn Belt Threatens Grasslands and Wetlands." *Proceedings of the National Academy of Sciences* 110 (10): 4134–39.

WTO. 1995. "Agreement on Agriculture." World Trade Organization.

———. 1996. "World Trade Organization GATT Bilateral Negotiating Material by Round." World Trade Organization.

Yang, Martin C. 1945. *A Chinese Village*. New York: Columbia University Press.

Zhang, Qing, and Lifeng Liu. 2014. "Perspective on the Trend of Soybean Production and Trade in China." In *Proceedings of Selected Articles of 2013 World Agricultural Outlook Conference*, edited by Xu, 19–28. Berlin: Springer.

Zoe Liu, Zongyuan. 2023. "China Increasingly Relies on Imported Food. That's a Problem." Council on Foreign Relations (CFR). Cf. https://www.cfr.org/article/china-increasingly-relies-imported-food-thats-problem (Accessed 2023-01-25).

6 Historicizing soy

Toward a new rupture?

We began this book with the observation that soy has become ubiquitous in the agrofood system. Processed soybeans are the world's largest source of animal protein feed and the second largest source of vegetable oil. But while these roles are clear, today's global food order is characterized by deep contradictions; growing meat consumption is a significant predictor for obesity, and yet, "meatification" coexists with rising rates of hunger, food insecurity and malnutrition (FAO et al. 2022; Khodayari et al. 2022). Moreover, soybean production is a major driver of biodiversity loss, deforestation, pollution, land degradation and significant greenhouse gas emissions (Nabuurs et al. 2022). While all agricultural activity changes the landscape and typically reduces levels of biodiversity, the pace and scope of current soybean expansion and intensification is historically unprecedented. As we argued in Chapter 5, contemporary patterns of soybean production and consumption raise deep concerns over social-ecological sustainability, especially as we move toward a future of increasing variability, and as foundational ecological processes are altered, perhaps most immediately, in the form of climate change (Steffen et al. 2018).

Counterintuitively, the more unstable and uncertain the present, the more important history is in making sense of it. The present is a product of history and the legacies of the past shape and constrain our choices today. One obvious way in which this happens is through the long history of agricultural frontier expansion. Notwithstanding all the important differences distinguishing its different cycles, the *longue-durée* structure of soy production has resulted in the loss of fertile land and few remaining areas with high ecological integrity. There is simply not much frontier left. The continuing frontier-based development model is thus losing its material foundations. As we write these lines, the world's population surpassed the 8 billion mark. Between 2013 and 2019 only, population increased by 21%, while world per capita cropland area decreased by 10%. The ratio between arable land and population has only increased in one continent, South America, but this new frontier expansion can hardly continue for much longer (Potapov et al. 2022). Even if the mechanics of soybean growth during the present era are historically unique, they still build on institutional, technological, social-ecological and cultural legacies. To repeat, the past lays down both the foundations and limitations of the future (Braudel and Wallerstein 2009). But history is also made

DOI: 10.4324/9780367822866-6

up of shifts and transformations: things have been different and soy has played other roles. There is nothing natural or spontaneous about the passage into the contemporary soy cycle when one takes a historical perspective.

For millennia, most soy production was small scale, labor-intensive and diversified. One of its main roles was to resupply agricultural soil with nitrogen, which was especially beneficial in crop rotation. Other legumes fulfilled a similar role in European agriculture, helping maintain a structural complementarity between crops and livestock (Gorman 2013a, 33). But, as we know, mixed-farming systems began to dissolve in the West around the time when the third soy regime was in its roots phase. A long-term process (structure) of specialization in the core of the world economy resulted in an increasing reliance on the import of distant nutrient sources (Manchurian soy cake shipped to fertilize fields in South China and Japan, for instance, or Peruvian guano shipped to fertilize Europe and the United States). A crucial catalyst in this trend was the invention and spread of cheap synthetic fertilizers. Mixed systems in advanced economies began to disintegrate and agriculture expanded and specialized, ending its dependence on the restorative power of legumes in crop rotations (Gorman 2013b, 61). While gradual, and still on-going, the shift has been deep: organic, circular agricultural systems moved toward industrial ones, relying on the cumulative use of external inputs, breaking the nutrient cycle. As a result of the lack of rotations, the use of synthetic fertilizers continues to increase significantly, leading to negative side effects such as eutrophication and greenhouse gas emissions. Since rotation also acts as a natural system of pest control (e.g. weeds and insects), the move away from it creates a need for even larger use of chemical pesticides (Chimonyo, Snapp, and Chikowo 2019). As we can see, then, a full discussion of the social-ecological challenges and contradictions of the current soy regime is brought into sharp focus by looking at the legacies and breaks of the last few years relative to the longer history of soy.

We have argued that each soy cycle's regime phase should be understood as a temporary constellation of formal and informal rules and relations of relative stability. But regimes are historical and, eventually, contradictions and limits trigger their rupture and end. The current soy regime is no different, and this chapter will consider the extent to which it has already entered its rupture phase. We will focus on three different types of identified challenges, which, we argue, could precipitate its coming apart. The first is a crisis of structure in the *longue durée*, which involves the idea that frontier expansion and ecological simplification are close to reaching more or less absolute biophysical limits. The second is a looming crisis in the logic of accumulation of soy production and commercialization, which could threaten its future profitability. The third and final type of challenge is political. It is manifested in a growing contestation and resistance to the soy model. This popular type of countermovement could compound the already noticeable crisis of legitimacy in the West. China's competition with the United States for the control of the global soy complex is yet another challenge considered below.

A crisis of structure? limits to frontier expansion and ecological simplification

Today's frontier-like soybean expansion in the Southern Cone has antecedents in world history. Humans have always sought to overcome scarcity – whether biophysical (poor soils) or political (wars) – by trying to exploit new spaces and resources, or "frontiers". A frontier can be described as an area of unusually abundant natural resources relative to economic capital and labor (Barbier 2011, 5, 7, 31). Ultimately all agricultural activities are about manipulating ecosystems: reducing their structural complexity, simplifying their functions and specializing all means to boost production (measured as yield). As agriculture expands, the variety of heterogeneous landscapes previously filled with diverse biological communities and abiotic structures, are homogenized and simplified (Friedman 2000, 481–83; Tscharntke et al. 2005).

There are, nevertheless, some important distinctions to bear in mind. Commodity-frontier expansion can happen in one of two ways. The first is extensive, or horizontal, when a piece of non-agricultural land (forest, savannah or wetland) is incorporated into any agricultural system. The consequence is a dramatic reduction of complexity, diversity and integrity, along with a disruption to the many ecosystem services they provide, such as water and climate regulation (IPBES 2020). The second type of frontier expansion, however, is intensive, or deep, and refers to the incorporation of already agriculturalized (commodified) spaces through the use of different technologies to "boost" yields (Langthaler 2020, 245–47). Intensive-frontier expansion includes developments such as high-yielding seeds, irrigation or synthetic fertilizers. This intensification of land use often causes severe social-ecological problems in the medium and long term. One of the most recurrent problems is soil degradation and erosion. While the decline of soil fertility and soil erosion do not need to be an irreversible process – the experience of the antebellum American South provides plenty of evidence to the contrary (Gray and Thompson 1933) – the harm is not always easy to correct. As economic historian Ernst Langthaler notes, commodity frontiers have expanded in both ways throughout history: "ranging between predominantly extensive and intensive modes of incorporation into regimes of capitalist accumulation and regulation" (Langthaler 2020, 246).

Soy and frontiers in ancient China

Long before the spread of capitalism, more than 2,000 years ago, soybeans were part of a high-yielding, crop-centered system in ancient China, where rotations with soy allowed for land-use intensification through continuous cropping. As we saw in Chapter 2, soybeans were produced in varied and complex crop rotations. Such a system included not only several different crops but many varieties of each cultivar, which allowed for a relatively high degree of biological complexity. Moreover, the nitrogen-fixation capacity of soybean plants (though still not understood in these terms) made them a popular way of enriching soils. The relatively small farming units, combined with a philosophy of product diversification and self-sufficiency which hindered specialization and monocultural production, promoted multifunctional farming. This multifunctionality diversified risks, reducing the danger of a comprehensive crop failure. It was also labor-intensive – particularly meticulous weeding was needed – and farmers had to develop multiple skills and knowledge about all the different species interactions. Soybeans in this system could be relied upon to yield between 5 and 10 bushels per *mu*, or approximately 2–4,000 kg/ha (Bray 1984, 514–15), similar to average soybean yields today, and many times higher than in Europe before the introduction of chemical fertilizers (Ibid, 7).

The limits of the Chinese frontier model

In broad strokes, from the 9th century CE, China followed a frontier expansion model to alleviate the recurring specters of scarcity and famine. The careful management of soils seems to have enabled high yields and the model seemed sustainable over time. However, as demographic pressure rose, the agricultural frontier expanded southwards, bringing massive deforestation, soil erosion and even the alteration of local precipitation patterns. This system eventually hit something like a Malthusian limit in the mid-18th century, as we saw in Chapter 3, when almost all land had been incorporated into agriculture and productivity (yields) stagnated. Interestingly, the model might have reached this limit much earlier but for the predominance of vegetarianism. All over East Asia, agrofood systems were centered on crops, with relatively little importance given to livestock. This value system seems to have been justified on multiple reinforcing grounds. For one thing, the dominant Confucian mentality extolled values anchored in tradition, "laws of nature" and humility; eating meat was associated with extravagance and greed; and soybeans were associated with austerity and contentment (Sterckx 2019, 283). Moreover, several records show an acute awareness of the large amount of resources (particularly land) required to sustain a meat-based diet. Emperors sought to maintain a growing population and avoid popular revolts, and thus also saw vegetarianism as a necessary ideal to make a scarce resource (land) produce enough to forestall further frontier expansions into remote areas where the state could not be sure it could impose stability and order. Western Europe, by contrast, already demanded access to vast, varied and cheap supplies of extra-European resources. Indeed, the exploitation of the world's "Great Frontier" was instrumental to the economic boom experienced in the new European metropoles (Barbier 2011, 7–9). However, transport costs were initially relatively high: partially constraining export-driven frontier expansion in the periphery until around the late-19th century, when steamships, railways and other innovations made long-haul freights rates decline sharply (O'Rourke and Williamson 1999, 33–37). For the first time in history, bulk agrofood commodities could be traded massively.

Frontier expansion at the turn of the century: the case of Manchuria

It was during the late-19th century wave of agrofood globalization that the first specialized and massive frontier expansion centered around soybeans emerged. Manchuria represented a "classic" case of frontier development – new land was incorporated into agriculture, creating a rapid increase in soy output, with decreasing yields per unit of land (Langthaler 2020, 247). While frontier soybean agriculture in Manchuria was a crucial step toward simplification, it was not yet a monoculture, as soy was rotated with other crops and still helped to restore some of the nutrients that the other cultivars (not least sorghum) mined from the soil. Population growth in China, coupled with the elimination of mobility and settlement restrictions, propelled a massive migration flow of Han-Chinese people into Manchuria. From 1860 to 1940, about 8 million Han Chinese settled in the region. During this period, the population rose 12-fold, while cultivated areas expanded by a factor of 9 (Federico 2010, 32). Undoubtedly, cash crop production supplying the world market with agrofood commodities, irrespective of tenure, has been associated with a constant search for new fertile land to be exploited. Indeed, this was the heyday of imperialism. The discovery and "development" of land and other natural resources often coincided with the emergence of new regional or global economic powers. Manchuria was subject to imperialist rivalries and extractive companies, explored in depth in Chapter 4. At the same time, Manchurian soy, in the form of bean cake, helped intensify Asian agriculture, enabling specialization, the simplification of systems and the disruption of the nutrient cycle. Japan completely conquered Manchuria in the early 1930s, but its power ambitions were ultimately smashed in Hiroshima and Nagasaki. While Manchuria ceased to be the world's soy frontier when World War II cut off global trade, the West's exploitation of new "Great Frontiers" continued.

The frontier within: soy in the United States

The next soy frontier, was not in the global periphery, however, but in the heartland of what had become the world's most powerful industrial nation – the United States. Soy's rise to a premier crop in the American mid-West also took the form of a frontier expansion, but a predominantly (though not exclusively) intensive one, marked by the logic of capitalist accumulation. As we saw in Chapter 4, during the roots phase of this soy cycle, soybeans first entered the United States as a forage crop within mixed systems, where most animal fodder and seeds were still produced on-farm. However, as synthetic fertilizers and commercial seeds spread and farming systems became increasingly specialized, agroecosystem simplification greatly accelerated. As the United States became the epicenter for global production and exports, the soy management model became capital-intensive, highly specialized – with just a few commercial crops (usually, corn) – and mostly reliant on Green Revolution technologies, such as improved seeds, irrigation, agrochemicals and machinery – thus depending on intensification to increase production.

Animal farmers also intensified and stopped producing their own fodder and instead bought protein-rich feed, most often with soy as a central ingredient. Thus, the industrialization of agriculture, including a mechanization boom, which had begun to take shape in the 19th century, exploded after World War II, moving the third soy cycle into its consolidated regime phase. The resulting rise in output came at a price, however. Besides the loss of biodiversity, this modern type of agriculture was also responsible for other significant social-ecological damage, such as water pollution, displacement of small farmers, soil erosion and the release of new, highly toxic substances.

The frontier moves South

While soy expanded continuously in the United States during the current regime phase of the Great Acceleration, the center of soybean production eventually gravitated south. Latin America became the world's largest soy producer. This shift took place in a context characterized by flexible currencies, debt crises and structural adjustment programs imposed on the region. In 1995, the WTO's *Agreement on Agriculture* removed many of the protections (trade barriers) previously allowed for agrofood trade (Baraibar Norberg 2022). Under this "free market" regime, hundreds of bilateral and multilateral investment agreements were signed, strengthening the hands of foreign businesses and making it easier for them to pursue legal action against governments. The state's ability to use and allocate land according to imperatives other than those of the market was further weakened (Ankersen 2006, 113–19; Romson 2012, 25–26, 359). Moreover, Latin American governments, motivated by debt-driven necessity and/or domestic political consent, moved to privatize land, initiated land-titling programs, facilitated foreign direct investments and adopted more flexible approaches to land leasing (Fearnside 2001). All this worked to dissolve the social function of land (Margulis and Porter 2013) and contributed to the continued expansion of soy.

Economic globalization encouraged producers to move sequentially across South American biomes, in search of new frontiers where fertile land was easily available (Lambin and Meyfroidt 2011). As a result, harvested areas in Brazil, Argentina and Paraguay more than tripled over the past 30 years, covering more than 50 Mha. In this way, soy-driven land use and land-cover change has caused forest degradation and fragmentation in the world's largest remaining resource frontiers in the Amazon and the Gran Chaco forest biome in Bolivia, Paraguay, Argentina and Brazil. Besides deforestation, soy has directly or indirectly also driven land-use conversion across other biodiverse zones like the Cerrado in Brazil and the Argentinian and Uruguayan Pampas (Garrett and Rausch 2016). Frontier expansion has come hand in hand with loss of biodiversity and natural carbon storage, the disruption of water cycles and enhancement of the risks of outbreaks of zoonotic diseases as humans encounter species to which we have no immunity (Ellwanger et al. 2020; Baraibar Norberg 2022, 110).

In addition, the soy-frontier model in Latin America has not only involved an extensive, horizontal expansion, but also an intensive, deep, frontier expansion by modifying practices through monoculturalization and increased use of agrochemicals as well as other technologies to boost crop yields in spaces that were already in cultivation or grazed. This has increased pressures on soil, water and other key resources. Today's high-tech, large-scale soybean production based on genetically-modified seeds has now been adopted almost exclusively throughout the Americas. The Gene Revolution has only further increased the level of ecological simplification from earlier soy models to the extreme. The fusion of frontier intensification and extensification has had a rapid, visible and essentially negative effect on ecosystem integrity and food security. It has perversely increased our use of resources, particularly more synthetic and chemical inputs, in order to compensate for the loss of natural ecosystem services – e.g. pest regulation, carbon sequestration and soil fertility. Thus, genetically engineered (GE) technology has not delivered on its promise of reducing resource use, but GE soybeans are instead one of the main drivers of increased environmental degradation and resource use (GRAIN 2013; Pendrill et al. 2019; Boanada Fuchs 2020; Ritchie and Roser 2021); although, of course, all "modern" agriculture is based on a similar logic (Friedman 2000). According to FAO's *Statistical Yearbook 2021*, between only 2000 and 2019, 127 Mha of land were lost, degraded and no longer suitable for cultivation – an area roughly the size of Niger (FAO et al. 2022).

As for forests, the global deforestation trend is undeniable – but it is also geographically uneven. While forest areas have expanded in Europe and Asia between 2000 and 2019 (FAO 2022, 49) – in the Americas – especially in Paraguay and Brazil – large areas of forest have been converted to agricultural lands. Latin America is likely to remain the central soy frontier for years to come. And as global crop outputs continue to supply growing demand from an intensified livestock industry (Nabuurs et al. 2022), arable land reserves will further shrink. Plainly, the extensive frontier expansion model cannot continue like this for much longer. The area of arable land available per person decreased from about 0.45 ha (1960) to about 0.25 ha (2010) and will continue to fall below 0.20 ha per capita after 2020 (Flachowsky, Meyer, and Südekum 2017). While the overall tendency is the result of a very long historical trend, the logic and forms of accumulation of the current soy cycle have accelerated the pace of the long-term structure of frontier expansion.

An important lesson is, then, that the approach to land areas as "Great Frontiers" has reached an end. Many biophysical processes crucial for functioning agroecosystems have limits that we are fast approaching (e.g. land use and biodiversity loss) and some that we have passed (e.g. nutrients and climate change) in a "race to the bottom" manner. Land degradation, broken nutrient cycles, biodiversity loss, losses of protein and energy in animal production and increasing use of toxic pesticides are just some of the problems we have created with this mindset. A fundamental change in worldview – away from Nature as an endless sink and source – is needed.

A crisis of profitability? The end of "cheap" soy

As we asked at the start of this book, what turned the soybean into the number one oilseed in the contemporary food system? Just like the peanut, the soybean is both an oilseed and a legume, but it has gained far more economic importance than all other oilseeds (e.g. sunflower, canola and cottonseed) and all other legumes (e.g. beans, peas, chickpeas, lentils, lupins, alfalfa and clover). Soy has several unique qualities which make it highly attractive: its nitrogen-fixation capacity, its hardiness and its high-protein and oil contents. Ironically, soy's capacity to bring nitrogen into the soil in a form that plants can assimilate actually ceased to be valued by farmers just as soy cultivation took off in earnest. Inexpensive synthetic fertilizers were used instead to supply nutrients to the soil, so the impressive growth of soy during the present regime phase (1950–today) had nothing to do with soil enrichment. Instead, it was a thirst for first industrial and then also edible oil that greatly raised the profile of soy oil, the extraction of which yielded soy cake or meal. Soon, soy's main use shifted from oil to a protein input for animal production. Whatever its advantages and qualities as crop, however, soy's "success" under the current regime phase depends on the fact that it is cheap. As we argued in Chapters 4 and 5, soy only entered a new market segment (whether industrial uses, margarine or feed) because it offered a cheaper way of producing the end product compared to other inputs. As mounting volumes of cheap soy meal from oil extraction became available in the United States, companies saw an opportunity to produce larger quantities of meat, dairy products and fish more cheaply than before. Under a capitalist mode of production, economic actors try to maximize profits by systematically cutting costs and selling their commodities at lower prices.

Modern agriculture is based on a model of boosting supply at low cost: this has essentially underwritten the tenfold agricultural output increase between 1800 and 2000 (Federico 2010, 5). During the latest phase of this period, 1950 to 2000, world agrarian production grew on average 2.3% yearly – thus, it more than tripled while the world population more than doubled (Ibid, 19). Moreover, agricultural prices decreased in relation to per capita income over the same years (Ibid, 24). "Accumulate, Accumulate! That is Moses and the Prophets!", writes Marx famously in the first volume of *Capital*. Cheap supplies are one of the "commandments" of contemporary agricultural production, particularly, in those commodities that are inputs in the value chains of other (higher-value) products. The whole soy model of the current regime phase rests on this fact: if soy were not produced and sold cheaply, we would not live in a "world of soy". But, as we have shown in this book, and particularly in Chapter 5, soy is only cheap because its full social-ecological costs remain hidden from its price. Carbon emissions, biodiversity loss, habitat degradation, land grabbing, polluted waterways and obesity are all "externalized": profits can be made while risks and costs are transferred to geographically distant locations and/or to the future (McMichael 2009, 160; Clapp 2014, 799–806). Under a system of capitalist accumulation, these negative externalities are concealed, but so are other positive externalities potentially generated when soy is used differently: for example, in rotation schemes with other crops and grasses.

The soy commodity chain is embedded in a political economy of globally institutionalized "free" markets, in which the largest agribusinesses have relentlessly pursued economies of scale to hold down prices. In fact, there are important economies of scale involved in all stages of the chain. On-farm production stages offer a strong competitive edge to large firms, putting small farmers out of business. A small producer from Uruguay, explained his decision to sell his land and give up farming thus: "[T]he logic today is that you have to be big, or you have to dedicate your time to something else" (Baraibar 2014, 243). The success of soy over other crops has spurred concentration of farmlands in the hands of a small number of large and powerful growers. Significant economies of scale are equally at play in the input (biotech, seeds and agrochemicals) and processing/trading stages. Upfront costs are very high and property rights are very strong. Moreover, the advancement of transnational agribusinesses has formed conglomerates active at several stages: from biotechnology, agrochemicals, finance and trade to land markets, infrastructure and processing, thus strengthening their hold over the entire value chain (Lang and Heasman 2015; Clapp 2016, 97–99; Baraibar Norberg 2020a). As we have seen in Chapters 4 and 5, large companies have specialized in using all kinds of strategic assets (financial, technological, "supply chain management" and organizational) to impose the specific sets of relations, techniques and quality standards which enable them to appropriate most of the surplus generated throughout the chain. But if soy is ubiquitous because it is cheap, cheap soy is now under threat on at least three different fronts: first, stricter environmental regulations and other attempts to internalize social-ecological costs into firms' costs; second, the technological treadmill; and third, lawsuits against biotech/agrochemical companies.

The challenge to profitability from stricter regulations and traceability requirements

If traders are to continue reducing costs, the ability to source soy anywhere is the *sine-qua-non* condition of the model. On the one hand, this requires that competition between farmers is high enough to maintain a downward pressure on prices. On the other hand, soy from "anywhere" must be easily mixed, to keep down logistics and transportation costs. Since storage and transport expenses weigh heavily in the cost structure of the soy complex, traders have developed a highly efficient supply chain practice, organized in accordance with the logic of *Just-in-Time* production, i.e., with a minimum amount of inventory on-hand ready to meet demand. Competing traders (say, Cargill and Louise Dreyfus Commodities) pay dearly for every minute their ships sit at port waiting to be loaded or unloaded. In order to avoid this, traders help each other load cargo as quickly as possible. In the Uruguayan port of Nueva Palmira, for instance, firms have found they can collaborate: they fill their incoming ships with whatever soy is at hand, independently of who owns it, and thus reduce freight rates and maritime transport costs. When the ship leaves port, traders square the accounts, paying, or getting paid, for the borrowed tons of soy (Baraibar 2014, 170–73). Such smooth supply-chain management might create overall efficiency gains, but it does so at the cost of product differentiation. Soybeans from a wide range of sources are mixed together; in the process, quality distinctions are almost completely lost, and so is any possibility of considering where the soybeans were grown, in which ways and by whom. In this sense, traders have made soy completely interchangeable, or, if one prefers, fully commodified.

This system, in turn, depends on a market driven purely by a cost-benefit logic, entirely indifferent to any social or environmental externalities. However, as climate change and biodiversity loss gain space on the international policy agenda, so does awareness about the damage wrought by agricultural extensification and intensification. About a third of all greenhouse gases come from agricultural production (Steffen et al. 2018). Between 2001 and 2015, cattle, palm oil, soybean and cocoa were the four main commodities driving deforestation. Climate change is expected to continue accelerating the rate of extinction of a large number of species, further undermining food security (IPCC 2014, 13; zu Ermgassen et al. 2022). The wealth of nature has been transferred to the private domain, while the effects of environmental destruction are being socialized in a geographically uneven way. The growing global appetite for food, feed and fuel is making itself felt mostly in anonymous and distant spaces, far removed from end consumers who are often unaware of their own soy consumption. But predictably, all of these trends generate points of resistance and countermovement.

A growing number of voices – in social movements but some even within policy circles – have begun calling for a reform in soy markets. The claim is that soy's contribution to climate change and biodiversity loss should be "paid" as baseline resource depletion, and that shareholders should not be put above the protection of biodiverse ecosystems or human rights, including indigenous peoples' rights to the land. While the same arguments can be made of many crops, soy's significant biotechnological amelioration and economic demand have turned it into a global-media villain of environmental destruction (Relly and de Majo 2022, 149). *Sojización* has been happening against the backdrop of rising environmental activism and the politicization of global ecological issues, and both media and policy circles have begun turning their focus on soy as both a direct and indirect driver of deforestation across the Brazilian Amazon and Cerrado, and the Paraguayan and Argentinian Chaco (zu Ermgassen et al. 2022; Lopes et al. 2021). In this context, several reformist initiatives aiming to "correct" the international soy market and make it more sustainable have emerged over the last years as pressure for tougher regulations mounted. We consider the more radical countermovements in the next section.

First and foremost, several Latin American states have strengthened environmental laws for forest protection, and some of them have announced their commitment to "zero deforestation" (de Castro, Hogenboom, and Baud 2016; Martinez-Alier, Baud, and Sejenovich 2016). These are promising, though insufficient, first steps. In some cases, environmental standards improved but, partly as a response, soy production moved to less-protected forest areas; in other cases, livestock moved into deforested areas first, to be replaced by soybean cultivation after only two or three years, making it difficult for regulators to discern what the original drivers of deforestation were and, thus, complicating the full implementation of environmental standards (Rocha et al. 2019). Moreover, as soy displaced ranching activities in grasslands – not least in the Pampas – livestock moved into forests. An important and illustrative example is Brazil, where soy cultivation constitutes the largest economic source of land use after cattle. Several efforts to reduce deforestation rates were implemented during Luiz Inácio Lula da Silva's first terms in office (2003–10), including the designation of indigenous lands, the demarcation of new conservation areas, tougher penalties in the enforcement of deforestation restrictions and sanctions directed at those local jurisdictions with the highest rates of deforestation. The state also implemented public property registries and a deforestation monitoring program to curb illegal logging. Deforestation directly related to soy production fell significantly up until 2019, although the total area planted with soy in the Amazon increased from 1.6 to 4.7 Mha in the same period (Rausch and Gibbs 2021). Critically, recent studies have revealed significant leakage effects, meaning that Amazon deforestation simply shifted to the Brazilian Cerrado. In fact, as of late summer 2022, this ecoregion – the world's most biodiverse savannah – already supported roughly half of the country's total soy production, suffering deforestation and conversion rates almost four times higher than in the Amazon over the course of only the previous year. Much of this is directly related to erstwhile President Jair Bolsonaro's policy of slashing funds for environmental agencies between 2019 and 2022. Much hope for a strengthened environmental mandate now rests on Lula's reelection as head of state.

Another focal point of regulatory reform comes from agribusinesses themselves. The so-called chain drivers have also announced a range of initiatives during the past decades, allegedly in order to improve soybean traceability and assist efforts to ensure soy does not come from deforested areas. One prominent and well-known example is the so-called "Roundtable on Responsible Soy" (RTRS), a global multi-stakeholder initiative in which all big traders and other soy-producing or managing agribusinesses take part (among the most salient: ADM, Bunge, Cargill, Dreyfus, Unilever, Bayer-Monsanto, Syngenta and COFCO). It involves a certification scheme based on a set of measures to prevent further soy expansion into native forests, ringfencing all areas which are not already in agriculture or pasture over the past 12 years (Ismail, Rossi, and Geiger 2011). Since the largest traders are estimated to manage more than 80% of international soybean trade and control 50% of installed crushing capacity (Oliveira 2022, 214), all reforms taken by the ABCD companies and the Chinese state-owned mega company COFCO should have a noticeable impact on the soy chain. However, the RTRS was established in 2006 and several studies using satellite imagery have shown that soy-driven deforestation has only continued to rise. As with so many initiatives and pledges emanating from the realm of Corporate Social Responsibility, the RTRS evidently lacks the capacity to ensure compliance with its own standards: for example, the procurement strategies of traders remain opaque (zu Ermgassen et al. 2022). Meanwhile, the industry has proven that traceability for soy products is impossible to implement fully for the time being. Another, more successful, industry-led initiative is *The Amazon Soy Moratorium* (Heilmayr et al.

2020). The *Moratorium* rests on an agreement by grain traders to refrain from buying soy grown on recently deforested land in the Brazilian Amazon. It is led by a partnership between soy traders, non-governmental organizations and government agencies called *Grupo de Trabalho da Soja*, which, using a combination of information from the Rural Environmental Registry, satellite data and field visits, prepares an annual list of farms which have violated its terms. The *Moratorium* is more robust than previous initiatives and is estimated to have had a substantial effect on deforestation rates in the Amazon, though it is hard to isolate it from other policy measures taken to protect the Amazon Forest under the same period (Ibid). Moreover, alternative media platforms (Bureau of Investigative Journalism, Unearthed and Repórter Brasil), have used satellite images to document how Cargill, Bunge and COFCO engaged in purchasing soybeans from companies buying soy from farmers fined and sanctioned multiple times after illegal deforestation (Wasley et al. 2021). Their investigations illustrated the way in which soy linked to illegal deforestation can find its way into supposedly "clean" international supply chains, despite apparently robust mechanisms (Ibid). The latest industry initiative to address deforestation associated with soy, beef and palm production is a plan announced at the 2021 United Nations (UN) Climate conference (COP-26). The world's ten largest commodity traders published a "shared commitment to halting forest loss" and announced that they would present a detailed plan in line with keeping deforestation rates with the 1.5 C warming target at the 2022 COP-27. So they did, but, according to a sobering report by WWF, while plans for palm and beef showed signs of progress, those for soy did not; and deforestation was left out of any strategy for habitat conversion, effectively excluding 74% of the Cerrado (WWF 2022).

Along with nation states and agrobusiness firms, a number of key international organizations have also moved toward regulatory reform. In October 2020, for instance, the European Parliament asked the Commission to introduce legislation to halt EU-driven global deforestation. EU consumption is estimated to have contributed to about 10% of global deforestation between 1990 and 2020, while European imports of soy and palm oil account for more than two-thirds of the totals. As we write these lines, EU legislators have reached an agreement to pass a new law, guaranteeing that a range of imports such coffee, cattle, cocoa, palm oil, timber and, indeed, soy, are not linked to the destruction or degradation of forests. In a so-called "due diligence" statement, the proposal would force companies to certify that their source of e.g. soy does not come from deforested land (European Parliament 2022).

The EU Parliament also secured a wider definition of forest degradation which includes the conversion of primary and naturally regenerating forests into plantation forests or into other wooded land. Brussels will have access to relevant information provided by companies, including geolocation coordinates, and will conduct traceability checks using satellite technology and DNA analysis. Moreover, companies will have to verify that the rights of affected indigenous peoples have been respected. The proposed legislation covers products which contain, have been fed or have been made with these commodities. Biodiesel (to a large extent produced from soy oil), however, is exempted from the law. Eventually, the Commission shall also evaluate extending the legislation to other ecosystems, including land with high carbon stocks and with a high biodiversity value, such as the Cerrado. The EU's initiative is the first law against deforestation in the world; setting "a global gold standard for due diligence requirements for deforestation-free supply chains", and environmental NGOs have expressed hope that many other countries will follow. At the same time, the infrastructure and organization required for traceability would significantly increase costs and soy would cease to be cheap.

Many other initiatives aiming to price market failures are emerging. One recent example comes from the International Maritime Organization, the UN body responsible for shipping. After years of discussion, a decision was reached in 2022 to put a price on greenhouse gas emissions from international shipping activities. If adopted, this would represent a major step in tackling global CO_2 emissions and internalizing this environmental cost could likely drive up the price of soy.

The challenge to profitability from the technological treadmill: broken nutrient cycles, weed resistance and soybean rust

Beyond these first regulative shifts and attempts to make the soy complex pay the full price for the real social-ecological costs it generates, a second threat to the profitability of the model comes from the visible ways in which agroecosystems react to the maniacal spread of standardized technological packages implemented in soybean production. As we saw in Chapters 2 and 3, the division between urban and rural areas in China triggered a disruption in metabolic interactions between humans and nature as soy cakes were traded to fertilize fields far in the South as well as in Japan. This rift widened significantly during the roots phase of the third cycle, in the wake of the second agricultural revolution, and further again with the Green Revolution's industrialization of agriculture. This model has led to the depletion of soil fertility, resulting in a huge increase in the demand for fertilizer (Bellamy Foster 1999, 373–76). Overuse and poor management of cheap synthetic fertilizers has, in turn, led to nutrient pollution on lands and in waterways and significant greenhouse gas emissions. Thus, this treadmill created a "metabolic rift" (Ibid) whereby cheap fertilizers were required to compensate for nutrient imbalances created by the economic need for specialization that resulted in the separation of animals that produce nutrients (manure) and the crops that need them for growth. The Gene Revolution – the latest step in this treadmill where big agrochemical companies moved into biotechnology and merged with seed companies – further added the challenge of an "arms race" to the "metabolic rift" in the logic of the treadmill.

As shown in Chapter 5, soy production today is conducted mainly under a technological package centered in herbicide-tolerant soybeans in combination with specific herbicides and/or insecticides. The package has allowed soy to expand dramatically over remarkably distinct ecologies, homogenizing diverse social-ecological relations in the process and has been central in keeping soy "cheap".

The use of herbicide-tolerant soy traits allowed for the use of cheap weed killers (glyphosate for Monsanto's soy varieties and glufosinate with Bayer's). Effective weed management is, of course, important; without it growers risk average losses close to 40% (Lucio et al. 2019) and the herbicide-tolerant soy model did the job in a very cost-efficient and labor-saving way (Trigo, Falck-Zepeda, and Falcon 2010). A high degree of "homogeneity in the agronomic techniques and technologies of production, from the smallest to the largest farms" set in, as soy expanded across varied landscapes (Oliveira 2022, 212). Glyphosate, in particular, became the most commonly used, since it controls more weed species than any other herbicide (Heap and Duke 2018, 1040–41). As Jennifer Clapp has noted: "glyphosate was widely considered to be a "perfect herbicide" because it was highly effective and widely deemed to be less toxic than other chemical herbicides" (2021, 202). At the same time, research and development into new active ingredients for herbicides dwindled: almost all soy seeds were designed to be combined with glyphosate, or if not, glufosinate (Ibid). This phenomenal success, however, carried within it the seeds of its own destruction, so to speak.

In fact, as GE soy spread over vast territories, the continuous application of non-selective herbicides – glyphosate for RR soybeans and glufosinate-ammonium for *LibertyLink* soybeans – led to a steady increase in glyphosate-resistant, as well as glufosinate-resistant, weeds. There are currently over 250 weed species that have evolved resistance to at least one type of herbicide (Lucio et al. 2019). Argentina, the United States and Brazil, the countries which most rapidly adopted glyphosate-tolerant soy, are now the ones with the largest infestation of glyphosate-resistant weeds by area – with 20.1 Mha in Brazil alone (Ibid). Thus began the "arms race" – farmers have responded with more herbicide applications, sometimes in combination with older chemicals, such as dicamba and 2,4-D (Benbrook 2016). The biotech industry then responded in a similar way, by combining several herbicide resistance varieties (stacked traits), including tolerance to the more toxic herbicide dicamba. While the USDA facilitated the regulatory process of these GE products and expedited their commercialization, today, many of these other herbicides (in addition to being more toxic) are also experiencing plant resistance. This is creating a combination of multiple herbicide resistance (glyphosate plus others) in weeds and taking herbicide-tolerant agriculture to the verge of a weed-management crisis. The issue is expected to worsen over the next 30 years (Heap and Duke 2018, 1040–41).

Thanks to soy's expansive success, the efficacy of herbicides has deteriorated significantly and might eventually be negligible. Monoculturalized and standardized production systems centered on a high-input/high-output paradigm, thus result in a technological treadmill requiring continuous reinvestment for new varieties and ever more chemical inputs. Also, as a consequence of soy monoculture, farmers need to apply increasing amounts of pesticides and fungicides to combat different kinds of plagues and infestations. The path is familiar: as the use of pesticides increases, populations of natural predators are reduced and selection pressures lead to pest populations with resistance to these compounds. This encourages applications of larger amounts of current pesticides or the substitution of more toxic pesticides. Selection pressures are, therefore, increased, and are only temporarily effective in reducing crop pests. Another example besides weeds is soybean rust, a fungal disease native to Asia which has spread to South America and the United States. The yield damage it can inflict is extreme, potentially causing losses of up to 80% (Ishikawa-Ishiwata, and Furuya 2021). Moreover, control is expensive, requiring fungicide applications which are gradually draining farmers' incomes. Farmers in Paraguay have more than doubled the application of fungicides over the last few years (Ibid). What has the response of the biotech industry been? To incorporate rust resistance into soybean varieties suitable for Latin American (ISAAA 2022).

While soybeans are a fundamental case (GRAIN 2013; Pendrill et al. 2019; Boanada Fuchs 2020; Ritchie and Roser 2021), the fact is that all industrial agriculture is based on ecological simplification, and on the cumulative use of synthetic and chemical inputs to substitute for the ecosystem services lost by these practices (Friedman 2000; MEA 2005). In short, as soy expands, farmers have to use more and more chemical applications, and constantly adopt new ones to keep output up, in an endless innovation/adoption loop that resorts to the same type of technologies that generated the problems they aim to solve (Oliveira 2022, 213). Ultimately, this feedback loop destabilizes profitability.

A way to slow the technological treadmill would be move industrial systems toward organic or agroecological farming practices that combine and rotate crops and rely on mechanical weed-control techniques (Sarandón and Flores 2014). Naturally, however, big agribusiness firms and their allies prefer chemical-dependent agriculture and the profits they earn. Other than developing new traits, many firms have also tied their color to the mast of new high-tech fixes, such as the so-called precision and smart agriculture, which rely on big data and remotely controlled machinery enabling farmers to apply chemical products surgically, taking into consideration the exact conditions of each square meter. Such techniques help avoid the over-use of agrochemicals, and in theory, reduce spraying to the minimum (Clapp 2021). But again, precision farming perpetuates the "get big or get out" model (Suppan 2020). Agribusiness firms also spread the Protocols on "Good Agricultural Practices" as a solution, offering courses to farmers in which they are taught not to over-apply agrochemicals, not to clean trucks in or near waterways and to rotate crops. Another response is the so-called Climate Smart Agriculture promoted by FAO over the last few years. This includes a wide range of practices such as crop rotation and intercropping, mulching, water management, the efficient use of pesticides and fertilizers and/or the use of high-quality seed varieties. Climate Smart Agriculture and organic agriculture – centered on rotation and inter-cropping – all include elements that resemble the ways soy was cultivated in ancient China millennia ago. Another possible solution would be to practice mixed systems based on integrating animals with crops. These systems were predominant in both the United States and Europe before the current regime, and were also common in Latin America before *sojización*. In mixed systems, weeds and insects can be controlled without overusing agrochemicals. Farmers' specialization in "pure" crop or livestock systems constitutes a kind of "rational" response to price relations, of course, despite the higher environmental costs and externalities they generate relative to mixed systems (Schut et al. 2021). A "return" to mixed systems and rotations would significantly increase labor costs and/or mechanization and depending also on the efficiency of fertilizers, cheap soy might just have run its course. While there are tradeoffs in either system, the fact is that the current regime phase is embedded in a wider agrofood system which is increasingly coordinated by large firms at input stages on one end of the chain and traders towards the other end of the chain. Clearly it is mainly the farmers that bear the costs of the treadmill (by having to put more money on inputs, as well as by their decreased independence and prospects for practicing renewable agriculture). For the producers of inputs (and owners of patents), however, farmers' adoption of new technologies should ideally be neither too fast nor too slow. Ideally, farmers adopt their technology and use it until the patents expire (most commonly after 20 years). Thus, as long as the treadmill does not move too fast, input producers can actually benefit from the constant need for farmers to change technologies.

The challenge to profitability from lawsuits

The wide use of glyphosate is controversial in many ways, not least for its toxicity for humans. In 2015, the WHO's International Agency for research on Cancer (IARC) classed glyphosate as "probably carcinogenic". In 2017, the European Chemicals Agency (ECHA) maintained its assessment that glyphosate is linked to serious eye damage and noted that it is toxic to aquatic life. At the same time, many industry-sponsored studies continue to challenge such results (Clapp 2021), though by 2020, Bayer had to pay settlements in lawsuits for over USD 10 billion and the company failed to win court approval for a proposed USD 2 billion settlement of future claims (Ibid). After enormous initial losses, Bayer won several consecutive trials in 2022 over claims that its weed killer, Roundup, caused cancer (Pierson 2022). In spite of the recent judicial victories, several European states have unilaterally banned glyphosate. France has not only prohibited the domestic application of glyphosate, but is also pushing the EU to follow suit. In 2016, Saudi Arabia, Kuwait, the United Arab Emirates, Qatar, Bahrain and Oman were among the first countries to put national bans in place on the use of glyphosate for any purpose, and Germany has announced it would ban the chemical by 2023. Several other countries have implemented partial or sub-national bans or restrictions on the use of glyphosate, though often with exceptions. Generally, then, the impact of legal challenges to glyphosate, and particularly their potential capacity to set in motion further regulative shifts, could have a strong impact, not only on the sales of pesticides and weed killers, but also on the trade of the particular seeds developed by firms to be used in tandem with those chemical products. Once more, this could constitute a major game changer for the soybean business. Moreover, legal challenges go beyond glyphosate.

In 2016, as Monsanto (Bayer) launched a new line of dicamba-resistant soybeans. Farmers and environmental organizations filed a string of lawsuits in the United States, arguing that the weed killer had drifted from soy fields onto adjacent areas, causing billions of dollars in crop damage (Rollins 2020; Turner 2022). Dicamba products are estimated to have caused drift damage to millions of hectares of soybeans. The lawsuits seek millions of dollars in compensation for the damage. On January 27, 2020, the first legal trial concerning dicamba-related products began in Cape Girardeau, Missouri. The lawsuit involved a peach farmer who claimed that dicamba-based herbicides had caused significant damage to his crops and trees. The jury ruled against Bayer, dicamba's owner, and its codefendant, BASF, in favor of the peach grower. The two companies were ordered to pay USD 15 million in damages, and an additional USD 250 million in punitive damages (Earls 2022).

A crisis of legitimacy?

Countermovements: alternatives and resistance

As we have seen, a regime typically rests on a particular global division of labor and sets of power relations between farmers, firms and governments (McMichael 2009, 140; 2016). The challenges to the current regime are not only economic, regulatory or legal: they also come from below, from petty producers and social movements in their fundamental opposition to the GE Soy Complex. Chapters 4 and 5 showed the ways in which Green Revolution technologies and the further development of industrial agriculture made economies of scale and specialization all-important in agriculture, chaining farmers to the necessity of perennially increasing output. Technology (new seeds and machines) replaced labor: between 1800 and 2000, the share of the total workforce engaged in agricultural production declined from more than 75 to less than 50% – with only 2.5% of the workforce in agriculture in rich countries (Federico 2010, 1). At Capitalism's core, the labor force that left agriculture became absorbed in other sectors of the economy. During the past decades, this trend is taking place in the global South, but with a crucial difference. While the majority of farmers have been involved in formal market structures for over a century now, many have continued relying on family labor and adopting diverse production strategies centered around pecuniary, but also non-pecuniary values (tradition, identity, experience) rather than specializing solely in line with the highest economic margins and wage labor. By the mid-20th century, the partial persistence of pre-capitalist elements in Latin American agriculture – i.e. high degree of subsistence farming, low use of wage labor and customary rights to land – was weakened by several rounds of "modernization" and by the partial adoption of Green Revolution technologies. However, traditional forms of agricultural labor were finally undermined with neo-liberalization and the adoption of export-oriented development strategies in line with a "comparative advantage" in soy and other primary products (Baraibar Norberg 2022). As a result, many people have been pushed off the land at a pace comparable to that of the European enclosures. The Gene Revolution further facilitated the rapid expansion of soy over small farms and into new lands (Botella-Rodríguez

2018; Kay and Vergara-Camus 2018). Areas that were previously considered as too remote, too difficult, too costly or too risky to exploit began to be commodified and brought into the soy chain (Mansourian et al. 2014; Baraibar Norberg 2020b, 93). In contrast to the case of industrializing Europe, then, those Latin American farmers who have been compelled to migrate to urban centers, find that there is no industry or economic sector to absorb them: they have become impoverished excess laborers and slum dwellers. Within this overall context, the GE Soy Model is particularly "labor-saving". As Latin America became the global hotspot for agribusiness-driven *sojización*, most petty producers – typically rich in (family) labor and poor in capital– have been outcompeted and displaced. At the same time, the neoliberal policies adopted in Latin America have made states less active or able to re-balance the relations of force between large producing firms, petty producers and rural laborers. In short, *sojización* has brought a model of agriculture "without farmers", where capital and technology replace labor (Baraibar Norberg 2020c). The few farmers who have remained "competitive" have nevertheless had to surrender control over what is produced, at what price and how, in favor of the transnational firms and their technological packages (Blum et al. 2008; Gomez 2008; GRAIN 2013; Lang and Heasman 2015). Since farmers need to specialize in the goods that give them the best return in the market, many have found they have no alternative but to grow soy (Baraibar Norberg 2022).

Local communities, small farmers and indigenous peoples have denounced the social-ecological hazards related to soybean monocultures and, indeed, social movements across Latin America have begun to speak up against the current soy regime (Relly and de Majo 2022, 146–47). One of the most influential voices here is *La Via Campesina* (LVC) – an international "movement of movements" with 182 member organizations, representing more than 200 million peasants, according to its website. LVC has produced innumerable texts and statements about and against the contemporary soy model, which it sees as corporate agriculture's ultimate threat to people and the planet. One expression of LVC's overall outlook is captured in the words of Perla Álvarez, a Paraguayan member of its Latin American steering committee (CLOC-LVC), while protesting against the WHO in Geneva in June 2022:

> These companies grab our territories to expand, because as they expand over our communities and forests, they accumulate capital to maintain this production model (…) [T]he agreements leveraged by these governments are part of our everyday lives; they are coming to our homes, our tables. Meanwhile, there is the destruction of the environment, climate change, and the impacts on our production, as they steal our seeds, reduce local varieties, and cause intense droughts, cold, rainfall, and heat.
>
> (Zelic and Pessoa 2022)

La Via Campesina and other peasant organizations see soy as the ultimate manifestation of corporate control, "free" trade, monoculture and life destruction, and as one of the prime movers behind continuous commodification through the privatization of seeds, plants and new species, imposed through patent laws. One battlefield is the right to grow, save, reproduce and use seeds, which is not compatible with the strong property-rights regime around GE soy covered in Chapter 5. Many small producers in Brazil and Argentina adopted GE soy, but continued to challenge Monsanto's (Bayer's) royalties by saving and reproducing seeds (Lapegna and Perelmuter 2020). In Ecuador, a long process of grassroots mobilization resulted in a seed law recognizing peasants' rights to save seeds (Ibid). In its struggle against commodification, LVC also disputes reformist "solutions" such as the aforementioned Climate Smart Agriculture framework, precision agriculture and "sustainable intensification", and counter that such initiatives and innovations are little more than a façade masking the ongoing corporatization of global agriculture. "[T]here is a trend boosting digital agriculture, which turns rural labor into data and turns data into commodities" (Zelic and Pessoa 2022). Instead of more commodification, these countermovements propose food sovereignty, defined as

> the right of peoples to define their own food and agriculture; to protect and regulate domestic agricultural production and trade in order to achieve sustainable development objectives; to determine the extent to which they want to be self-reliant; [and] to restrict the dumping of products in their markets.
>
> (Rosset 2003)

This explicit ideal of LVC was first formulated toward the end of the 1990s. It came out of the conviction that

> hunger will only be ended if governments and nation-states espouse public policies for access to land and the promotion of local food production by family farmers and peasants. The development, growth, and progress of humanity have always been associated with food that is produced in the territory where it is consumed.
>
> (Stedile 2019)

Beyond and beneath the second-tier organization of LVC, hundreds more peasant organizations across South America are mobilizing against soy. Instead, they advocate "agroecology" and/or "food sovereignty". These ideals propose a radical transformation of the power relations shaping the agrofood system, from corporations, to farmers and local communities and defends the importance of national food security (Altieri 2009; Altieri and Nicholls 2017; Desmarais 2007; Gómez, Ríos-Osorio, and Eschenhagen 2015; Padilla and Guzmán 2009). From their vantage point, export-oriented soy is a factor behind increasing food insecurity in the region, and *sojización* is cast as a direct threat to the farming activities that produce for local markets and contribute to national food supply (Leguizamon 2014).

In several places this has led to violent conflicts over land. Peasants and indigenous groups have had sharp confrontations with soy farmers and firms.

In Paraguay, specifically, this has involved the so-called *brasiguayos*, or the big Brazilian soy farmers who control most soy production in the Eastern part of the country. Soy has become synonymous with "agribusiness", "corporate control," "foreignization of land" and "land grabbing" (Domínguez and Sabatino 2006; Dominguez and Sabatino 2008; Human Rights Council 2015; Irala and Cardozo 2016; Lapegna and Perelmuter 2020; Baraibar Norberg 2022). Besides struggles over land, increasing local resistance against the widespread use of agrochemicals in soy production is also evident. Producers' organizations argue that the large landholdings of soy do not want agricultural workers, but prefer to replace them with agrochemicals that poison water and land (La Vía Campesina 2020). For their part, peasant organizations have struck alliances with many environmental and community organizations in both big urban centers and smaller rural towns, e.g. where agrochemical drifts have become a recurring problem. Mobilization across several local provinces and departments in Argentina and Uruguay, for instance, have succeeded in launching moratoria on the application glyphosate and other agrochemicals (Baraibar Norberg 2020b).

Argentina is probably the country where soybeans have been politicized most sharply. The center-left Peronist party under Cristina Fernández de Kirchner has worked with urban environmentalists, peasant movements and indigenous groups, articulating a highly antagonistic program against the soy model and attacking it on the grounds that it is socially and ecologically exploitative, it displaces family farmers and indigenous communities and that it is destroying the land and the water. At the same time, producers' organizations, firms and a wider range of other actors identifying themselves with the "rurality" (*el campo*) have rallied in defense of the country's top crop. When the government increased export taxes on soybeans in 2008, producers' organizations responded with a nationwide lockout. The level of conflict between the Peronist governments (2003–15) and the farm sector has been very high (particularly during Cristina Fernandez de Kirchner's two presidential terms (2007–15). However, the export tax on soy was an important revenue source for the state, financing debt payments as well as rural development programs in support of family farmers. Given Argentina's dependence on soy's hard-currency revenues, the government has also adopted policies that strengthened the soy business and has not managed to curb the rapid process of land concentration (Vergara-Camus and Kay, 2017).

In short, the soy model is heavily contested by a wide range of actors, not leastly the well-organized peasant organizations. While these radical movements have mobilized against the soy model, they are still too weak, fragmented and uncoordinated to pose any real challenge to the larger soy regime. Alliances with other social movements are still growing, however, and it would be rash to write off the challenge they could mount against the soy complex in years to come.

The rise of the East (… and the fall of the West?)

As we have seen throughout our study, shifts between regime and rupture phases in previous soy cycles have typically followed in tandem with wider geopolitical shifts. The first soy regime, entered its rupture phase around 200 CE with the fall of the Han Dynasty and the gradual disintegration of the North. The second soy regime began to unravel in the late-18th century and finally collapsed with the geopolitical shift brought by the economic subordination of Qing China by British imperialism during the Opium Wars. Notwithstanding the potential contradictions and challenges which could take the third soy regime into its own rupture phase, and which we have just discussed, it is impossible to ignore the signs of a new hegemonic shift, away from the selectively free trade policies under Washington's hegemonic oversight. Even if the United States continues to play the dominant role within the organization of the world economy, its agricultural model is now also internationally contested. "Free trade" is, as we just argued, increasingly under fire from local communities and peasant organizations in producing countries. The notion of an ecological debt owed them by high-income countries for ecological damage wrought by their overconsumption has likewise gained considerable traction, also in the advanced economies. The United States' commitment to free trade has always been stronger in foreign policy, and weaker and more partial with regard to its own economy. The Trump administration, however, has explicitly turned away from the traditional free trade discourse in favor of the "America First" slogan and this has had momentous effects on the soy business.

Soybeans are the United States' number one export to China. They are worth almost three times more than the second most important export item, which is corn. Both are mainly used for feed (Greenwood 2022). In 2018, China imposed tariffs on US-grown soy in response to those levied by the United States on a number of Chinese goods. As a result, US soybeans lost more than 75% of their export value to China 2018 to 2019. More specifically, soybeans accounted for the largest share of total trade loss, making up nearly 71% – USD 9.4 billion – of annual losses (Morgan et al. 2022). China instead moved even closer to South America and accepted Argentina's demand for access to soy oil and meal markets (Jingxi 2022). After the so-called Phase One Agreement of 2020, China committed to buying US soy again, but while sales are almost back to pre-conflict levels, the US share of China's soy imports has fallen and trade tensions between the two countries remain high (Morgan et al. 2022). According to a recent analysis in *China Daily*, "the diplomatic, trade, financial, technological and ideological wars launched by the United States against China have continued to escalate" (Siu-kai 2022). In this context, China has pushed ahead with its strategy of decreasing its dependence on global trade for food supplies, especially US soy dependence.

Beijing has made several attempts to increase domestic production (Yimeng 2022). Favorable soy policies and higher self-sufficiency targets have led to more Chinese farmers planting soy. However, at the same time, China's arable lands are decreasing due to urbanization and pollution from pesticide and fertilizer use (Greenwood 2022). Also, farms remain unmechanized and their size is small. Despite this, China seems to hope that as the rural population declines, industrial farming approaches will increase productivity. A second approach has been to reduce the total demand for soy by lowering the soybean meal content in animal feed. In 2021, the share of soybean meal in the feed industry dropped to 15.3%, down 2.5% from 2017, saving an equivalent of 11 MMT of soybean meal and 14 MMT of soybeans (Morgan et al. 2022). Moreover, in its latest Five-Year Plan for Crop Farming, China promoted the development of alternative proteins. A third strategy has been to invest abroad, acquiring foreign land, technology and other resources to build up the domestic agricultural system. China has thus increasingly invested in Latin America, Africa and, in recent years, also in the United States. One seminal example is the Chinese WH Group's purchase of the big US hog-company, Smithfield – the largest Chinese acquisition of an US company to date. This turned the WH Group into one of the largest overseas owners of American farmland, and Smithfield became the world's largest pork processor and hog producer (Wang 2020). The hog herds are still based in the United States since feeding them there is cheaper, and the environmental impacts (mostly water pollution from manure) are far away from China's borders.

According to the *US-China Economic and Security Review*, since the United States is the global leader in agricultural innovations and GE, China has turned an eye to the United States in its quest to decrease its long-term US dependence. According to the report, China recognizes that

> the United States has something China does not, whether that is infrastructure, land mass, IP [intellectual property], or an already built out supply chain. By accessing these areas through capital, or in some cases theft, China gains useful intelligence for achieving agricultural self-sufficiency.

The report goes to note that "Chinese nationals have smuggled seeds out of the United States to China to be replicated" (Greenwood 2022). No doubt, readers of this book will immediately call to mind the thousands of seeds that USDA took from Manchuria under a period of many years around a century ago, and which were vital in the long and laborious efforts to make soy thrive in the United States. This exercise in historical analogy seems out of reach for many current analysts, though. The *US-China Economic and Security Review* instead argues that "China's illicit acquisitions of GM [GE] seeds provides a jumpstart to China's own development of such seeds, deprives US companies of revenue, and offers an opportunity to discover vulnerabilities in US crops" (Ibid). China's actions, according to the same source, involve several risks to US economic and national security and might even dampen China's need for US farm products. More importantly, it argues, China might gain leverage over US supply chains and erode America's competitiveness, as well as potentially developing the capacity to create bioweapons:

> Beijing could easily hack the code or DNA of U.S. GM seeds and conduct biowarfare by creating some type of blight that could destroy U.S. crops. For example, biotechnology experts have recognized that fungal spores could be used as biological warfare agents to target staple crops.
>
> (Greenwood 2022)

While the United States' fears of biological warfare could be taken as exaggerated, what is clear is that China is engaging more proactively in all stages of the value chain. While Chinese state-owned companies in trading and input stages are latecomers, they have rapidly advanced their market shares in the past few years. In this way, the world's four largest agribusinesses traders referred to as "the ABCD companies" have been challenged by COFCO in global purchasing, processing, warehousing and transportation. In fact, COFCO has become the largest grain exporter in Argentina, Paraguay, Brazil and Uruguay. However, the company has struggled to plan its industry chains in North America (Xingqing, Yu, and Xu 2022). While it is too early to say whether the rise of China as an economic super power and its strategies to decrease dependence on US soy will result in a geopolitical shift that will bring the current soy regime to an end, the nonlinearity of capitalist development across the unequal state-system and successive political-economic hegemonic orders are at the very foundation of subsequent food regimes.

Reflections on the soybean through world history: lessons learnt for sustainable agrofood systems

As we have shown in this book, soy plays a key role in today's increasingly vulnerable and unsustainable agrofood system. But, as we have also shown, things have been, and could still be, different. The way the agrofood system is formed and functions today is not the way it has always been – nor is it the way it will always be. Instead, it is change that is natural – it is change that is the norm. For an example of a profound historical discontinuity wrought by our current regime, let us remember that soybeans were an important food source during the first soy cycle (cooked, fermented, milled and sprouted) and a key food staple during the second soy cycle (tofu, soy sauce, soy milk, miso soup and tempeh). While these roles have remained important especially in East Asian cuisine, only 7% of global soy production today is consumed directly by humans with an additional 13% consumed as oil: the bulk is used for animal feed (Ritchie 2021). Ironically, perhaps absurdly, eating soy-fed animal products results in a substantial protein conversion loss: we lose 90% of the protein value in soy when it is fed to beef cattle; 80% when fed to pigs and 60% when fed to chickens. In this way, meat-based diets require huge amounts of land. According to Ritchie's calculations, in fact, producing a kilocalorie of beef or lamb takes up to a 100 times more acreage than a kilocalorie of peas or tofu (Ibid). Further, soy produces between four and eight times as much usable protein per unit of land as milk or eggs (Lander and DuBois 2022, 32). Again, according to Ritchie (2021), global land use for agriculture could be reduced by 75% by shifting to a plant-based diet. The production of meat also causes significant greenhouse gas emissions and has a large freshwater footprint. From this vantage point, the current agrofood system is clearly irrational and inefficient – no matter how profitable it may prove for some of its economic agents. A more vegetarian diet would not only decrease pressures on nature, but also have health benefits. The nutritious properties and remarkable culinary diversity of soyfoods could potentially play a positive role in such a shift.

In short, there is nothing inevitable in the current role of soy in the agrofood system. If "agroecology", "food sovereignty" or other lines of thought challenging the current agrofood system gain traction, it would for example be possible to imagine the development of locally adapted soy varieties planted in labor-intensive, diverse and "place-based" agricultural systems, controlled by independent small farmers. These, in turn, would make a better fit with the "golden rule" of natural resource management which is to "strive to retain critical types and ranges of natural variation in ecosystems [...and] facilitate existing processes and variabilities rather than changing or controlling them" (Holling and Meffe 1996).

Soy can evidently take on a wide variety of roles. As a versatile crop which is easy to store and transport, it can be seen as a mere reflection of the power relations of the contemporary food order. As we have shown, governance in the commodity chain is mainly driven by giant agrofood corporations engaged mostly in the input, processing and trading stages, and prioritizing short-term profit. Moreover, the international institutional context mostly reinforces this arrangement (although sometimes disguising it as food security, development and food-aid strategies). While it is difficult to imagine any radical change as long as existing power relations stay fixed, all regimes reach an end, and so, eventually, the current regime phase will disintegrate. As we have discussed in this last chapter, "cheap" soy depends on its full social-ecological costs remaining hidden from its market price and on opportunities for extensification (frontier expansion) and intensification. The continued economic extraction from social-ecological systems is not sustainable and this model cannot continue for much longer. Thus, the question is not so much if this will come to an end, but more importantly, whether there will be a deliberate and, necessarily, planned transformation, or if business as usual will continue until it hits biophysical limits with potentially nonlinear and chaotic responses. While we balance our options, the world history of soy shows that there is much to be learnt from past experiences, including the roads not taken.

References

Altieri, Miguel. 2009. "El Estado Del Arte de La Agroecología: Revisando Avances y Desafíos." In *Vertientes Del Pensamiento Agroecológico: Fundamentos y Aplicaciones*, edited by Altieri, 69–90. Medellín: Sociedad Científica Latinoamericana de Agroecología (SOCLA).

Altieri, Miguel A., and Clara I. Nicholls. 2017. "Agroecology: A Brief Account of Its Origins and Currents of Thought in Latin America." *Agroecology and Sustainable Food Systems* 41 (3–4): 231–37.

Ankersen, Tom and Thomas Ruppert. 2006. "Tierra y Libertad: The Social Function Doctrine and Land Reform in Latin America." *Tulane Environmental Law Journal*; 19 (69): 70–120.

Aurélie Desmarais, Anette. 2007. *La Vía Campesina – Globalization and the Power of Peasants*. Winnipeg, Manitoba: Fernwood Publishing.

Baraibar, Matilda. 2014. "Green Deserts or New Opportunities? - Competing and Complementary Views on the Soybean Expansion in Uruguay, 2002–2013." Monograph, Stockholm: Stockholm University.

Baraibar Norberg, Matilda. 2020a. "Agrofood Globalization: The Global Soybean and Beef Commodity Chains." In *The Political Economy of Agrarian Change in Latin America*, edited by Norberg, 117–63. Cham: Palgrave Macmillan, Springer Nature.

———. 2020b. "Regulative Shifts and Agrarian Change of the Twenty-First Century." In *The Political Economy of Agrarian Change in Latin America*, edited by Norberg, 209–326. Cham: Palgrave Macmillan, Springer Nature.

———. 2020c. "Agrofood Globalization: The Global Soybean and Beef Commodity Chains." In *The Political Economy of Agrarian Change in Latin America*, edited by Norberg, 117–63. Cham: Palgrave Macmillan, Springer Nature.

———. 2022. "Sojización as a New First Movement: A Polanyian Analysis of the South American Soybean 'Boom.'" In *The Age of the Soybean: An Environmental History of the Soyacene during the Great Acceleration*, edited by da Silva and de Majo, 91–114. Winwick, Cambridgeshire: The White Horse Press.

Barbier, Edward. B. 2011. *Scarcity and Frontiers: How Economies Have Developed through Natural Resource Exploitation*. Cambridge: Cambridge University Press.

Bellamy Foster, John. 1999. "Marx's Theory of Metabolic Rift: Classical Foundations for Environmental Sociology." *American Journal of Sociology* 105 (2): 366–405.

Benbrook, Charles M. 2016. "Trends in Glyphosate Herbicide Use in the United States and Globally." *Environmental Sciences Europe* 28 (3):1–15.

Blum, Alfredo, Ignacio Narbondo, Gabriel Oyhantcabal, and Diego Sancho. 2008. *Soja Transgénica y Sus Impactos En Uruguay – La Nueva Colonización*. Montevideo: RAP-AL.

Boanada Fuchs, Vanessa. 2020. "Chinese-Driven Frontier Expansion in the Amazon: Four Axes of Pressure Caused by the Growing Demand for Soy Trade." *Civitas - Revista de Ciências Sociais* 20 (1):16–31.

Botella-Rodríguez, Elisa. 2018. "La cuestión agraria en América Latina: desafíos recurrentes y nuevas preguntas para la historia rural." In *Del pasado al futuro como problema: La historia agraria contemporánea española en el siglo XXI*, edited by David and Berasáin, José-Miguel, 285–311. Zaragoza: Prensas universitarias de Zaragoza.

Braudel, Fernand, and Immanuel Wallerstein. 2009. "History and the Social Sciences: The Longue Durée." *Review (Fernand Braudel Center)* 32 (2): 171–203.

Bray, Francesca. 1984. "Science and Civilisation in China. Volume 6, Biology and Biological Technology. Part 2, Agriculture", In Science and Civilisation in China, edited by Needham. Cambridge: Cambridge University Press.

Chimonyo, Vimbayi Grace Petrova, Sieglinde S. Snapp, and Regis Chikowo. 2019. "Grain Legumes Increase Yield Stability in Maize Based Cropping Systems." *Crop Science* 59 (3): 1222–35.

Clapp, Jennifer. 2014. "Financialization, Distance and Global Food Politics." *The Journal of Peasant Studies* 41 (5): 797–814.

———. 2016. *Food*. 2nd ed. Cambridge: Polity Press.

———. 2021. "Explaining Growing Glyphosate Use: The Political Economy of Herbicide-Dependent Agriculture." *Global Environmental Change* 67 (March): 102239.

de Castro, Fabio, Barbara Hogenboom, and Michiel Baud. 2016. *Environmental Governance in Latin America*. Hamphire: Palgrave Macmillan.

Domínguez, Diego, and Pablo Sabatino. 2006. "*Con La Soja al Cuello: Crónica de Un País Hambriento Productor de Divisas.*" In *Los Tormentos de La Materia, Aportes para una ecología política latinoamericana*, edited by Alimonda, 249–274, Buenos Aires: CLACSO.

———. 2008. "La Conflicitividad En Los Espacios Rurales de Argentina." *Lavboratorio, Cambio Estructural y Desigualdad Social (CEyDS)/Facultad de Ciencias Soicales, UBA*. 1m0 (22): 38–52.

Earls, Maya. 2022. "BASF Scores New Trial Over $60 Million Dicamba Damages Award (1)." *Bloomberg Law*, July 7, 2022. https://news.bloomberglaw.com/environment-and-energy/basf-scores-new-trial-over-60-million-dicamba-damages-award.

Ellwanger, Joel Henrique, Bruna Kulmann-Leal, Valéria L. Kaminski, Jacqueline María Valverde-Villegas, Ana Beatriz G. Da Veiga, Fernando R. Spilki, Philip M. Fearnside, et al. 2020. "Beyond Diversity Loss and Climate Change: Impacts of Amazon Deforestation on Infectious Diseases and Public Health." *Anais Da Academia Brasileira de Ciências* 92 (1): e20191375.

European Parliament. 2022. "Deal on new law to ensure products causing deforestation are not sold in the EU" Press release 06-12-2022. *Press room*, Cf. https://www.europarl.europa.eu/news/en/press-room/20221205IPR60607/deal-on-new-law-to-ensure-products-causing-deforestation-are-not-sold-in-the-eu.

FAO, WHO, UNICEF, IFAD, and WFD. 2022. *The State of Food Security and Nutrition in the World 2022: Repurposing Food and Agricultural Policies to Make Healthy Diets More Affordable*. Vol. 2022. Rome: Food & Agriculture Organization.

Fearnside, P. M. 2001. "Soybean Cultivation as a Threat to the Environment in Brazil." *Environmental Conservation* 28 (1): 23–38.

Federico, Giovanni. 2010. *Feeding the World: An Economic History of Agriculture, 1800-2000*. Princeton: Princeton University Press.

Flachowsky, Meyer, and Südekum. 2017. "Land Use for Edible Protein of Animal Origin – A Review." *Animals* 7 (3): 25.

Friedman, Harriet. 2000. "What on Earth Is the Modern World-System? Foodgetting and Territory in the Modern Era and Beyond." *Journal of World-Systems Research* 6 (2): 480–515.

Garrett, Rachael D., and Lisa L. Rausch. 2016. "Green for Gold: Social and Ecological Tradeoffs Influencing the Sustainability of the Brazilian Soy Industry." *The Journal of Peasant Studies* 43 (2): 461–93.

Gomez, Marcelo. 2008. "La Soja de La Discordia. Los Sentidos y Estrategias En La Movilizaci6n de La Pequeña Buruesía." *Lavboratorio, Cambio Estructural y Desigualdad Social (CEyDS)/Facultad de Ciencias Soicales, UBA*. 10 (22): 22–35.

Gorman, Hugh. 2013a. "Innovation within an Ecological Limit." In *The Story of N: A Social History of Hte Nitrogen Cycle and the Challange of Sustainability*, edited by Gorman, 33–42, New Brunswick, NJ: Rutgers University Press.

———. 2013b. "N and the Rise of Science." In *The Story of N: A Social History of the Nitrogen Cycle and the Challenge of Sustainability*, edited by Gorman, 52–64. New Brunswick, NJ: Rutgers University Press.

GRAIN. 2013. *La República Unida de La Soja Recargada*. Barcelona: Grain.

Gray, Lewis Cecil, and Esther Katherine Thompson. 1933. *History of Agriculture in the Southern United States to 1860*. Washington DC: Carnegie Institution of Washington.

Gómez, L Fernando, Leonardo Ríos-Osorio, and M Luisa Eschenhagen. 2015. "Las Bases Epistemológicas de La Agroecología." *Agrociencia* 49 (6): 679–88.

Greenwood, Lauren. 2022. "China's Interests in U.S. Agriculture: Augmenting Food Security through Investment Abroad." Staff Research Report. Washington DC: U.S.-China Economic and Security Review Commission.

Heap, Ian, and Stephen O. Duke. 2018. "Overview of Glyphosate-Resistant Weeds Worldwide." *Pest Management Science* 74 (5): 1040–49.

Heilmayr, Robert, Lisa L. Rausch, Jacob Munger, and Holly K. Gibbs. 2020. "Brazil's Amazon Soy Moratorium Reduced Deforestation." *Nature Food* 1 (12): 801–10.

Holling, C. S., and Gary K. Meffe. 1996. "Command and Control and the Pathology of Natural Resource Management." *Conservation Biology* 10 (2): 328–37.

Human Rights Council. 2015. *Report of the Special Rapporteur on the Rights of Indigenous Peoples, Victoria Tauli-Corpuz: addendum.* Human Rights Council, Thirtieth Session, Agenda Item 3. A/HRC/30/41 United Nations-General Assembly, Geneva: UN.

IPBES. 2020. *Workshop Report on Biodiversity and Pandemics of the Intergovernmental Platform on Biodiversity and Ecosystem Services.* Daszak, P., Amuasi, J., das Neves, C. G., Hayman, D., Kuiken, T., Roche, B., Zambrana-Torrelio, C., Buss, P., Dundarova, H., Feferholtz, Y., Földvári, G., Igbinosa, E., Junglen, S., Liu, Q., Suzan, G., Uhart, M., Wannous, C., Woolaston, K., Mosig Reidl, P., O'Brien, K., Pascual, U., Stoett, P., Li, H., Ngo, H. T., Bonn, Germany: IPBES Secretariat.

IPCC. 2014. *Climate Change 2014: Synthesis Report.* Fifth Assessment Report (AR5). Geneva: IPCC.

Irala, Abel Enrique, and Hugo Javier Pereira Cardozo. 2016. "Violencia Armada y Avance de La Soja En El Norte Del Paraguay." *Conflicto Social* 9 (16): 180–208.

ISAAA. 2022. "GM Approval Database." International Service for the Acquisition of Agri-Biotech Applications. https://www.isaaa.org/gmapprovaldatabase/.

Ishikawa-Ishiwata, Yuki, and Jun Furuya. 2021. "Fungicide Cost Reduction with Soybean Rust-Resistant Cultivars in Paraguay: A Supply and Demand Approach." *Sustainability Science* 13 (2): 887.

Ismail, Rossi, and Geiger. 2011. *A Compilation of Bioenergy Sustainability Initiatives: Update.* Rome: FAO.

Jingxi, Mo. 2022. "Xi Calls for Deepening Ties with Argentina." *China Daily*, September 29, 2022, Global Edition edition, sec. Xi's Moments. https://www.chinadaily.com.cn/a/202209/29/WS63347f0ea310fd2b29e7a48f.html.

Kay, Cristóbal, and Leandro Vergara-Camus. 2018. *La Cuestión Agraria y Los Gobiernos de Izquierda En América Latina.* Buenos Aires: CLASCO.

Khodayari, Shaghayegh, Omid Sadeghi, Maryam Safabakhsh, and Hassan Mozaffari-Khosravi. 2022. "Meat Consumption and the Risk of General and Central Obesity: The Shahedieh Study." *BMC Research Notes* 15 (1): 339.

La Vía Campesina. 2020. "Opinion: Agrarian Reform Today and the Challenges We Face." September 18, 2020. https://viacampesina.org/en/opinion-agrarian-reform-today-and-the-challenges-we-face/.

Lambin, Eric F., and Patrick Meyfroidt. 2011. "Global Land Use Change, Economic Globalization, and the Looming Land Scarcity." *Proceedings of the National Academy of Sciences* 108 (9): 3465–72.

Lander, Brian, and Thomas David DuBois. 2022. "A History of Soy in China: From Weedy Bean to Global Commodity." In *The Age of the Soybean: An Environmental History of the Soyacene during the Great Acceleration*, edited by da Silva, and de Majo, 29–47. Chapter 2. Winwick, Cambridgeshire: The White Horse Press.

Lang, Tim, and Michael Heasman. 2015. *Food Wars: The Global Battle for Mouths, Minds and Markets.* 2nd ed. New York: Routledge.

Langthaler, Ernst. 2020. "Broadening and Deepening: Soy Expansions in a World-Historical Perspective." *Historia Ambiental Latinoamericana y Caribeña (HALAC) Revista de La Solcha* 10 (1): 244–77.

Lapegna, Pablo, and Tamara Perelmuter. 2020. "Genetically Modified Crops and Seed/Food Sovereignty in Argentina: Scales and States in the Contemporary Food Regime." *The Journal of Peasant Studies* 47 (4): 700–719.

Leguizamon, Amalia. 2014. "Roundup Ready Nation: The Political Ecology of Genetically Modified Soy in Argentina." Doctoral Dissertation, Adviser: Kenneth A. Gould. CUNY Academic Works: City University of New York.

Lindborg, Regina, Jan Bengtsson, Åke Berg, S. A. O. Cousins, Ove Eriksson, Tomas Gustafsson, Knut Per Hasund, et al. 2008. "A Landscape Perspective on Conservation of Semi-Natural Grasslands." *Agriculture, Ecosystems and Environment* 125 (1–4): 213–22.

Lopes, Gabriela Russo, Mairon G. Bastos Lima, Tiago N.P. dos Reis. 2021. "Maldevelopment revisited: Inclusiveness and social impacts of soy expansion over Brazil's Cerrado in Matopiba." *World Development*, 139, 1–17.

Lucio, Felipe Ridolfo, Augusto Kalsing, Fernando Storniolo Adegas, Caio Vitagliano Santi Rossi, Núbia Maria Correia, Dionisio Luiz Pisa Gazziero, and Alexandre Ferreira da Silva. 2019. "Dispersal and Frequency of Glyphosate-Resistant and Glyphosate-Tolerant Weeds in Soybean-Producing Edaphoclimatic Microregions in Brazil." *Weed Technology* 33 (1): 217–31.

Mansourian, Stephanie, Lucy Aquino, Thomas Erdmann, and Francisco Pereira. 2014. "A Comparison of Governance Challenges in Forest Restoration in Paraguay's Privately-Owned Forests and Madagascar's Co-Managed State Forests." *Forests* 5 (4): 763–83.

Margulis, Matias E., and Tony Porter. 2013. "Governing the Global Land Grab: Multipolarity, Ideas, and Complexity in Transnational Governance." *Globalizations* 10 (1): 65–86.

Martinez-Alier, Joan, Michiel Baud, and Héctor Sejenovich. 2016. "Origins and Perspectives of Latin American Environmentalism." In *Environmental Governance in Latin America*, edited by Castro, Hogenboom and Baud, 29–57. London: Palgrave Macmillan, Springer.

McMichael, Philip. 2009. "A Food Regime Genealogy." *The Journal of Peasant Studies* 36 (1): 139–69.

———. 2016. "Commentary: Food Regime for Thought." *The Journal of Peasant Studies* 43 (3): 648–70.

MEA. 2005. *Millennium Ecosystem Assessment Ecosystems and Human Well-Being: Synthesis.* Washington, DC: Island Press.

Morgan, Stephen, Shawn Arita, Jayson Beckman, Saquib Ahsan, Dylna Rusell, Philip Jarell, and Bart Kenner. 2022. *The Economic Impacts of Retaliatory Tariffs on U.S. Agriculture.* ERR-304. Economic Research Report. Washington DC: Economic Research Service (ERS) – United States Department of Agriculture (USDA).

Nabuurs, Hatab, Bustamante, Clark, and Havlik. 2022. "Agriculture, Forestry and Other Land Uses (AFOLU)." In *Climate Change 2022: Mitigation of Climate Change. Contribution of Working Group III to the Sixth Assessment Report of the Intergovernmental Panel on Climate Change*, edited by Skea, Shukla, Reisinger, Slade, Pathak, al Khourdajie, et al., IPCC Report (AR6), Worlk Group III. Chapter 07, 1–186, Cambridge and New York: Cambridge University Press.

Oliveira, Gustavo de L. T. 2022. "Political Ecology of Soybeans in South America." In *Political Ecology of Industrial Crops*, edited by Ahmed and Gasparatos, Chapter 9, 201–20. New York: Routledge.

O'Rourke, and Williamson. 1999. *Globalization and History: The Evolution of a Nineteenth-Century Atlantic Economy*. Cambridge, MA: The MIT Press.

Padilla, Mamen Cuéllar, and Eduardo Sevilla Guzmán. 2009. "Aportando a La Construcción de La Soberanía Alimentaria Desde La Agroecología." *Ecología Política*, no. 38: 43–51.

Pendrill, Florence, U. Martin Persson, Javier Godar, Thomas Kastner, Daniel Moran, Sarah Schmidt, and Richard Wood. 2019. "Agricultural and Forestry Trade Drives Large Share of Tropical Deforestation Emissions." *Global Environmental Change* 56: 1–10.

Pierson, Brendan. 2022. "Bayer on Winning Streak in Roundup Litigation after Huge Initial Losses." *Reuters*, September 2, 2022. https://www.reuters.com/legal/litigation/bayer-winning-streak-roundup-litigation-after-huge-initial-losses-2022-09-02/.

Potapov, Peter, Svetlana Turubanova, Matthew C. Hansen, Alexandra Tyukavina, Viviana Zalles, Ahmad Khan, Xiao-Peng Song, et al. 2022. "Global Maps of Cropland Extent and Change Show Accelerated Cropland Expansion in the Twenty-First Century." *Nature Food* 3 (1): 19–28.

Rausch, Lisa, and Holly K. Gibbs. 2021. "The Low Opportunity Costs of the Amazon Soy Moratorium." *Frontiers in Forests and Global Change*, no. Tropical Forests, 4:621685.

Relly, Eduardo, and Claudio de Majo. 2022. "A Soyacene of Fire:" In *The Age of the Soybean*, edited by de Majo and da Silva, 143–63. An Environmental History of Soy during the Great Acceleration. White Horse Press.

Ritchie, Hannah. 2021. "If the World Adopted a Plant-Based Diet We Would Reduce Global Agricultural Land Use from 4 to 1 Billion Hectares." *Our World in Data*. March 4, 2021. https://ourworldindata.org/land-use-diets.

Ritchie, Hannah, and Max Roser. 2021. "Forests and Deforestation." *Our World in Data*. February 9, 2021. https://ourworldindata.org/soy.

Rocha, Juan, Matilda Baraibar, Lisa Deutsch, Ariane de Bremond, Jordan S. Oestreicher, Florencia Rositano, and Cecilia Gelabert. 2019. "Toward Understanding the Dynamics of Land Change in Latin America: Potential Utility of a Resilience Approach for Building Archetypes of Land-Systems Change." *Ecology and Society* 24 (1): 17.

Rollins, Brigit. 2020. "The Deal with Dicamba: An Overview of Dicamba-Related Litigation." The National Agricultural Law Center, National Agricultural Library, Agricultural Research Service, U.S. Department of Agriculture. https://aaes.uada.edu/research-highlights/economics-law/understanding-dicamba-related-litigation/.

Romson, Åsa. 2012. "Environmental Policy Space and International Investment Law." Doctoral Dissertation, Stockholm: Acta Univeritatis Stockholmiensis, Stockholm University.

Rosset, Peter. 2003. "Food Sovereignty: Global Rallying Cry of Farmer Movements." *Food First Institute* 9 (4): 1–4.

Sarandón, Santiago Javier, and Claudia Cecilia Flores. 2014. *Agroecología*. La Plata, Argentina: Editorial de la Universidad Nacional de La Plata (EDULP).

Schut, Antonius, Emily Cooledge, Marc Moraine, Gerrie Van de Ven, Davey Jones, and David Chadwick. 2021. "Reintegation of Crop-Livestock Systems in Europe: An Overview." *Frontiers of Agricultural Science and Engineering* 8 (1): 111–29.

Siu-kai, Lau. 2022. "Head-of-State Diplomacy Broadens China's Global Space." *China Daily*, November 22, 2022. https://epaper.chinadaily.com.cn/a/202211/22/WS637c114ca31009d7c3da528f.html.

Smil, Vaclav. 2002. "Worldwide Transformation of Diets, Burdens of Meat Production and Opportunities for Novel Food Proteins." *Enzyme and Microbial Technology* 30 (3): 305–11.

Stedile, Joao Pedro. 2019. "The Struggle against Hunger Is an Enduring Challenge; Food Sovereignty Is Essential." October 30, 2019. https://viacampesina.org/en/

october-16th-the-struggle-against-hunger-is-an-enduring-challenge-food-sovereignty-is-essential/.

Steffen, Will, Johan Rockström, Katherine Richardson, Timothy M. Lenton, Carl Folke, Diana Liverman, Colin P. Summerhayes, et al. 2018. "Trajectories of the Earth System in the Anthropocene." *Proceedings of the National Academy of Sciences* 115 (33): 8252–59.

Sterckx, Roel. 2019. *Chinese Thought: From Confucius to Cook Ding.* Pelican: Penguin Books.

Suppan, Steve. 2020. "Revisiting Crisis by Design: The Agricultural Technology Treadmill." Institute for Agriculture & Trade Policy, April 2020, Minneapolis: IATP.

Trigo, Eduardo, José Falck-Zepeda, and César Falcon. 2010. "*Biotecnología Agropecuaria Para El Desarrollo En América Latina: Oportunidades y Retos.*" LAC/01/10. Washington DC: FAO – BID.

Tscharntke, Teja, Alexandra M. Klein, Andreas Kruess, Ingolf Steffan-Dewenter, and Carsten Thies. 2005. "Landscape Perspectives on Agricultural Intensification and Biodiversity – Ecosystem Service Management." *Ecology Letters* 8 (8): 857–74.

Turner, Terry. 2022. "Dicamba Lawsuits." Consumer Notice. September 30, 2022. CN https://www.consumernotice.org/legal/dicamba-lawsuits/.

Vergara-Camus, Leandro, and Cristobal Kay. 2017. "New Agrarian Democracies: The Pink Tide's Lost Opportunity." *Socialist Register* in Rethinking Democracy, edited by Panitch and Albo, Vol. 54: 224–243, London: Merlin Press.

Wang, Jennifer. 2020. "The Chinese Billionaire Whose Company Owns Troubled Pork Processor Smithfield Foods." *Forbes*, April 16, 2020. forbes.com/sites/jennifer-wang/2020/04/16/the-chinese-billionaire-whose-company-owns-troubled-pork-processor-smithfield-foods/?sh=28213d962c55.

Wasley, Andrew, Alexandra Heal, André Campos, and Daniel Camargos. 2021. "Loophole Lets Soya Farmers Tear down the Amazon." Entreprises et Les Droits de l'Homme. Centre de Ressources sur les. https://www.thebureauinvestigates.com/stories/2021-05-19/loophole-lets-soya-farmers-tear-down-the-amazon.

Whitton Clare, Diana Bogueva, Dora Marinova and Clive J. C. Phillips 2021. Are We Approaching Peak Meat Consumption? Analysis of Meat Consumption from 2000 to 2019 in 35 Countries and Its Relationship to Gross Domestic Product. *Animals* 11(12): 3466.

WWF. 2022. "WWF Statement on the COP27 Agri-Commodity Sector Roadmap." https://www.worldwildlife.org/press-releases/wwf-statement-on-the-cop27-agri-commodity-sector-roadmap.

Xingqing, Ye, Cheng Yu, and Zhang Xu. 2022. "Grain Teaser." *China Daily*, October 12, 2022, Global Edition, sec. National Affairs.

Yimeng, Zhao. 2022. "Action Plan Puts Measures in Place to Ensure National Food Security." *China Daily*, April 12, 2022, Global Edition, sec. National Affairs.

Zelic, Helena, and Bianca Pessoa. 2022. "End Hunger, End the WTO: The Peasant Caravan to Geneva against Free Trade." Translated by Aline Scátola. La Via Campesina. June 22, 2022. https://viacampesina.org/en/end-hunger-end-the-wto-the-peasant-caravan-to-geneva-against-free-trade/.

zu Ermgassen, Erasmus K. H. J., Mairon G. Bastos Lima, Helen Bellfield, Adeline Dontenville, Toby Gardner, Javier Godar, Robert Heilmayr, et al. 2022. "Addressing Indirect Sourcing in Zero Deforestation Commodity Supply Chains." *Science Advances* 8 (17): 1–15.

Index

Note: *Italic* page numbers refer to figures and page numbers followed by "n" denote endnotes

ABCD companies 144, 186–88, 192, 193, 223, 235
acquisitions *see* horizontal integration; vertical integration
adaptive cycle 9
ADM (Archer Daniels Midland) 13, 140, 144, 145, 178, 187–92, 223; *see also* ABCD companies; traders
adzuki beans 72, 85
Agent Orange 169
agreements *see* trade agreements
agribusiness 156, 158, 168, 177, 188–89, 191–92, 198, 200, 220–21, 223, 227, 232, 235; *see also* ABCD companies; traders
agricultural output 36, 147, 165, 218, 220, 229; Brazil 168; China 71–78, 93–95, 105, 163; cotton 95; demand 77; demographic growth 73; Europe 105, 117; feeding the population 93, 95; in the market (marketization) 74, 75; labor 95; Manchuria 76–77, 217; mulberry 95; per capita 72, 73; prices 75; rice 94; soy 73, 75–78, 124, 125, 145, 147, 168, 217, 227; United States 117; world (global) 117, 147, 195, 219–20
agricultural policies 24, 163; agroecology 231, 236; Farm Bill 165, 199; Food for Peace Act 175; protectionism 11, 175, 178; Public Law 480 165, 175; subsidies 162, 165, 177, 192; surplus 31; *see also* Confucianism; division of labor; trade policies
agricultural revolution (Europe) 105; second agricultural (or agrarian) revolution 165, 225
agricultural treatises, Chinese 22, 32, 43, 66, 67, 98; *see also Fan Shengzhi Shu; Qimin Yaoshu; Simin yueling*
agriculture: animal husbandry (livestock) 65, 126, 214; biodiversity loss 6, 29, 218; books 58–59, 116; capitalist relations 105, 230; Chinese civilization 31; Climate Smart (renewable) 227–28, 231; Confucian 51, 52n4, 67, 73–74; deforestation 48, 223; economic history of 147; economies of scale 105, 220, 229; European 214; expansion 51, 100, 214, 215; export 175; Fan Shengzhi shu 24, 37; financialization of 156; frontier expansion 1, 7, 30, 34, 37, 47, 61–62; "from nowhere" 11; globalization of 6, 175; "great leap forward" 163; green revolution 11, 165, 167, 200, 225; Han rulers 25, 40; herbicide tolerant 226–27; human settlement 29; industrialization of 5, 11, 148, 165, 171, 200, 218–20, 225, 227, 229; intensive (intensified) 15, 42, 48–49, 51, 106, 128; international trade in 136, 175; land degradation 48, 219, Manchuria 126–29, 217; north Chinese 31–49; population growth 34, 70, 220; precision 34, 227, 231; productivity 77, 167; Qimin Yaoshu 25; root of prosperity 24, 34, 60; school of 138; science 131; SMR 128; South Chinese 50, 61, 107; stagnation 102, 216; subsistence 107, 229; technology (and innovation) 99–102, 132, 147, 199; transnational 156; value of 38; "without farmers" 230; work (labor) 29, 64, 95, 229, 231; WTO 175, 178, 218; Yearbook of (USDA journal) 139; *see also* farming
agrofood system 1–17, 57, 70, 108, 134, 147, 157, 178, 189, 201, 235–36
agronomy 5, 15, 26, 107, 128, 162, 166, 226

alternatives 11, 14, 229–35; challengers 11, 174; counterpoints 11, 173, 198; *see also* countermovement; roles, uses and functions
Amazon 173, 200, 218, 222–24; Soy Moratorium 223–24
animal feed *see* feed
animal power *see* power
Aoyagi, Akiko 26, 52n1, 117
Argentina 32, 133, 138, 167–69, 172–74, 181, *183*, *185*, 186–87, 200, 218, 231–33
Arthur (port) 122–23
ASA (American Soybean Association) 13, 117, 141–44, *142*, 146, 175–77, 192, 199

Bacon, L. B. 136
Barbier, Edward 102–03
barley 24, 28, 30, 38, 40, 42
BASF 174, 229; *see also* agribusiness
Bayer (including Monsanto) 2, 169, 173, 174, 187, *201*, 223, 226, 228, 229, 231
Bencao Gangmu 92
Bennett's Law 194; *see also* nutritional transition
Beresford, Lord 120–21
Bergius, Peter Jonas 100–01
biodiversity 29, 166, 173, 223, 225; loss 1, 6, 147, 166, 201, 213, 218, 219, 220, 222
biofuel 2, 186, 188, 190, 195–97, 224–25; biodiesel 2, 186, 196, 224
Bolivia 80, 162, 169, 172, 173, 200, 218
Bolsonaro, Jair 223
botany 33; botanical 101; botanical garden 105–06
bovine spongiform encephalopathy (BSE) 192
Bowen, Samuel 106, 138
Boxer Rebellion 121; *see also* rebellions
Bozhong, Li 75, 95
Bramah, Joseph 134
Brasiguayos 168, 232
Braudel, Fernand 7
Bray, Francesca 25, 26, 94
Brazil: ABCD companies 193; Bahia School of Agriculture 138; biodiesel 196; brasiguayos 168, 232; Bunge 166, 168, 178, 198; Cargill 167, 178, 218; Cerrado 173, 200–201, 218, 222–25; COFCO 235; deforestation 168, 173, 200, 218–19, 222, 224; displacement 166, 173; land-use change 166, 172–73; mixed farming 165; pesticide use 173; resistance 173, 226; rotation with soy 166; RR soy (GE) 169–71, 226, 231; soy expansion 218; soy export 167, 178, *179–85*, 186; soymeal 193; soy production *159–160*, 162, 164, 181; soy promotion 162, 166, 168, 176; soy specialization 166, 186; soy taxes 186; soy yields 165; sugar trade 82; value-added 186, 193
breeding, seeds 101, 105, 116, 139, 141, 172
Bretton Woods system 156, 176, 199, 200
Britain (including England and British) 80, 83–85, 103, 104, 106, 115, 156; agriculture 105; balance of trade 58, 83; Manchurian soybeans 124, 133, 134; capital 99; colonies 106, 164; consulate 117–21; cookbooks 99; diplomacy 104; government report 136; great divergence 95; hegemony 11; imperialism 233; imperial trade 133; port of Hull 124; soap maker 137; soy sauce 82–83, 99, 106
broomcorn millet *see* millet
Buddhism: China 48; monks 63–64, 65, 66, 87, 90; soyfoods 63–65, 66, 87, 89
Bunge 144, 145, 166–68, 178, 187–92, 198, 223, 224; *see also* ABCD companies; traders
Butterfield & Swire 120
by-products of soy 90, 93, 94, 97, 98, 107, 127, 134, 139, 147, 149, 189, 192, 195, 196, 199

cake: beancake 72, 75, 77, 78, 93, 95, 98, 101, 102, 105, 107–08, 115, 119–21, 124, 125, 127, 129, 136; pulse-cake 118; sesame pancake 66; soybeans 44, 59, 64, 67, 71, 73, 74, 86, 92–98, 107, 117, 137–38, 162–63, 225; tofu 88
Canada 138, 140, 144, 169, 185
canola (rapeseed) 67, 132, 135, 169, *170*, 195, 196, 220
Canton (Guangdong) 68, 79, 82, 84, 103, 121; system 103–04, 121; traders 103
capitalism 9, 10, 11, 97, 108, 126, 201, 215, 229
capitalist accumulation 215
Cargill 13, 144–46, 155, 167, 178, 187–93, 195, 198, 224; *see also* ABCD companies; traders
Carletti, Francesco 89–90

cash crop 26, 40, 51, 61, 70–74, 77, 93, 96, 107, 127, 142, 143, 155, 162–64, 167, 177, 201, 217
Cerrado see Brazil
chain drivers 9, 13, 64, 173–75, 188–91, 223
cheap: fertilizers 214, 225; food 11, 133, 138, 200; meat 5, 6, 141, 198, 199, 201; soy 146, 192, 219–29, 236; *see also* labor, cheap
ChemChina 174–75, 188–89
China 4–6, 27–42, 47–50, 57–59; agriculture 25, 38, 51, 61, 107; ancient 215–16; ChemChina 174–75, 188–89; farming system 23; five-year plan 163, 234; food 58, 60, 75; frontier expansion 11, 30, 34–35, 47, 51, 57, 216; land 70, 101–02; population 70, 72, 125; roots 60–63; soybeans 70–73, 85–89; soyfoods evolution 63–67; soy sauce 89–91; trade 69, 102, 103, 107, 120; unequal treaties; weakened 104; WH Group 234; *see also* Han Chinese
Chinese Eastern Railway (CER) *see* railways
Chinese Maritime Customs Service (CMCS) (or Imperial Chinese Maritime service) 104, 115, 117, 121
civil unrest 47, 62; *see also* rebellions
Clapp, Jennifer 189, 226
Climate Smart Agriculture 227
COFCO (China Oil and Foodstuffs Corporation) 188–89, 223, 224, 235; *see also* traders
coffee 1, 82, 83, 136, 166, 224
colonies 80, 103, 105–06, 128, 131–38, 164
colonization 15, 66, 122, 126
Columbian Exchange 57, 69, 94
commercialization 4, 12, 14, 26, 49, 58, 74, 76, 90, 107, 187–91, *190*, 201, 214, 226
commodification 230–31
commodity chain *see* Global Commodity Chain (GCC)
Commodity Credit Corporation 143
Common Agricultural Policy (CAP) 199
complexity 6, 7, 13, 15, 215–16
concessions 104–05, 122, 130
Confucian 23, 26, 35, 38, 50, 51, 58, 67, 73–74, 87, 216; confucianist 22, 72; Confucianism 23
congee *see* soyfoods
conventions theory 10
cookbooks 5, 58–60, 64, 85, 87, 99
core areas (regions) 91, 107, 127, 128, 214
corn (maize): adoption in China 71–72; biofuels 196; biotech 169–70; Argentina 166; Brazil 166; Columbian Exchange 69, 71, 94, compared to soy 143, 161; feed (fodder) 146, 192, 195; foodstuff (staple) 127; Green Revolution 165; intensive production 69, 94, 107; Manchuria 73, 127; prices 176; rotation with soy and other crops 139, 164–66; trade 133, 188, 233; in Ukraine 164; in the United States 6, 116, 139, 146, 161, 165–66, 192, 217, 233; world production 1, 160
Corn Belt (US Midwest) 6, 116, 143, 158–75
Corn Laws (UK, 1846) 133
corporate 158, 169; agriculture 230; Award for Corporate Excellence 195; control 232; domination 13; food 198; food regime 156; hegemony 11; Social Responsibility 223
cotton (cottonseed) 8, 58, 69–71, 74, 82, 94, 95, 127, 133, 137, 140, 142, 160, 166–69, *170*, 172, 195, 220; cottonseed oil 135, 136, 140
countermovement 214, 222, 229; alternatives 229–35; challengers 11, 174; contestation 214; counterpoints 11, 173, 198; resistance (social movements) 214, 222, 229–35; *see also* alternatives
Crook, Fredrick 158
crop rotations *see* rotation schemes
Crosby, Alfred W 69
crushing 66, 67, 128, 129, 133, 135, 136, 140–42, 144, 166–67, 179, 181, 182, 186, 188, 190–92, 195, 223; mills 124, 129, 133, 136, 137, 140

Dairen (today Dalian, Manchuria) 122–25, 129
deforestation 1, 6, 37, 51, 102, 107, 172, 173, 213, 216, 218, 219, 222–25
Denmark 130, 135, 136, 137, 138, *185*, 192
Dennys, Nicholas Belfield 117–18
Deshima (Japan) 60, 80, 81, 83, 91
development aid 175
Diderot, Denis 83
diversity 6, 10, 50, 147, 148, 174, 215, 236
division of labor 8, 11, 29, 31, 74, 77, 127, 147, 156, 229

domestication of soybeans 4, 12, 14, 27–28
Dominican 87–88
Dongguan Hanji 41, 44
Dorsett, Howard 139
draft animals 49, 77, 127; *see also* labor, animal
Dreyfus (Louis Dreyfus Commodities or LDC) 144, 187, 190, 221, 223; *see also* ABCD-companies
drought 31–33, 35, 37, 39, 43, 47, 60–62, 118, 163, 174, 189, 230
due diligence 224

East Asian cuisine 85–93
ecological simplification 6, 7, 29, 167, 214–19, 227
economies of scale 105, 165, 166, 188, 199, 201, 221, 229
ecosystem service 7, 127, 148, 204, 215, 219, 227
Ecuador 231
edamame *see* soyfoods
Edo period 59–60, 90
EIC (East India Company) 17, 60, 79, 80, 84, 91, 100, 103, 106
Ekeberg, Carl Gustaf 91, 100–01
Ekken, Kaibara 72
elites 35, 41, 47, 48, 50, 62, 69, 70, 84, 97
England *see* Britain
erosion, soils: in ancient China 48, 51 in contemporary Latin America 166; in contemporary US 166; in early modern China 73, 102, 216; intensification of land use 215, 218
eurocentrism 15–16
Europe: agricultural productivity 105; agrochemicals 167; biodiesel 196; Canton system 103; contrast to China 22, 35, 50, 51, 61, 68, 84, 95–96, 101, 216; cookbooks 99; cultivation of soybeans 105, 128, 131–38, 146, 161; GE traits 174, 187; feed 137, 192–93; food imports 133, 135; industrializing 11, 133; Manchuria 124, 128, 131, 133–38; margarine 5, 13, 135–36; meat consumption 149; medieval 61; oil-industry 129–30, 134–37; solvent extraction techniques 134; soy commodity chain 191; soy enthusiasm 58, 99–101, 132, 138; soy imports (demand) 124–25, 128, 134–37, 161, 175–76, *180*, 187, 192, 224; soy sauce 82–85, 89–90, 99–101, 107; trade 68–72, 78–82, 83–85, 145; rise of the west 11, 57, 68, 82, 101; United States 103, 138, 140, 142, 144–45, 176, 192
European: agriculture 131–33, 165, 171, 214, 227; agro-industry 105; 116; 131, 133–38; colonies 105, 128, 131; conquest of the Americas 71; demand for Asian goods 68–69, 84; family farmers 132; force 82; imperialism 79, 84, 216; literature on soy 60, 100, 116; maritime power 68; mechanization (of agriculture) 105; merchants (companies) 5, 69–70, 79–80, 104, 106, 174; nationalisms 16; parliament 187, 224; policies 136, 145, 175, 199; powers 80, 83–85; scientists 12, 99–101, 106, 131–32; shipbuilding 68; soils 101; soy demand 124–37, 175, 187; soyfood 87–92; tariffs on soy oil 131; trade 57, 79–86; travelers 5, 106; Treaty of Nanjing 103
excrement (night soil) 37, 50, 93, 94, 96, 97
expansion/extensification *see* land
experiment station 138
export: boom 128; costs 5, 7, 11, 41, 57–58, 61, 82, 101, 122, 126–28, 133, 147, 216, 221; driven 216; oriented 126, 145, 167, 168, 229, 231; regions 128; structure 125, 130, 176, 178–88
externalities 215, 220, 222, 228; environmental 222; negative 220; positive 220; social 222

factory 80, 81, 91, 136, 193
fallow (land) 36, 37, 70, 77, 93–94
famine 4, 34, 35, 39, 40, 42, 47, 57, 62, 76, 86, 98, 118, 163, 216
Fan Shengzhi shu 23–24, 25, 32, 33, 37, 43
Fan Shengzi (Fan Shen) 24, 32, 33, 36, 43
fandou see soyfoods
FAO (Food and Agriculture Organization of the United Nations) 157–58, 176, 219, 227
Farm Bill (US) *see* agricultural policies
farm machinery 105, 165, 174, 217–18, 227
farm size 35–36, 126, 163, 233
Federico, Giovanni 147
feed 2, 5, 12, 35, 57–58, 107, 142–48; animal 97–99, 138, *197*, 200, 213, 233–35; concentrates 148, 177, 191, 193, 199; fodder 58, 62, 97, 98, 123,

135–37, 165, 175–76, 194, 217, 218; Manchurian soybeans 137; nutritional value (heated) 141; on-farm use 142; prices 176; soybean meal, flour or cake 136, 191–95
feedlot 193–94
fermentation 4, 45, 63–65, 90; fermented paste 41, 45
fertilizers 5, 12, 13, 50, 52, 57–59, 67, 69–75, 77, 98, 107–08, 127, 139, 148, 160, 165–67, 174, 214, 227, 228, 233; beancake 97; cakes as fertilizer 93–97; cheap 214, 225; chemical 162, 200, 216; green manure, soybeans as 37, 50, 63, 71, 73, 93, 94, 115, 121, 139, 165, 166; organic 37; revolution 95; soy cakes in Japan 124; synthetic 148, 165, 166, 214, 215, 217, 220, 225
financialization 13, 156, 157, 189–90
fish 2, 42, 45, 86, 89, 90, 91, 108, 194, 220; oil 137; meal 194
fodder *see* feed
food 25, 40–41, 51–52, 76–77, 127–28, 130–36, 187–91, 200–01, 231, 235–36; agriculture 147; Chinese 58, 60, 75; crops 167, 169, *170*; cultures 148–49; East Asian cuisine 85–93; human consumption 41–47; industries 137, 143, 145; nutritious 197–99; security 124, 129, 169; soybean oil 195–97; surplus 29, 72; subsidies 156; *see also* soyfoods
Food for Freedom program 145
food preparation (processing) techniques 4, 52, 134, 136
food regime 7, 10–14, 31, 133, 136, 156, 235; first 11, 133, 136; second 11, 156; third (corporate) 11, 13–14; *see also* soybean cycles
foxtail millet *see* millet
France 34, 104, 120, 126, 136, 145, 228; French 7, 80, 98, 104, 128, 135
Frank, Andre Gunder 69
free trade *see* trade policies
Friedmann, Harriet 10, 136
frontier: agricultural 1, 35, 30–31, 51, 213; expansion 34–35, 37, 47, 57, 117–20, 214–19; frontier-based development 61, 62, 126, 213; Gene Revolution 200; Great Frontier 216, 219; Latin American 158–75; Manchurian 6, 117–20, 124–28; settler 11; soy 11, 217–18; transfer of nitrogen 124
fuel *see* biofuel
futures/futures market/futures contracts 70, 134, 141, 145, 177, 189, *190*, 201, 213, 214, 220, 228

GE (Genetically Engineered) 1–2, 5, 6, 168, 169, 219
GE Soy Model 173, 230
Gene Revolution 5, 11, 156, 157, 164, 169–74, 200, 219, 225, 229–30
geopolitical: aims 138; history 10; shifts 106, 232–33, 235; tensions (Manchuria) 105, 118–19, 121
Gereffi, Gary 9
Germany: BASF 174; Bayer 174; crush (mill) 131, 144; feed 138; Kaempfer 91; glyphosate-ban 228; inoculum (Nitragin or Germ Fertilizer) 132; lecithin 137, 195; Manchuria 104, 120, 126, 130–32; Rumphius 88; soy-based substitute for rubber 135; soy for food 138; soy export *180, 185*, 186; soy import 130, 131, 135–38, *180, 185*; vessels 118
Gills, Barry K. 69
global biotech crop 169, *170, 171*
Global Commodity Chain (GCC) 7–10, 14, 196; agrofood chains 190; chain drivers 9, 13, 173, 187, 223; commodity chain 8–9, 108, 116, 129, 144, 157, *157*, 164–66, 187, 188, 109–91, 196, 200–01, 220–21, 236; soybean chain (including soy chain) 4, 128, 143, 144, 157, 158, 196, 200, 201, 223, 230; supply chain 133, 156, 188, 189, 221, 224, 225, 234; Global Value Chain (GVC) 8, 17n1; value chain 188, 195, 197, 220, 221, 235
globalization 6, 177, 191, 218; agrofood 4–7, 14, 217; neo-liberal 156
glufosinate 226
Glycine max 1, 27
glyphosate 2, 169, 170, 172–74, 201, 226, 228, 232
gold standard 11, 128, 136, 225
Goldsmith, Peter D. 168
governance 9, 23, 49, 58, 129, 178, 236
Grain Robbery 176
Gran Chaco 200, 218
granary 40, 76
Grand Canal 50, 75
Great Acceleration 2, 5, 6, 13, 156
Great Britain *see* Britain
Great Depression 142

Great Divergence 95
Great Leap Forward 163, 176
green manure *see* fertilizer
Green Revolution 11, 165, 167, 173, 200, 217–18, 225, 229
greenhouse gas emissions (GHG) 1, 181, 196, 197, 201, 213, 214, 225, 235
Gu, Bang 41
Guangdong (Canton) 68, 79, 84, 105
Guangwu 41
Guanzi (Kuan Tzu) 33, 44
guild *see* merchant organization
gunboat diplomacy 103

Haber-Bosch method 148
Han Chinese 23, 24, 30, 31, 49, 61, 62, 72, 73, 76, 78, 103, 119, 129, 217
Han Dynasty 4, 12, 22–27, 29–31, 34–42, 44–51, 62, 63, 67, 73, 233
Harbin 122, 125
Hart, Robert 104
hegemonic 4, 10, 11, 52, 69, 81, 101, 106, 176, 233, 235
herbicide tolerant (HT) 169, 173, 225, 226
Hirado 60
Hitsudai, Hitomi 89
Holland *see* Netherlands
Hopkins, Terrence 8
horizontal integration (including acquisitions) 157, 169, 173–75, 187–92, 201, 234
horse 28, 62, 71, 92, 97, 98, 123, 124
Hosie, Alexander 117, 119, 121, 122
Hou Hanshu 25, 40, 41
Huang He see Yellow River
Huang, Hsing-Tsung 26, 27, 45, 58

imperialism 79, 105, 116, 120–25, 155, 217, 233
Import Substitution Industrialization 168
import tariffs (taxes) 131, 136, 178, 182, 186, 187, 194
India 63, 66, 81, 82, 100, 103, 128, 162–64, 169, 181, 182, *183*
Indonesia 66, 81, 128, 184, *185*, 186
industrial agriculture (modernization of; industrialization of) 11, 165, 171, 200, 218–20, 225, 227, 229
Industrial Revolution 155
industrialized food (system, supply) 127
inputs (biotech, seeds and agrochemicals) 174, 221
intensification, of agriculture 29, 36, 37, 97, 107, 156, 157, 164, 166, 167, 170–72, 194, 215, 217–19, 231
International Institute of Agriculture (IIA) 107
international political economy 9
International Soybean Growers Alliance (ISGA) 187
intra-Asiatic commerce *see* trade
investment agreements 218
Isett, Christopher 47, 97, 102

Japan 4–5, 28, 63–66, 72–73, 79–87, 89–93, 176, 225; bean cake 97; Manchuria 115–17, 120–22, 126, 128, 130, 131; soybean functions 106–08
Jardine & Matheson Steamship Co 120
Jesuit 79, 87–88, 90, 98
Ji Ran 39
Jia Sixtie 25–26, 32–33, 36–37, 49; *see also Qimin youshou*
jiang see soyfoods
Jinizi 38–40
Jonsson, Ulf 17
Jurchen (tribe) 61
Just-in-Time 221

Kaempfer, Engelbert 91, 100
Kangxi (emperor) 81, 84, 102
ketchup (soy sauce)(other names: catsup, ketjap, katsiap, ke-tsiap (Chinese pickled-fish sauce)) 81, 82, 99; catch-up 27, 58, 81, 83, 99
Keynesian 39
Khan, Gengis 61
Khan, Kublai 61
Kishlar, Lamar 145
Korea 4–5, 28, 29, 63, 65, 66, 81, 92, 120, 128, 130
Kuhn, Thomas 9

labor: animal power 105, 129; cheap (low wages) 77, 126, 127, 129, 135, 147; Han-Chinese 129; human power 105, 127, 129; division of 124, 147, 156; intensive 33, 37, 45, 50, 126, 129, 214, 216, 236; saving (including sparing and reduced use of labor) 62, 94, 165–66, 168, 170, 172, 226, 230; substitution with fossil fuel power 105, 127, 129; substitution with capital 105, 164
land: cover (use) change 164, 172, 218; China 70, 101–02; degradation 6, 166,

213, 219; expansion/extensification 34, 126, 160, 164, 219, 222, 236; fertile 213, 217; foreignization 232, 234; grabbing 232; Great Frontiers 219; Japan 97; prices 194; private 30, 218; redistribution 53n7; reform 35, 48, 163; scarcity 106, 216; soy boom 76–78
landowners 32–33, 47, 48, 59, 167
Langthaler, Ernst 138
latifundio-minifundio structure 167; latifundistas 194
Latin America
lecithin 137, 195, 196
legitimacy 229–35; longue-durée structure 213;
legumes 26, 28, 29, 36, 37, 42, 62, 93, 94, 107, 132, 141, 147, 148, 192, 214, 220
Li, C.N. 26, 58
LibertyLink 226
Linné, Carl von 100
livestock 1, 49, 51, 65, 94, 98, 107, 132, 137, 142, 145–47, 165, 176, 187, 191, 193–96, 199, 214, 216, 219, 222–23, 227–28
Locke, John 99
Lockyer, Charles 82
Loewe, Michael 23, 26, 34, 35
longue durée 7, 14, 213, 214
Lula (Luiz Inácio da Silva) 223
LVC (*La Vía Campesina*) 230, 231
Lyons, Thomas 122

Maddison, Angus 147
Malthusian (Malthus, Thomas) 73, 102, 105, 216
Manchuria 5, 13, 62, 73–78, 98, 104–05, 115–17, 134, 138–39, 141, 144–47, 234; extraction and control 126–31; frontier expansion 117–20, 217; imperial rivalries and railway imperialism 120–25; migration 126; Manchukuo 130
Manchus 70, 73, 76, 118
Manila 65–66, 68–69, 71, 79
margarine 136
marketization 59, 61, 74, 107
Marx, Karl 220
Masuda, Tadayoshi 168
Ma-wang-tui (tombs) 46
McMichael, Philip 10–13, 136
McNeill, JR 7, 93
meat 5, 6, 11, 12, 14, 35, 42, 45, 46, 51, 63, 65, 87, 89, 94, 98, 136, 137, 140, 141, 143, 145, 146, 149, 155, 175, 176, 177, 186, 187, 190, 191, 192, 194, 195, 197, 199, 200, 201, 214, 216, 220, 235; *see also* cheap meat; US Soy-Meat Complex
meatification 11, 149, 213
mechanization 167, 200, 218, 228
Meiji 120
Mengzhu, Ye 78
mercantilism *see* trade policies
merchant organization/house (*liangzhan*) 76; *Dounqi* merchants 78; guilds 77–78, 107, 119; Shanghai soybean guild (*douhang*) 78; Shantung guild 119; trader
mergers *see* horizontal integration
metabolic rift 148, 225
military 4, 12, 31, 44, 51, 57, 62, 72, 82, 101, 103, 130, 131
Miller, Stephen 47, 97
millet (including foxtail and broomcorn) 24, 28, 29–33, 35, 36, 38, 40, 42, 44, 46, 49, 50, 52, 60, 70, 72, 74, 126, 127
Ming: bureaucracy 62–62; Confucian revival 67; corruption 84; fermented soy 45; second soy cycle 57; revival of the North 63, 70; soy fertilizer 94–95; trade restrictions 67–68; Yuan defeat 62
Mintz, Sidney W. 52
miracle bean 11–12, 143, 145, 163
miso *see* soyfoods
mission: missionaries (Christian/Jesuit/Dominican) 5, 79, 87–88, 90, 98, 104, 106; trading missions 79; tributary missions 66–67, 68
Mitsui Bussan Kaisha 124, 128
mixed farming systems 51, 137, 139, 149, 155, 165, 167, 172, 191–92, 214
mixed feed *see* feed
modernization of agriculture (or agrarian modernization) 129, 171
Mongols 61–63, 67
monoculture 127, 147, 164, 166, 217–19, 226, 230
Monsanto 169, 173, 174, 187
Moronobu, Hishikawa 88
Morse, William 139, 143
mou (Chinese unit of land measurement) 31–32
multifunctional farming 216; *see also* rotation schemes
mutual trade *see* trade policies
Myrdal, Janken 17, 22, 38

Nagasaki 68, 79–82, 87, 131, 217
Nakajima, Gakusho 68
Napoleonic Wars 136
natto see soyfoods
Navarette, Fernández 88
Needham, Joseph 27
neoliberal 11, 156, 168, 200, 229, 230
Neolithic Revolution 27, 28
Netherlands (including Holland and Dutch): archives 17, 60, 83; Britain 84; cultivation of soy 100, 128; Dutch East Indies 81, 138; imperial trade 133; importer of soy 130, 138, *180*, 181; Japan (including Nagasaki and Deshima) 80, 83; margarine 135, 137; merchants 80; newspapers 60; ships 82; soy oil 130, 181–82; soymeal trade 185–86, 192; soy sauce trade 83; VOC 17, 60, 79, 81, 83, 84
New Deal 142
new global food order 128
New World 69, 76, 80, 82, 128
Newchwang (or Niu-chwang - today's Yingkou) 105, 115, 117–21, 124, 125
Nigeria 1, 128, 164
night soil *see* excrements
nitrogen (including nitrogen-fixation) 36, 37, 50, 70, 93, 94, 96, 127, 132, 139, 147, 148, 166, 201, 214, 215, 220
Nixon, Richard 176
nomadic 42, 48, 49, 51, 61, 98
North America 5, 84, 105–06, 124, 138, 235
North-Chinese agriculture 31–38
nutrient cycle 147, 148, 214, 217, 219, 225–28
nutritional transition 155, 194, 195; *see also* Benett's Law

occupation 88, 120, 126, 168
oil press, including wedge 66–67, 86, 92–93, 134, 149
oilseeds 5, 188, 195, 196, 219–20
okara 88, 93, 98, 107, 198
opium 99, 103; First Opium War 103; Opium Wars 115, 233; Second Opium War 104; trade 103, 104
organic fertilizers 37

paddy-rice production 71; *see also* rice
palm oil 2, 195, 222, 224
Pampas 165, 167, 200, 218, 223
Paraguay 162, 164, 168, 169, 172, 173, 181, 188, 196, 200, 218, 219, 227, 232, 235
pastoral land-use (pastoralism): history as a linear process 15; inefficient and wasteful 37, 42; non-Han (barbaric) 42, 48–49; pre-Han dynasty 37; "pastoral turn" 49, 62, 63
peanuts 69, 71, 72, 94, 107, 135, 166, 195, 219
peasant: agriculture 106–07; family 35, 231; farms 48, 74; food 85; household 12, 58, 73, 74; labor 77, 147; land 35–36; movements 232; organizations 230–33; peasantry 34, 35, 47–49, 62, 70
periphery (including hinterland) 8, 68, 72, 78, 118, 127, 128, 216, 217
pesticides *157*, 160, 165, 173, 174, 200, 214, 219, 226–28, 233
Petyt, William 83
Ping, Su 64
political economy 1–17
Pomeranz, Kenneth 70, 72, 74, 80, 95, 96, 101
population 4, 26, 31, 34–35, 40, 42, 60–61, 93, 95, 96, 102, 195, 217; China 70, 72, 125; European and American 149; Han Dynasty 47–50; production 59, 73; value system 51
ports: Amsterdam 135; Cargill (Minnesota) 144; Copenhagen 135; Hull 124, 133–34, 135; Liverpool 133, 135; Los Angeles 139; Nueva Palmira 221; port cities along the North Atlantic coast 84; Rotterdam 135; treaty port 105; Yangjiang 188; *see also* Darien port; Newchwang port; Port Arthur; Vladivostok; Yinkou
Portuguese 4, 68, 79–81, 87
potato 69, 71, 169, *170*
power 2, 9, 13, 202n12; ASA 141, 177, 199; China 5, 12, 29–31, 34, 36, 38, 44, 47–48, 67–70, 73, 101, 106, 188, 235; Chinese merchants 41, 77–78, 107; European 80, 83–85, 101; fossil fueled (e.g. steam, hydraulic) 105, 127, 129; Japan 73, 120, 129, 217; large US growers 166, 221; maritime power 68; Portuguese 68; power relations (configurations) 8, 9, 11, 14, 69, 101, 104, 116, 120, 129, 136, 155, 217, 229, 231, 236; private corporations 155, 159, 188, 191; purchasing 76, 195, 199; Russia 122; soybean's restorative power for soils 148, 214; technology (agrochemical-biotech companies) 188;

traders 188; US 217; US government (state) in GCC 141, 148, 156, 168, 175, 177, 188, 195–96, 199–200; *see also* concentration
precision agriculture 8, 9, 11, 14, 34, 229, 231, 236
price convergence 128
privatization 231, 168
processing techniques 4, 41, 45, 63, 64, 133
Prodöhl, Ines 128–29
production 59, 71, 73, 158, 194
property-rights regime 221, 231
protectionism *see* trade policies
protein 5, 28, 42, 52, 65, 87, 91, 97, 136, 140, 141–45, 161, 197–99, 213, 234–35; animal 29, 145; conversion loss 245; fodder (feed) 137, 142–47, 178, 187, 191–95, 199, 213, 218, 220; source of 89, 94, 123, 146, 149, 187, 191–94, 197

Qi, Wang 92
Qianlong (emperor) 75, 84, 102, 103
Qimin yaoshu 25–26, 32, 33, 36, 37, 45, 49, 98
Qing Dynasty (including Manchu): agricultural marketization 74, 76; bean cake 96–99; British 104, 233; collapse of 125; corruption 79, 84, 103; frontier expansion 70; gazetteers 59; granary system 76; Grand Canal 75; 101–02; Japan 120–23; Manchuria 73–75, 118–21; migration 72, 118; neo-Confucian 73; opium 104, 233; regime phase of the second soy cycle 57, 67, 69–101; rupture phase of the second soy cycle 101–06; Russia 120–22; settlement 73, 147; specialization 74; soy sauce 89; soy trade 75–78; Taiping Rebellion 104; taxes 76, 103, 118; trade protectionism (regulations, restrictions) 73, 77–78, 79; tribute trade 79; uprisings 78

railways (railroads) 122, 127–29, 147, 178, 216; Chinese Eastern Railway (CER) 122, 124, 130, 131; imperialism 120–25, *123*; South Manchurian Railway (SMR) 123–25, 128–29
Rath, Eric 59–60
rebellions, upheavals and uprisings: 44, 47–48, 50, 78; Boxer Rebellion 121; Taiping 104, 118
recipes 58, 63–64, 86, 87, 89, 99
regulatory 2, 108, 169, 200, 223, 224, 226, 229; environmental 221; regulation 75–79, 119, 192–93, 199–200, 215, 219
resilience 7–10, 14
resistance (ecological) 33, 226–27; drought 33; pesticide 173, 214, 225–27; weed 173, 225–27
revolution: agricultural (agrarian) 105, 155, 165, 225; Chinese 125; fertilizer 95–96; Gene 5, 11, 156, 157, 164, 169, 173, 200, 219, 225, 229; industrial 155; National Revolutionary Army 130; scientific 9, 102; *see also* Green Revolution
rice: biotech 169–70; Champa 60; Chinese crop system 29, 30, 33, 50; communist China 163; domesticated in China 28; foodstuff (staple) 38, 42, 60, 85, 87, 88, 125; green revolution 165; in Japan 72, 85, 90, 125; intercropped with soy and other crops 28, 72, 85; intensive production 58, 69, 93–95, 107; Ma-wang-tui tomb 46; "necessities of life 64; paddy 70–71, 93–94; price 40; rotated with soy 71; sacred/superior crops 24, 70, 71; in South China 12, 42, 50, 61, 93, 95, 96; soy fertilizer 94–96, 107, 125, 129; soy sauce 80; tax 94; trade 41, 79; world production 1, 160
Richards, John 70
Rihua, Li 86
roles, uses and functions of soybean: 2, 3, 4–6, 14, 26, 71, 131, 136, 139; alternative/relief crop 139; food aid 11, 156, 175; forage 139; military rations 4, 12; rituals and medicine 45–47; *see also* cash crop; feed; fertilizers; biofuel; soyfoods
roots-regime-rupture phases of soybean cycle 14
rotation schemes (crop) 36–38, 50, 70, 71, 93, 127, 132, 142, 164–67, 214–15, 227, 228; abandonment of long, diverse 127
Roundtable on Responsible Soy (RTRS) 223–24
Roundup Ready (RR) 169, 172, 173, 226
Royal Swedish Academy of Agriculture and Forestry (KSLA) 100
rupture phase of the soy cycle 12–14, 22, 47–50, 57, 58, 101–06
Russia: The Chinese Eastern Railway (CER) 122–24; Grain Robbery 176; immigrants 126; imperialist rivalry 119–21; invasion of Ukraine 164; investment

147; Manchuria 104, 116–17, 119–24, 126, 147; margarine 135; merchants 5, 124; profitability 219–29; Russo-Japanese War 122–24; Sino-Russian alliance 122; soldiers 123; soy exporter 180–81, 185; soy importer 136, 178, 180, 185; soy production 164; Treaty of Aigun 104; Treaty of Portsmouth 123; Vladivostok 124; *see also* Soviet Union

Schloemer, FC 136
self-sufficiency 12, 34, 40, 61, 131, 132, 148, 163, 178, 192, 200, 216, 233, 234
Shang Dynasty 37
Shen Nung 24
sheng (shēng – ancient chinese unit of measurement) 31–32, 33, 42, 97
Shennong Bencaojin 24
shi (basic unit of weight in ancient China) 32, 75, 76, 95, 105
shi (fermented black soybeans) *see* soyfoods
Shih Hsien Hung Mi 87
Shih Huang Ti 23
Shih, Shêng-han 36
Shiji (Shih Chi) 23, 25, 39, 41
Shijing 23, 29, 30
Shiliao bencao 46
ship 76, 79, 79, 81, 82, 91, 105, 124, 139–40, 144, 178, 191, 194, 221, 225
Shizheng, Li 79
Shônin, Nichiren 66
shoyu see soyfoods
Shurtleff, William 26, 52n1, 117
silver 68–69, 76, 79, 80, 84, 101, 103, 104, 107, 120, 133
Simin yueling 24, 26, 32, 33
Sinocentric soy regime 58, 101
sinocentrism (sinocentric) 5, 16
Sino-Japanese War 115, 120, 131
Sixtie, Jia 25, 26, 32–33, 36–37, 49
soap 116, 123, 134, 135, 136, 137, 140, 195
social function of land 218
social-ecological system (SES) 2, 9–10
socioeconomic complexity 6–7
sojización 1, 172, 194, 222, 227, 230–31
Song Dynasty (rule) 24, 52, 57, 58, 61, 63, 64, 65; era (period) 57, 62; frontier-based development model 61; officials 60; navy 65; soyfoods 64, 65
sorghum (*kaoliang*) 70, 74, 126, 127, 147, 165
South Africa 136, 164, 169
South America 148, 161, 162, 164–68, 172, 181, 196, 200, 201, 213, 231, 233; *see also* Latin America
Southeast Asia: Columbian Exchange 69, 71; commerce (trade) 5, 65–66, 68, 78–79, 81–82; domestication of soy 4, 27–28; European demand 68–69, 82–85; fertilizer 93; Japanese conquest plans 131; meat consumption 194; soyfoods 63–67, 69, 88, 106–07, 198; women 69
Soviet Union 126, 130, 131, 135, 147, 164, 176; *see also* Russia
soy cycle: first 12, 22–53; second 12–13, 57–108; third 13–14, 115–49, 155–202
soy meal (flour): Argentina 186, 233; Brazil 186, 193; byproduct 141, 149; crush spread 189; EU imports 187, 192; exports 184–86 export taxes 186; feed (fodder) 136–38, 141–45, 147, 155, 157, 159, 187, 191–93, 233, 234; feedlot 193–94; fishmeal 194; foods 136; futures market 189; imports 184–85; mills 140–41; oil 181, 220; promotion abroad (from the US) 192; protein 191, 192, 194, 199; tariffs 141, 181, 186; trade 130, 136, 146, 175, 181, 186–87; traders 189–90, 193; US exports 176–77, 182
soy oil 2, 5, 66, 67, 69, 75, 86, 89, 92–94, 105, 107, 121, 123, 125, 130, 131, 135–37, 140, 141, 143–47, 149, *159*, 181–84, *182–83*, 186–87, 195, 196, 199, 220, 222, 233; hydrogenation 134; processing technology crushing 134, 139
soy output 124, 126, 145, 147, 168, 194, 217–20
soyabean oil *see* soy oil
soyfoods: boiled soybeans (congee, gruel, porridge, "sucking soybeans") 4, 20, 43–44, 63, 64, 92; edamame 86, 197, 198; *fandou* 43; *jiang* 45–46, 63, 64, 66, 86, 89, 92; miso 64, 72, 81, 85–87, 89–92, 143, 198, 235; *natto* 66, 86, 87, 88, 92, 98; *shi* 41, 45–46, 77, 95, 105; *shu* 4, 30; soy sauce (*jiangyou*, *shōyu*, ketchup) 5, 12, 45, 47, 57–58, 60, 63–67, 77, 79–92, 94, 99–101, 106, 107, 129, 133, 198, 199; sprouted soy 66, 86, 92, 198, 235; tofu 5, 45–47, 63–67, 69, 77, 85–89, 91–94, 98, 107, 123, 129, 143, 197, 198, 235; *see also* ketchup; margarine; soy oil
Soyinfo: Center 16, 58, 60, 117; Database 27, 52n2

Spanish 4, 66, 68, 80, 87–88
specialization 4, 40, 74, 77, 93, 95, 107, 127, 142, 147, 148, 155, 165–67, 171, 181, 201, 214, 216, 217, 225, 227, 229; *see also* monoculture; rotation schemes
sprouted soy *see* soyfoods
state: commission 23, 26, 37, 62; nation-16, 231; official 25, 59, 94, 148; owned 175, 188–89, 201, 235; policies 24; statesmen 23; system 10, 235
Stroessner, Alfredo 168
sugar 74, 75, 79, 82, 84, 127
Sui Dynasty 49
sunflower 167, 195, 196, 220
Swatow 121
Sweden 17, 60, 83, 84, 91, 99–101, 130, 136
sweet potato 69, 70, 89
Syngenta 174–75, 188–89, 223

T'ang Dynasty 49–50
Taiping Rebellion 104, 118
tax 35, 47–49, 59, 61, 68, 74–76, 78, 79, 94, 97, 103, 144, 181, 186, 232
technological complex, agrarian 38; technical and theoretical innovation in agriculture 5, 12, 57, 61, 66, 96–97, 101, 227
technological treadmill 173, 221, 225–28
Thunberg, Carl Peter 83, 91, 100
timber 41, 72, 74, 91, 224
tofu *see* soyfoods
Tokugawa shogunate 72, 73, 80, 85, 106, 120
tombs 43, 45–46
Tongye 43
Topik, Steven 80
traceability 221–25
tractors 155
trade agreements 10, 224; Agreement on Agriculture of WTO 178, 218; Dillon Round of GATT 175; free trade 104; General Agreement on Tariffs and Trade (GATT) 175; Germany 131; investment 218; Luso-Chinese Trade Agreement 68; Phase One Agreement 2020 233
trade policies: America first 233; barriers 178, 187; Canton system 103; comparative advantage 229; deregulation of agricultural markets 177; duties 104; export promotion 177, 239; free trade 11, 103–04, 133, 136, 139, 176–78, 192, 230, 233; liberalization 5, 77, 168, 177–78, 200, 218; mercantilism 80, 136; mutual trade 67–68; protectionism 11, 136, 141, 175, 178, 186; quota 76, 175, 178, 187, 200; restrictions 67, 74–78, 80, 84, 104, 130, 175, 177, 187; tariffs 68, 94, 103, 117, 131, 136, 141, 178, 182, 186, 187, 233; tariff-free 139, 175
trade: agricultural (agrofood) 1, 5, 12, 75, 116–17, 136, 145, 175, 178, 216, 218; ancient Chinese 23, 38–41, 69; Asia (Asiatic) 5, 65, 68, 84, 106, 107; balance 58, 83, 101, 103, 176; Chicago Board of 145; distrust (aversion) 40, 66, 74–75, 79; early-modern Chinese 82, 101–04, 107; EIC 80; Europe 83–84, 101, 103, 131, 145, 161, 177, 192; food regime 11–12; fertilizer 77, 93, 96, 225; geographical complementarity/specialization 4, 12, 61, 107; imperialism 105; increase in (expansion of) 2, 41, 61, 69, 107, 156; infrastructure 74, 76, 126; international (global) 1, 6, 67, 82, 84, 116, 117, 136, 145, 157, 175–91; intra-Asiatic 5, 12, 78–82, 106; Japan 80–82, 83, 91, 115, 120–22, 124, 129–30; long-distance (transoceanic) 57, 75, 82, 89, 106–07, 133, 200; Manchuria 73, 104, 115, 117–21, 122–29, 130, 217; Manila 69; maritime 65–66, 68, 79, 103, *108*; meal 181–84, *184*–85, 186–87; modern China 177, 186, 233; monopoly 79; mutual (tribute) 67–68, 79, 85, 103; networks 6, 79, 84; opium 103–04; price 40; processing techniques 41, 137; Russia 123–24; slave 80; smuggling (piracy) 68, 76, 78, 82, 84, 103–04; soy 1, 5, 12, 38–41, 73, 74, 75, 76, 78, 81, 105, 115, 118, 145, 147, 175–91; soy sauce 12, 57, 60, 80–81, 82–85, 88; standaridization 147; statistics (accounts) 16, 59, 116–17, 121, 158, 163, 187; transport costs/prices 41, 61, 66, 128; United States 145, 176–78, 181, 186, 192, 201, 233; unproductive 40; VOC (dutch) 79–80, 81, 91; World War I 136; *see also* trade agreements; trade policies; traders
traders: ADM 13, 145, 178, 187, 189–92, 223; Asian 81, 82; British 104, 119–20; Bunge 145, 166–68, 178, 187–90, 223–24; Cargill 13, 144–46, 155, 167,

178, 187–8, 190–93, 221, 223–24; Chinese tradesmen/merchants 39–40, 65, 77–78, 103, 119; COFCO 188, 223–24, 235; crushers 13; Dreyfus (LDC) 187, 190, 221, 223; financial services 189; horizontal integration of 157, 190–91; European 106; international 6; Japanese (e.g. Mitsui, Mitsubishi and SMRl) 123–24, 128; RTRS 223; Russian 124; USDA 13; *see also* EIC; VOC
transport/transportation: costs 5, 7, 11, 41, 57–58, 61, 75, 82, 101, 122, 126–28, 133, 147, 216, 221; freight (infrastructure, roads, ports, railways, steamships) 40, 41, 60, 68, 69, 79, 80, 83, 103–07, 117, 118, 120–25, *123*, 127–29, 131, 133, 139, 146, 147, 167–68, 178, 188, 216; Han-Chinese carriers 129
Trans-Siberian Railway 122
Treaty: Treaty of Aigun 104; Treaty of Nanjing 103–04; Treaty of Peking 104; Treaty of Portsmouth 123–24; Treaty of Shimonoseki 120; Treaty of Tientsin 104, 117
tribute 53n7, 68, 80; system 67, 78–79, 85; trade 103
Trump 186, 233
Tsarist Russia *see* Russia

Ukraine 164, 180, 185, 189
Unilever 137, 223
United Nations Environment Programme (UNEP) 195
United Nations International Maritime Organization 225
Uruguay 162, 169, 172, 181, 196, 221, 232, 235
US (United States) 2, 5, 6, 11, 104–06, 161–72, 177–78, *180*, *185*, 233–35; Agricultural Trade Expansion Act 177; Commodity Credit Corporation 177; Corn Belt 116, 143, 158–75; Farm Bill 165, 199; farming model 148; 192; food aid 11, 156, 175; "Food for Freedom" program 145; Food for Peace Act 176; Public Law 480 175; Smithfield 234; in South America 166–68; US Bureau of Manufactures 119, *123*, 124, 125, 127, 129; US soy-meat complex 157, 191–95, 199, 200; US Soy Model 156, 164–73, 199, 201
USDA (United States Department of Agriculture) 13, 17, 117, 137–41, 143, 158, 175, 176, 199–200, 226; division of Plant Exploration and Introduction 123; officials 138–139; Economics, Statistics and Market Information System (ESMIS) 158; Foreign Agricultural Service (FAS) 175, 199
uses of soybeans *see* roles, uses and functions

Van Goen, Rijckloff 81
vegetarian/vegetarianism 51, 66, 87, 198, 199, 216, 235
vertical integration 124, 128, 157, 173–75, 187–91, 200–01, 202n12
vessels 76, 80, 118, 121, 134, 141
Vietnam 65, 81, 85, 172, *180*, 184, *185*, 186
Vladivostok 122, 124, 137
VOC (Dutch East India Company) 17, 60, 79–81, 83, 84, 88, 91, 100, 103, 108n4

Wallerstein, Immanuel 8
Warg, Cajsa 99
Washington Consensus 168
ways to mill/grind/crush/extract oil and meal (flour): hydraulic power mill 129; oil mills 133, 134, 136–37, 140; rotary stone mill 44, 86, 93, 129; screw press 129, 134, 144; solvent extraction 144; steam-powered mills 129; wedge press 66, 86, 93, 129, 134
wheat: Chinese multi-crop system 29–30, 33, 50, 70, 93–96; compared to soy 44, 143, 161–62, 163, 176; domesticated outside China 28; European protectionism 175, 192; feed (fodder) 98, 99; foodstuff (staple) 42, 44–46, 127; grains 29–33, 42; 36, 42, 176; Grain Robbery 176; Green Revolution 165; intensive production 93–95, 107; in Japan 85; in Manchuria 127; Ma-wang-tui tomb 46; in Paraguay 162; population growth 7; rotation with soy and other crops 33, 36, 71, 162, 165; sacred/superior crops 24, 44, 70, 71; soy fertilizer 94–96, 107, 125, 129; tax revenues 94; trade 79, 137, 176, 188; in the United States 142, 161–62, 165; world production 1, 160
whole soybeans 92, 124, 125, 136–38, 145, *179*, 179–81, *180*, 186, 190, 198–99
Widgren, Mats 15

woodblock 88
world history 4–7, 10, 14–16, 104, 201, 215, 235–36
World War I 132, 136
World War II 116, 131, 132, 138, 144, 145, *146*, 155, 160
WTO (World Trade Organization) 168, 175, 178, 200, 218; Agreement on Agriculture 178, 218
Wu emperor 40, 64, 71

Xiangjin, Wang 95, 98
Xun Kuang 35

Yangtze River 12, 60; Yangtze Delta 74, 93–96
Yellow River (*Huang He*) 4, 12, 22, 27–32, 37, 39, 40, 42, 47–48, 51, 52, 61, 75
Yingkou (Ying-tz') 115, 117, 125; *see also* Newchwang
Yingxing, Song 71, 93, 98
Yizhi, Fang 93
Yuan: army (military) 61, 62; dynasty (rule) 57, 61, 62, 63, 65; era (period) 62; Northern 62

Zedong, Mao 158, 163
zero deforestation 222
Zhen, Liu 41, 42
Zhen, Wang 62
Zhou Dynasty 22, 23, 29, 30
Zimu, Wu 64

www.ingramcontent.com/pod-product-compliance
Lightning Source LLC
LaVergne TN
LVHW021122300926
849160LV00010B/823
9781032509358